Home

THE SCHOOL MATHEMATICS PROJECT

When the SMP was founded in 1961, its main objective was to devise radically new secondary-school mathematics courses (and corresponding GCE and CSE syllabuses) to reflect, more adequately than did the traditional syllabuses, the up-to-date nature and usages of mathematics.

This objective has now been realized. SMP *Books 1–5* form a five-year course to the O-level examination 'SMP Mathematics'. *Books 3T, 4* and *5* give a three-year course to the same O-level examination (the earlier *Books T* and *T4* being now regarded as obsolete). *Advanced Mathematics Books 1–4* cover the syllabus for the A-level examination 'SMP Mathematics' and five shorter texts cover the material of the various sections of the A-level examination 'SMP Further Mathematics'. Revisions of the first two books of *Advanced Mathematics* are available as *Revised Advanced Mathematics Books 1* and *2*. There are two books for 'SMP Additional Mathematics' at O-level. All the SMP GCE examinations are available to schools through any of the Examining Boards.

Books A–H, originally designed for non-GCE streams, cover broadly the same development of mathematics as do the first few books of the O-level series. Most CSE Boards offer appropriate examinations. In practice, this series is being used very widely across all streams of comprehensive schools, and its first seven books, followed by *Books X, Y* and *Z*, provide a course leading to the SMP O-level examination. An alternative treatment of the material in *SMP Books A, B, C* and *D* is available as *SMP Cards I* and *II*.

Teachers' Guides accompany all series of books.

The SMP has produced many other texts, and teachers are encouraged to obtain each year from the Cambridge University Press, Bentley House, 200 Euston Road, London NW1 2DB, the full list of SMP books currently available. In the same way, help and advice may always be sought by teachers from the Director at the SMP Office, Westfield College, Hampstead, London NW3 7ST, from which may also be obtained the annual Reports, details of forthcoming in-service training courses and so on.

The completion of this first ten years of work forms a firm base on which the SMP will continue to develop its research into the mathematical curriculum, and is described in detail in Bryan Thwaites's *SMP: The First Ten Years*. The team of SMP writers, numbering some forty school and university mathematicians, is continually evaluating old work and preparing for new. But at the same time, the effectiveness of the SMP's future work will depend, as it always has done, on obtaining reactions from a wide variety of teachers – and also from pupils – actively concerned in the class-room. Readers of the texts can therefore send their comments to the SMP in the knowledge that they will be warmly welcomed. *1974*

ACKNOWLEDGEMENTS

The principal authors, on whose contributions the SMP texts are largely based, are named in the annual Reports. Many other authors have also provided original material, and still more have been directly involved in the revision of draft versions of chapters and books. The Project gratefully acknowledges the contributions which they and their schools have made.

This book – *Book Z* – has been written by

Professor J. V. Armitage
T. Easterbrook
D. Hale
E. W. Harper
Joyce Harris

and edited by Mary Tait.

The Project owes a great deal to its Secretaries, Mrs Jacqueline McDonald and Mrs Julie Whatton, for their careful typing and assistance in connection with this book.

We would especially thank Professor J. V. Armitage and P. G. Bowie for the advice they have given on the fundamental mathematics of the course.

Some of the chapter openings in this book are by Ken Vail.

We are grateful to the Oxford and Cambridge Schools Examination Board and the Southern Regional Examinations Board for permission to use questions from their examination papers. Our thanks to the Royal Western Yacht Club of England for supplying the information about the trans-Atlantic single-handed yacht race, and to the Australian News and Information Bureau for the picture of Sydney Harbour Bridge on page 182.

We are very much indebted to the Cambridge University Press for their cooperation and help at all times.

THE SCHOOL MATHEMATICS PROJECT

Book Z

CAMBRIDGE UNIVERSITY PRESS

Published by the Syndics of the Cambridge University Press
Bentley House, 200 Euston Road, London NW1 2DB
American Branch: 32 East 57th Street, New York, N.Y. 10022

ISBNS: 0 521 20194 2 hard covers
0 521 08622 1 limp covers

First published 1974
Reprinted 1974

Printed in Great Britain by
William Clowes & Sons Ltd.,
London, Colchester and Beccles

Preface

This is the third of three books designed for O-level candidates who have previously followed the *A–H* series of books. *X*, *Y* and *Z* follow on from *Book G* and cover the remainder of the course for the O-level examination in 'SMP Mathematics'. The books will also be found suitable for students following a one year revision course for O-level and for those who have previously taken the CSE examination.

Many of the topics introduced in *Books A–G* are extended in *X*, *Y* and *Z* and several new topics are also introduced.

Book Z commences with a chapter on Vectors which continues the work developed in *Book X*. The chapter on Growth and Logarithm Functions follows on from the Looking for Functions chapter in *Book Y* and introduces the growth function and its inverse, the logarithm function.

The chapters on Speed and Acceleration, Invariance and Use of Combination Tables review the work introduced in earlier books *A–G*, *X* and *Y*, on these topics and then extend the topics further. Chapter 6 not only introduces Latitude and Longitude but also considers the surface area and volume of the sphere, thus completing the work on mensuration required for the O-level course.

There are comprehensive review chapters on Matrices, Geometry, Statistics, Probability, Graphs, Algebra and Computation with many revision examples. These chapters are separated from the other chapters by an Interlude on Proof.

The three books will each be accompanied by Teacher's Guides to be published shortly after publication of *Book Z*. The Teacher's Guides will contain answers, teaching suggestions and ideas.

Contents

Glossary of Symbols and Units

Sets

$\{x : x < 5\}$	Set of all x such that x is less than 5
$\{1, 2, 3, 4\}$	Set with listed elements
$\mathscr{E}$	Universal set
$\varnothing$	Empty set
$\in$	Is an element of, belongs to
$\notin$	Is not an element of
$A \cap B$	Intersection of A and B; set whose elements are elements of both A and B
$A \cup B$	Union of A and B; set whose elements are elements of either A or B or both
A'	Complement of A, set whose elements are elements of the universal set but not of the set A
$n(A)$	The number of elements in the set A

Relations, functions and transformations

$x = y$	x is equal to y
$x \neq y$	x is not equal to y
$x \approx y$	x is approximately equal to y
$x > y$	x is greater than y
$x \geqslant y$	x is greater than or equal to y
$f: x \to y$	The function f which maps x onto y
$x \xrightarrow{f} y$	x is mapped onto y by f
$f(x)$	The image of x under the function f
f^{-1}	The function that is inverse to the function f
fg	The composite function, 'f operating on the result of g'; the function mapping x onto $f(g(x))$
$f . g$	The function mapping x onto the product of $f(x)$ and $g(x)$
$\mathbf{M}: P \to P'$	The transformation $\mathbf{M}$ which maps P onto P'
$\mathbf{M}: \begin{pmatrix} x \\ y \end{pmatrix} \to \begin{pmatrix} x' \\ y' \end{pmatrix}$	The transformation $\mathbf{M}$ which maps $\begin{pmatrix} x \\ y \end{pmatrix}$ onto $\begin{pmatrix} x' \\ y' \end{pmatrix}$
$\mathbf{M}(P)$	The image of P under the transformation $\mathbf{M}$
$\mathbf{MH}$	The combined transformation, '$\mathbf{M}$ operating on the result of $\mathbf{H}$'; the function mapping P onto $\mathbf{M}(\mathbf{H}(P))$

Vectors

i	Unit vector in direction of x increasing, $\begin{pmatrix}1\\0\end{pmatrix}$
j	Unit vector in direction of y increasing, $\begin{pmatrix}0\\1\end{pmatrix}$
$\overline{PQ}$	Directed line segment from P to Q
PQ	Sometimes used as (i) the infinite line through P and Q and sometimes as (ii) the length of the segment from P to Q
PQ	The displacement from P to Q or the vector which describes this displacement

Logical symbols

$p \Rightarrow q$	If p, then q
$p \Leftrightarrow q$	p if and only if q; p is equivalent to q

s, h, min	Time: seconds, hours, minutes. (Days and years are written in full)
m, mm, cm, km	Length or distance: 1000 mm = 1 m; 100 cm = 1 m; 1000 m = 1 km
l	Volume: litre (≈ 1000 cm^3)
kg, g	Mass: kilogram, gram
N	Force: newton. This is a force which will give a mass of 1 kg an acceleration of 1 m/s^2. The 'weight' of an object mass m kg is approximately $10m$ N
m/s, km/h, kg/m^3	Rate: metres per second, kilometres per hour, kilograms per cubic metre

$x \to \sin x$
$x \to \cos x$
$x \to \tan x$
$x \to \log x$
$x \to 2^x$

} Some special functions mentioned

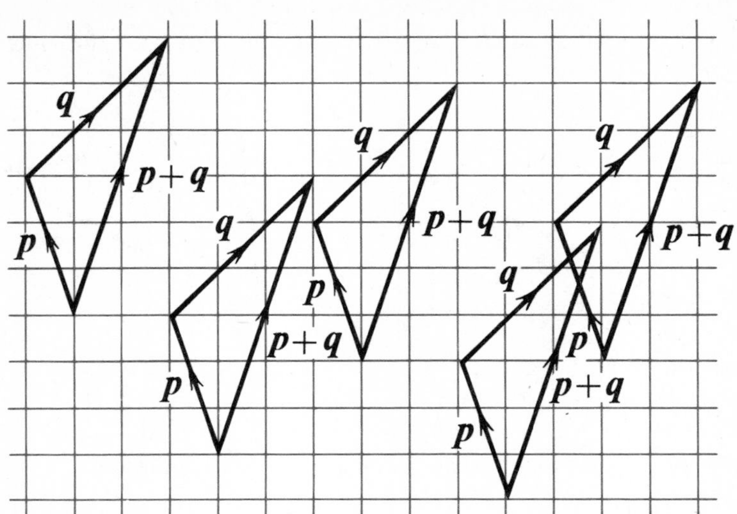

1 Vectors again

1 Combining vectors

(*a*) The diagram above reminds us that the sum of the vectors $\mathbf{p} = \begin{pmatrix} ^-1 \\ 3 \end{pmatrix}$ and $\mathbf{q} = \begin{pmatrix} 3 \\ 3 \end{pmatrix}$ may be found by a triangle method. What is $\mathbf{p} + \mathbf{q}$ written as a column vector?

(*b*) If **a** and **b** are representatives of two vectors (Figure 1), how can we find $\mathbf{a} - \mathbf{b}$?

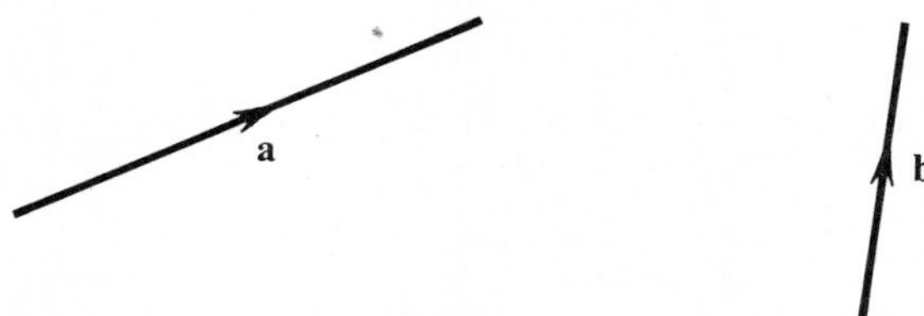

Fig. 1

Suppose $\mathbf{a} - \mathbf{b} = \mathbf{x}$. Do you agree that we can write this as $\mathbf{a} = \mathbf{b} + \mathbf{x}$? In this form we want to know the vector which, when *added* to **b**, will give the answer **a**.

In Figure 2 we take representatives of **a** and **b** which have the same starting point.

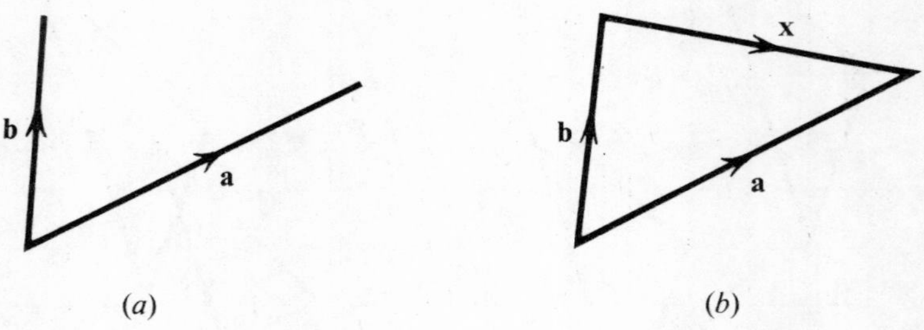

Fig. 2

We see in Figure 2(*b*) that $\mathbf{b}+\mathbf{x}=\mathbf{a}$ and so **x** is a representative of $\mathbf{a}-\mathbf{b}$.

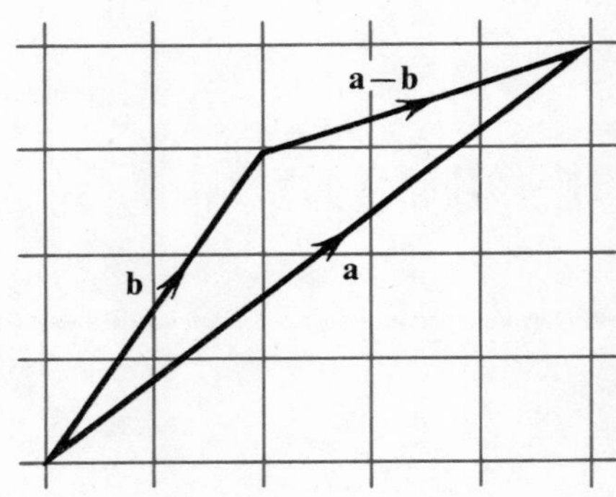

Fig. 3

Look at the vectors in Figure 3:

$$\mathbf{a}=\begin{pmatrix}5\\4\end{pmatrix} \quad \text{and} \quad \mathbf{b}=\begin{pmatrix}2\\3\end{pmatrix}.$$

What is $\begin{pmatrix}5\\4\end{pmatrix}-\begin{pmatrix}2\\3\end{pmatrix}$? Is this column vector a description of the directed line segment labelled $\mathbf{a}-\mathbf{b}$?

(*c*) If a translation has a vector $\mathbf{a}=\begin{pmatrix}5\\4\end{pmatrix}$, then the inverse translation has vector $\begin{pmatrix}^-5\\^-4\end{pmatrix}$ which we may write as $^-\mathbf{a}$. What is $\begin{pmatrix}5\\4\end{pmatrix}+\begin{pmatrix}^-5\\^-4\end{pmatrix}$?

Does your result confirm that the two vectors describe inverse translations? Describe the translation which has the vector $\begin{pmatrix}0\\0\end{pmatrix}$.

(*d*) Another way of looking at vector subtraction is to regard it as *adding* the inverse vector. So $\mathbf{a} - \mathbf{b}$ can be thought of as $\mathbf{a} + {}^{-}\mathbf{b}$.

We take

$$\mathbf{a} = \begin{pmatrix}5\\4\end{pmatrix} \quad \text{and} \quad \mathbf{b} = \begin{pmatrix}2\\3\end{pmatrix}$$

again and show the vectors $\mathbf{a}$, ${}^{-}\mathbf{b}$ and $\mathbf{a} + {}^{-}\mathbf{b}$ in Figure 4.

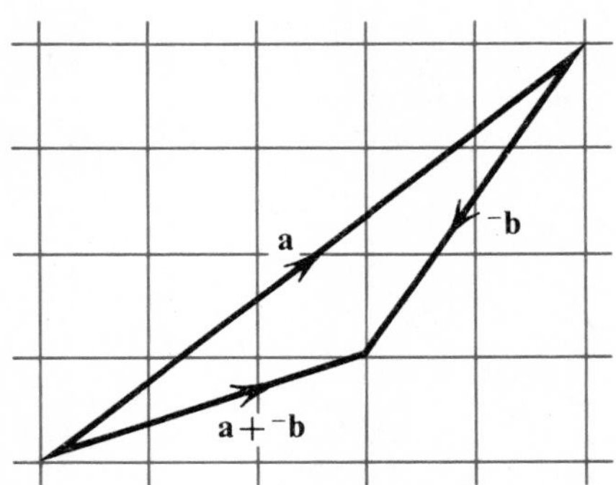

Fig. 4

What is $\mathbf{a} + {}^{-}\mathbf{b}$ as a column vector? Does your result agree with that obtained for $\mathbf{a} - \mathbf{b}$? Figure 5 shows Figures 4 and 3 combined.

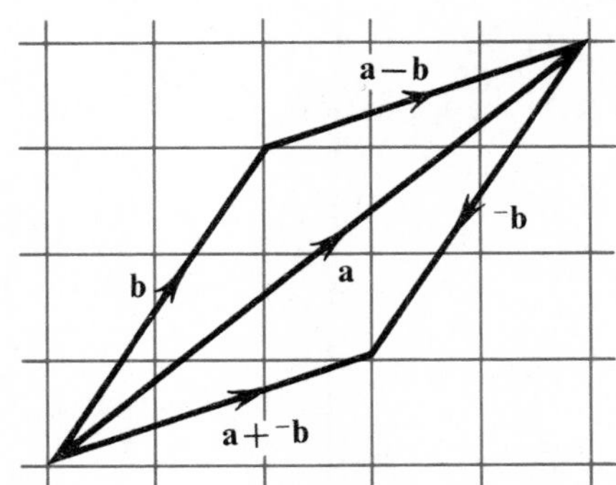

Fig. 5

Do you agree that $\mathbf{a} - \mathbf{b}$ and $\mathbf{a} + {}^{-}\mathbf{b}$ are representatives of the same vector? This confirms again that our two ways of looking at vector subtraction:

(i) $\mathbf{a} - \mathbf{b} = \mathbf{x} \Rightarrow \mathbf{a} = \mathbf{b} + \mathbf{x}$, (ii) $\mathbf{a} - \mathbf{b} = \mathbf{a} + {}^{-}\mathbf{b}$

are equivalent.

Vectors again

Exercise A

1

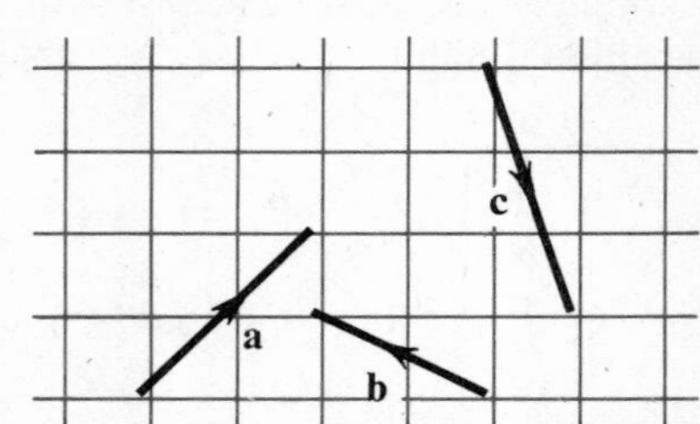

Fig. 6

Representatives of three vectors are shown in Figure 6. Copy them onto squared paper and construct representatives of:

(*a*) $\mathbf{a}+\mathbf{b}$; (*b*) $\mathbf{a}+\mathbf{b}+\mathbf{c}$; (*c*) $2(\mathbf{b}+\mathbf{c})$;
(*d*) $\mathbf{a}+2\mathbf{c}$; (*e*) $\mathbf{a}-\mathbf{b}$; (*f*) $\mathbf{b}-\mathbf{c}$.

2 If $\mathbf{p}=\begin{pmatrix}4\\3\end{pmatrix}$, $\mathbf{q}=\begin{pmatrix}2\\1\end{pmatrix}$ and $\mathbf{r}=\begin{pmatrix}3\\1\end{pmatrix}$, express as column vectors:

(*a*) $\mathbf{p}-\mathbf{q}$; (*b*) $\mathbf{q}-\mathbf{r}$; (*c*) $\mathbf{p}+\mathbf{q}-\mathbf{r}$; (*d*) $\mathbf{p}-\mathbf{q}+\mathbf{r}$;
(*e*) $\mathbf{p}-(\mathbf{q}+\mathbf{r})$; (*f*) $\mathbf{p}-(\mathbf{q}-\mathbf{r})$.

3 P is the point with coordinates (3, 2) and $\mathbf{a}=\begin{pmatrix}1\\5\end{pmatrix}$. Give the coordinates of the image of P under a translation with vector:

(*a*) $\mathbf{a}$; (*b*) $^{-}\mathbf{a}$; (*c*) $2\mathbf{a}$; (*d*) $^{-}2\mathbf{a}$.

4 If $\mathbf{a}=\begin{pmatrix}2\\3\end{pmatrix}$, $\mathbf{b}=\begin{pmatrix}4\\-1\end{pmatrix}$, write in column vector form:

(*a*) $^{-}2\mathbf{a}$; (*b*) $^{-}3\mathbf{b}$; (*c*) $\mathbf{a}-2\mathbf{b}$; (*d*) $\mathbf{a}+{}^{-}2\mathbf{b}$;
(*e*) $2\mathbf{b}-\mathbf{a}$; (*f*) $2\mathbf{b}+{}^{-}\mathbf{a}$; (*g*) $3\mathbf{a}-3\mathbf{b}$; (*h*) $3(\mathbf{a}-\mathbf{b})$.

5 In Figure 7, *ABCDEF* is a regular hexagon and *O* is its centre. Write in terms of **a**, **b**, **c**:

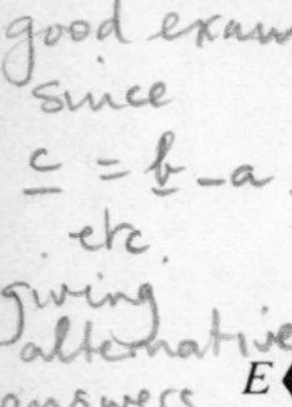

(*a*) **AB**; (*b*) **BC**; (*c*) **CD**; (*d*) **DE**; (*e*) **EF**; (*f*) **FA**.

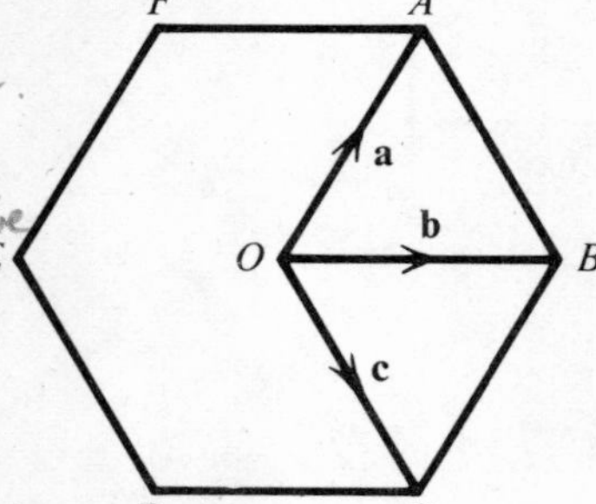

Fig. 7

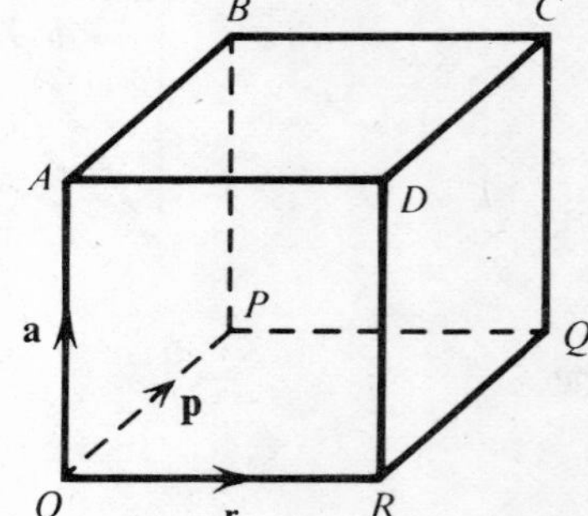

Fig. 8

6 Express the following (see Figure 8) in terms of **p**, **a**, **r**:

(*a*) **AP**; (*b*) **PR**; (*c*) **OB**; (*d*) **OQ**; (*e*) **BQ**; (*f*) **BR**.

7 Two trains, one going north at 100 km/h and the other travelling south at 85 km/h pass on parallel tracks.

(*a*) Describe the velocity of the northbound train as it appears to a passenger in the southbound train.

(*b*) Describe the velocity of the southbound train as it appears to a passenger in the northbound train.

8 A passenger train travelling east at 90 km/h overtakes a goods train travelling in the same direction on an adjacent track at 65 km/h.

(*a*) Describe the velocity of the passenger train as it appears to the guard of the goods train.

(*b*) Describe the velocity of the goods train as it appears to the guard of the passenger train.

(*c*) If $\mathbf{p} = \begin{pmatrix} 90 \\ 0 \end{pmatrix}$ and $\mathbf{g} = \begin{pmatrix} 65 \\ 0 \end{pmatrix}$ write $\mathbf{p} - \mathbf{g}$ and $\mathbf{g} - \mathbf{p}$ as column vectors.

(*d*) Write your answers to (*a*) and (*b*) in terms of $\mathbf{p}$ and $\mathbf{g}$.

2 Distance and direction

(*a*) A ship leaves a port P and some time later its position Q is given as 3 nautical miles east and 4 nautical miles north of the port. Look at Figure 9 and see if you agree that the distance PQ is 5 nautical miles.

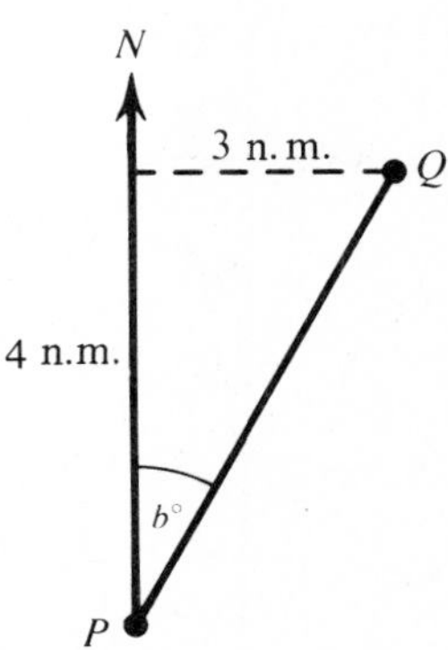

Fig. 9

(*b*) A translation is described by the vector $\mathbf{a} = \begin{pmatrix} 3 \\ 4 \end{pmatrix}$. How far does each point move under this translation? Figure 10 shows a point A and its image A' under the translation.

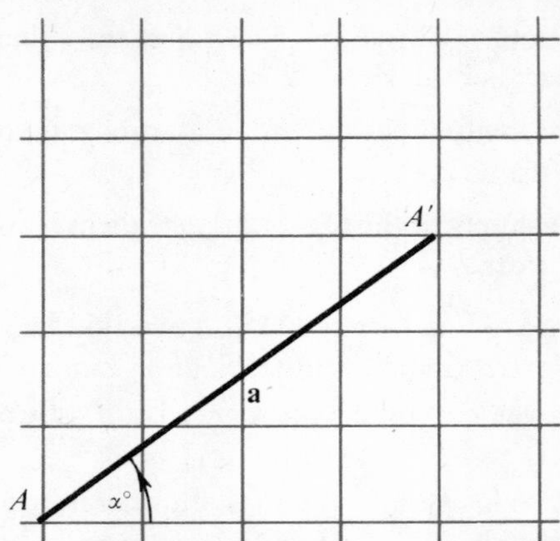

Fig. 10

Do you agree that AA' is 5 units long? We say that the *magnitude* of the vector **a** is 5. What are the magnitudes of the vectors

$$\mathbf{b} = \begin{pmatrix} 5 \\ 12 \end{pmatrix}, \quad \mathbf{c} = \begin{pmatrix} ^{-}3 \\ 4 \end{pmatrix} \quad \text{and} \quad \mathbf{d} = \begin{pmatrix} 3 \\ 2 \end{pmatrix}?$$

(*c*) On what bearing is the ship sailing in Figure 9?

Do you agree that $\tan b^\circ = \frac{3}{4}$? Use the tangents table to check that $b \approx 36{\cdot}9$.

(*d*) What is the direction of the vector $\begin{pmatrix} 3 \\ 4 \end{pmatrix}$? From Figure 10 we obtain

$$\tan \alpha^\circ = \tfrac{4}{3}$$
$$\approx 1{\cdot}33,$$

and from a tangents table $\alpha \approx 53{\cdot}1.$

So the vector $\begin{pmatrix} 3 \\ 4 \end{pmatrix}$ has a direction which makes an angle of $53{\cdot}1^\circ$ with the x-axis.

Without using trigonometry, find the direction of the following:

(i) $\mathbf{p} = \begin{pmatrix} 3 \\ 3 \end{pmatrix}$; (ii) $\mathbf{q} = \begin{pmatrix} 4 \\ 0 \end{pmatrix}$; (iii) $\mathbf{r} = \begin{pmatrix} 0 \\ 1 \end{pmatrix}$.

(*e*) Figure 11 shows the point B and its image B' under the translation with vector $\mathbf{c} = \begin{pmatrix} ^{-}3 \\ 4 \end{pmatrix}$.

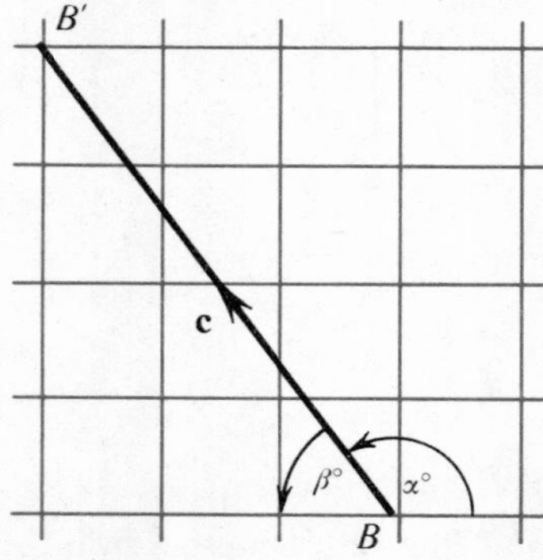

Fig. 11

Using the results of 2(*b*) and 2(*d*) we see that the magnitude of **c** is 5 and that $\beta \approx 53{\cdot}1$. Do you agree that $\alpha = 126{\cdot}9$? We usually give the direction of a translation by referring to the angle which it makes with the positive direction of the x-axis.

(*f*) Under a translation **T**, points of the plane are displaced a distance 10 units in a direction making an angle of 30° with the x-axis. Figure 12 shows a point C and its image C' under this translation.

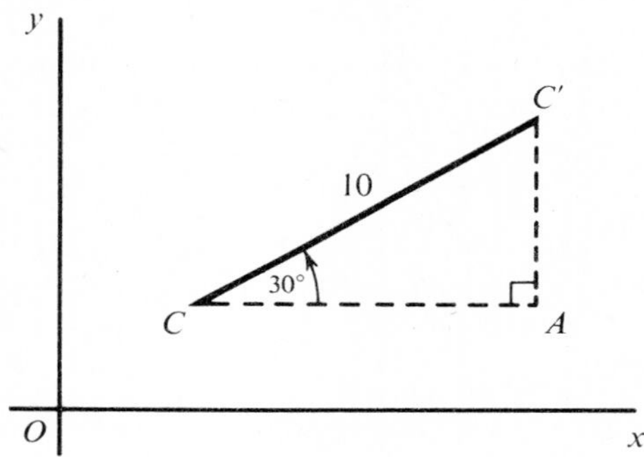

Fig. 12

What is the vector which describes **T**? To answer this we need to know the lengths of AC and AC'.

Do you agree that $AC = 10 \times \cos 30°$
and that $AC' = 10 \times \sin 30°$?

Using tables we obtain:

$$AC \approx 10 \times 0{\cdot}866 = 8{\cdot}66$$

and

$$AC' = 10 \times 0{\cdot}5 = 5.$$

The vector describing **T** is $\begin{pmatrix} 8{\cdot}66 \\ 5 \end{pmatrix}$. We say that the vector has *components* 8·66 and 5.

Summary

1 The magnitude of the vector $\mathbf{a} = \begin{pmatrix} u \\ v \end{pmatrix}$ is $\sqrt{(u^2 + v^2)}$ and is obtained by using Pythagoras' rule.

2 The direction of the vector $\mathbf{a} = \begin{pmatrix} u \\ v \end{pmatrix}$ makes an angle α with the positive direction of the x-axis, where $\tan \alpha = (v/u)$.

3 A translation in which points are displaced a distance p units, in a direction making an angle α with the positive direction of the x-axis is described by a vector **a** having components $p\cos\alpha$ and $p\sin\alpha$.

That is:
$$\mathbf{a} = \begin{pmatrix} p\cos\alpha \\ p\sin\alpha \end{pmatrix}.$$

Exercise B

1 How far and on what bearing should a ship sail to pick up the occupants of a lifeboat reported 10 miles east and 15 miles north of her current position?

2 Find the distance and direction of the translation represented by each of the following vectors:

(*a*) $\mathbf{a} = \begin{pmatrix} 6 \\ 8 \end{pmatrix}$; (*b*) $\mathbf{b} = \begin{pmatrix} 5 \\ 12 \end{pmatrix}$; (*c*) $\mathbf{c} = \begin{pmatrix} 3 \\ 3 \end{pmatrix}$; (*d*) $\mathbf{d} = \begin{pmatrix} ^-5 \\ 3 \end{pmatrix}$;

(*e*) $\mathbf{e} = \begin{pmatrix} 2 \\ ^-1 \end{pmatrix}$; (*f*) $\mathbf{f} = \begin{pmatrix} ^-8 \\ ^-2 \end{pmatrix}$.

3 Find the components of the vectors in Figure 13 (the dotted lines show, in each case, the direction of the x-axis).

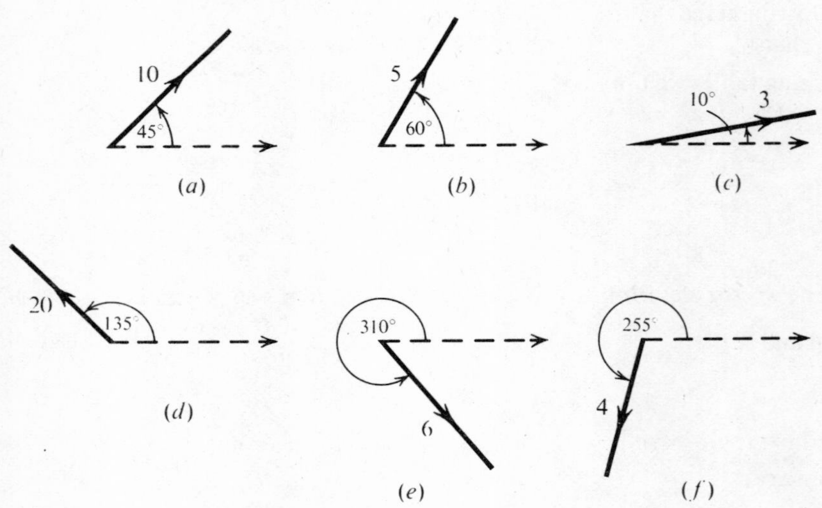

Fig. 13

4 A ship leaves port on a bearing of 150° with a speed of 20 knots. How far (i) east and (ii) south of the port is it after two hours sailing?

3 A two-part journey

(*a*) A ship sails from a port P, a distance of 5 nautical miles on a bearing of 070°, then alters course and sails the next 10 nautical miles on a bearing of 030° (Figure 14).

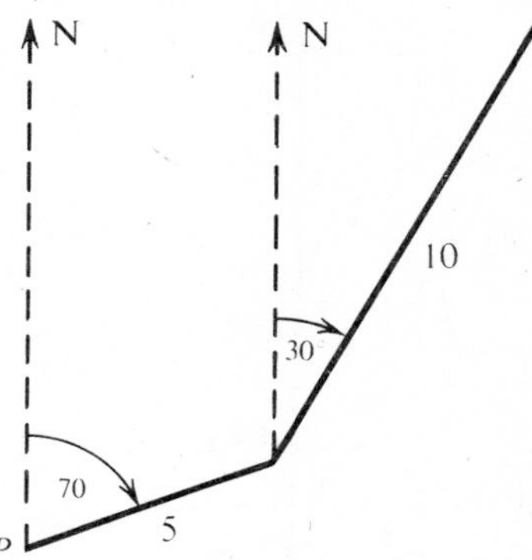

Fig. 14

The first part of its journey can be written as a column vector

$$\begin{pmatrix} 5\cos 20° \\ 5\sin 20° \end{pmatrix}.$$

Why does 20° appear in the components of this vector? Figure 15(*a*) shows the first part of the ship's journey and should help you to answer this question.

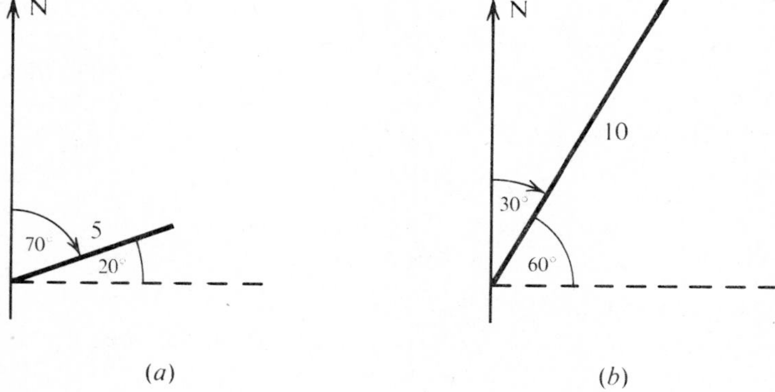

Fig. 15

Using tables we obtain:

$$\begin{pmatrix} 5\cos 20 \\ 5\sin 20 \end{pmatrix} \approx \begin{pmatrix} 4{\cdot}70 \\ 1{\cdot}71 \end{pmatrix}.$$

Do you agree that the vector describing the second part of the journey is

$$\begin{pmatrix} 10\cos 60 \\ 10\sin 60 \end{pmatrix}?$$

Use tables to find approximate values for the components of this vector. The components of the two vectors are shown in Figure 16(*a*).

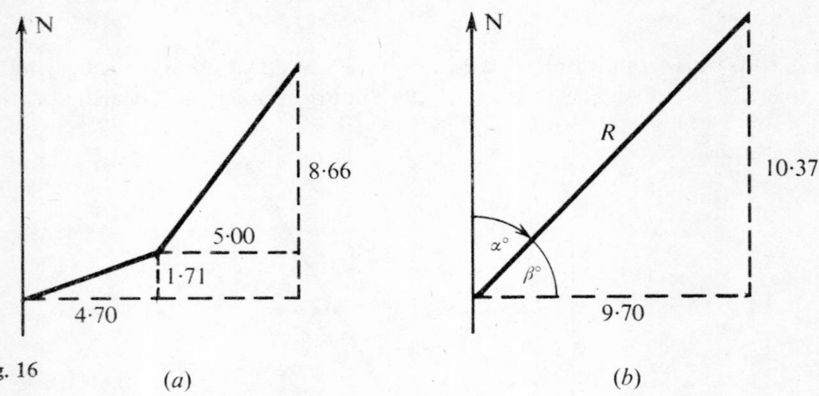

Fig. 16 (*a*) (*b*)

The vector which gives the ship's total displacement from the port is

$$\begin{pmatrix}4{\cdot}70\\1{\cdot}71\end{pmatrix}+\begin{pmatrix}5{\cdot}0\\8{\cdot}66\end{pmatrix}=\begin{pmatrix}9{\cdot}70\\10{\cdot}37\end{pmatrix}.$$

We see that at the end of its journey the ship is 9·7 nautical miles east and 10·37 nautical miles north of the port.

(*b*) Figure 16(*b*) shows the distance R and bearing $\alpha°$ of the ship from its port. We can find the value of R by Pythagoras' rule as before.

$$\begin{aligned}R^2 &= 9{\cdot}70^2 + 10{\cdot}37^2\\ &\approx 94{\cdot}1 + 108\\ &= 202{\cdot}1.\\ \text{So } R &\approx 14{\cdot}2.\end{aligned}$$

To find the value of α we first observe that

$$\begin{aligned}\tan\beta° &= \frac{10{\cdot}37}{9{\cdot}70}\\ &\approx 1{\cdot}07 \text{ by slide rule.}\end{aligned}$$

From tangent tables we obtain $\beta \approx 47$. Do you agree that $\alpha \approx 43$?

The ship's final position is 14·2 nautical miles from the port on a bearing of 043°.

Exercise C

1 (*a*) A hockey ball is passed from A to B (see Figure 17) where the distance AB is 20 metres at an angle of 20° to the sideline. Find the components of the vector **AB**.

(*b*) The ball is now hit from B to C where BC is 15 metres at 75° to the sideline. Find the components of the vector **BC**.

(*c*) Hence find the components of the vector **AC**.

(*d*) Find the distance AC and the angle it makes with the sideline.

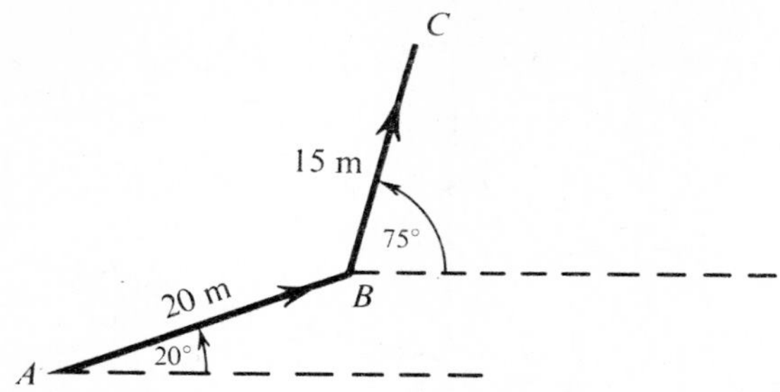

Fig. 17

2 A by-pass QR is being constructed to avoid town P (see Figure 18). Calculate the length and direction of the by-pass. Assuming an average speed of 40 km/h on the town route and an average speed of 64 km/h on the by-pass, how much time would be saved by taking the new road?

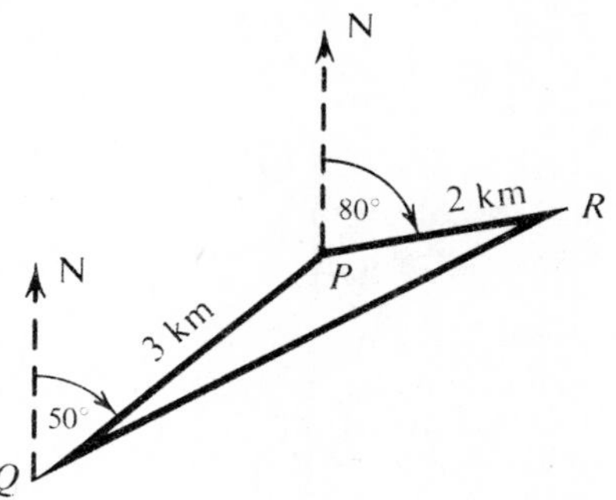

Fig. 18

3 A plane flies 350 km on a bearing of 030°, followed by 200 km on a bearing of 140°. Draw a sketch to show its route. Find the distance and bearing of the shortest route back to the starting point.

Exercise D (Miscellaneous)

1 If $\mathbf{p} = \begin{pmatrix} 2 \\ -1 \\ 1 \end{pmatrix}$, $\mathbf{q} = \begin{pmatrix} 0 \\ 2 \\ -3 \end{pmatrix}$ and $\mathbf{r} = \begin{pmatrix} 5 \\ 1 \\ 1 \end{pmatrix}$, write in column vector form:

(*a*) $\mathbf{p} + 2\mathbf{q}$; (*b*) $\mathbf{p} - \mathbf{q}$; (*c*) $2\mathbf{p} + \mathbf{q} - 3\mathbf{r}$; (*d*) $5(\mathbf{q} - 2\mathbf{r})$.

2 $ABCD$ is a parallelogram with diagonals intersecting at E. M is the midpoint of CD and $\mathbf{AB} = \mathbf{x}$, $\mathbf{AD} = \mathbf{y}$. Write the following in terms of $\mathbf{x}$ and $\mathbf{y}$.

(*a*) **AE**; (*b*) **BD**; (*c*) **DM**; (*d*) **AM**; (*e*) **MB**.

3 If $\mathbf{a} = \begin{pmatrix} 4 \\ 1 \end{pmatrix}$ and $\mathbf{b} = \begin{pmatrix} {}^{-}1 \\ 2 \end{pmatrix}$, find $\mathbf{x}$ in the following equations:

(*a*) $2\mathbf{x} - \mathbf{a} = \mathbf{x} + \mathbf{b}$; (*b*) $3(\mathbf{x} + \mathbf{a}) = \mathbf{b} - \mathbf{x}$.

4

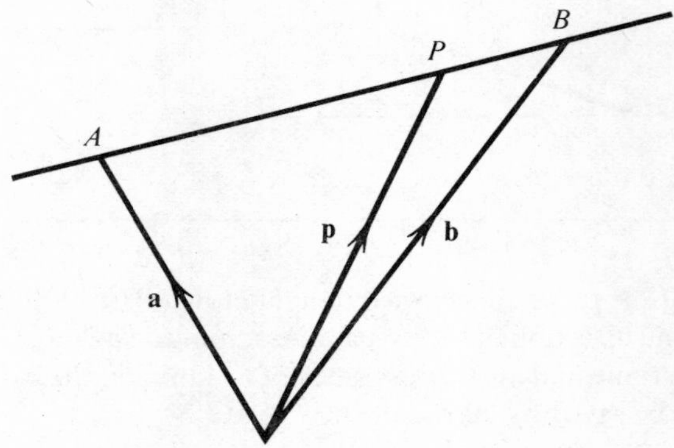

Fig. 19

In Figure 19, $\mathbf{AP} = 3\mathbf{PB}$.

(*a*) Write **AB** in terms of **a** and **b**.
(*b*) Write **AP** in terms of **a** and **b**.
(*c*) Find an expression for **p** in terms of **a** and **b**.

5 Find the velocity of a jet with an airspeed of 1000 km/h attempting to fly due south and being blown off course by a north-westerly gale of 150 km/h.

6 If **OA** is a vector of unit length in the direction Ox, and it is rotated about O through an angle of $\theta°$ to the position OP, what are the coordinates of P?

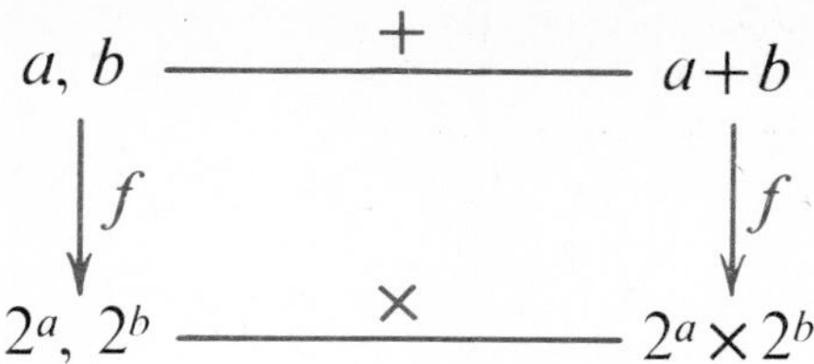

2 Growth and Logarithm Functions

1 Properties of growth functions

(*a*) We consider again the example of a growth function introduced in *Book Y*, Chapter 12. The following table shows the number of cells of a frog's egg at unit intervals of time.

Time	0	1	2	3	4	5
Number of cells	1	2	4	8	16	32

If the doubling pattern continues:

(i) how many cells will there be when the time is 8 units;
(ii) at what time will there be 512 cells?

(*b*) Let us look at three pairs of the mapping time $\rightarrow$ number of cells.

Time		Number of cells
2	$\rightarrow$	4
3	$\rightarrow$	8
5	$\rightarrow$	32

The times have been *added*,

$$2+3=5,$$

and the corresponding numbers of cells have been *multiplied*,

$$4 \times 8 = 32.$$

Now copy and complete the following:

$1 \rightarrow ?$			$2 \rightarrow ?$			$0 \rightarrow ?$
$3 \rightarrow ?$			$1 \rightarrow ?$			$5 \rightarrow ?$
$4 \rightarrow ?$			$3 \rightarrow ?$			$5 \rightarrow ?$

Do you agree that, in each case, adding the times corresponds to multiplying the numbers of cells?

(*c*) Now look at these three pairs:

Time		Number of cells
5	→	32
1	→	2
4	→	16

We have *subtracted* the times,

$$5 - 1 = 4,$$

and the corresponding numbers of cells have been *divided*,

$$32 \div 2 = 16.$$

Copy and complete the following:

3 → ?		4 → ?		5 → ?
1 → ?		3 → ?		5 → ?
2 → ?		1 → ?		0 → ?

Do you agree that, in each case, subtracting the times corresponds to dividing the numbers of cells?

(*d*) In the preceding sections we have been considering the growth function $f: x \to 2^x$ with domain {0, 1, 2, 3, 4, 5} and range {1, 2, 4, 8, 16, 32} and the results we have obtained suggest that it has the following properties:

(i) addition in the domain corresponds to multiplication in the range;
(ii) subtraction in the domain corresponds to division in the range.

Make a table of corresponding pairs for the growth function $g: x \to 3^x$ with domain {0, 1, 2, 3, 4, 5}. Test carefully to see if it has properties (i) and (ii) above.

(*e*) These properties of growth functions indicate that their inverses have the following properties:

(i) multiplication in the domain corresponds to addition in the range;
(ii) division in the domain corresponds to subtraction in the range.

Copy and complete the table of pairs for the inverse relation of the function $x \to 5^x$.

5^x	1	5	?	125	?	?
↓	↓	↓	↓	↓	↓	↓
x	0	?	2	3	4	5

Exercise A

1 Copy and complete the table of pairs for $f: x \to 2^x$.

x	$^-5$	$^-4$	$^-3$	$^-2$	$^-1$	0	1	2
↓	↓	↓	↓	↓	↓	↓	↓	↓
2^x	?	?	?	$\frac{1}{4}$	?	1	2	4

2 If h is the growth function $h: x \to 6^x$, what are the values of:

(*a*) $h(2)$; (*b*) $h(3)$; (*c*) $h(0)$;
(*d*) $h(^-2)$; (*e*) $h^{-1}(6)$; (*f*) $h^{-1}(1296)$?

3 j is the inverse of the growth function $h: x \to 6^x$.
(*a*) Copy and complete:

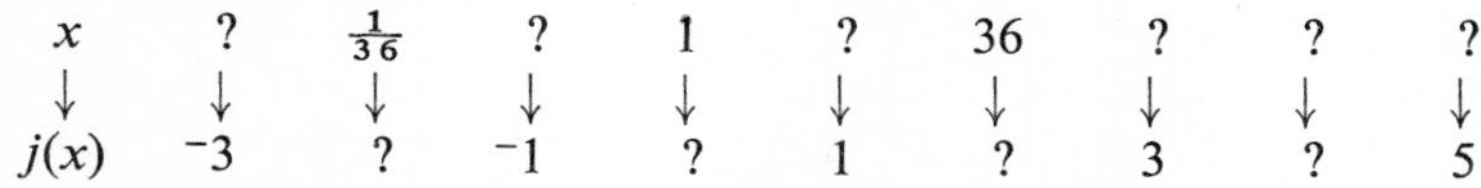

x	?	$\frac{1}{36}$	?	1	?	36	?	?	?
↓	↓	↓	↓	↓	↓	↓	↓	↓	↓
$j(x)$	$^-3$	?	$^-1$	?	1	?	3	?	5

(*b*) Complete the following *pairs* of statements:

(i) $216 = 36 \times ?$, $j(216) = j(36) + j(\quad)$;
(ii) $1296 = ? \times 36$, $j(1296) = j(\quad) + j(36)$;
(iii) $\frac{1}{216} = \frac{1}{6} \times ?$, $j(\frac{1}{216}) = j(\frac{1}{6}) + j(\quad)$;
(iv) $216 = 1296 \div ?$, $j(216) = j(1296) - j(\quad)$.

4 The images of 1, 2, 3, ..., 12 under the function l are as follows:

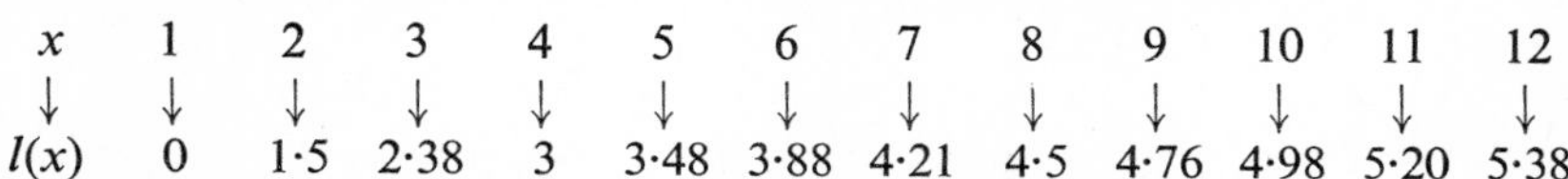

x	1	2	3	4	5	6	7	8	9	10	11	12
↓	↓	↓	↓	↓	↓	↓	↓	↓	↓	↓	↓	↓
$l(x)$	0	1·5	2·38	3	3·48	3·88	4·21	4·5	4·76	4·98	5·20	5·38

(*a*) What are: (i) $l(3)$; (ii) $l(4)$; (iii) $l(12)$?
(*b*) Write down a relation connecting $l(12)$, $l(3)$ and $l(4)$.
(*c*) Copy and complete:

(i) $l(12) = l(2) + l(\ )$; (ii) $l(6) = l(\ ) + l(3)$;
(iii) $l(8) = l(2) + l(\ )$; (iv) $l(9) = 2 \times l(\ )$;
(v) $l(2) = l(10) - l(\ )$.

(*d*) Do you think that l is the inverse of a growth function?
(*e*) Use relations similar to those in (*c*) to find:

(i) $l(14)$; (ii) $l(15)$; (iii) $l(36)$; (iv) $l(4{\cdot}5)$.

5 Figure 1 shows part of the D scale of a slide rule.

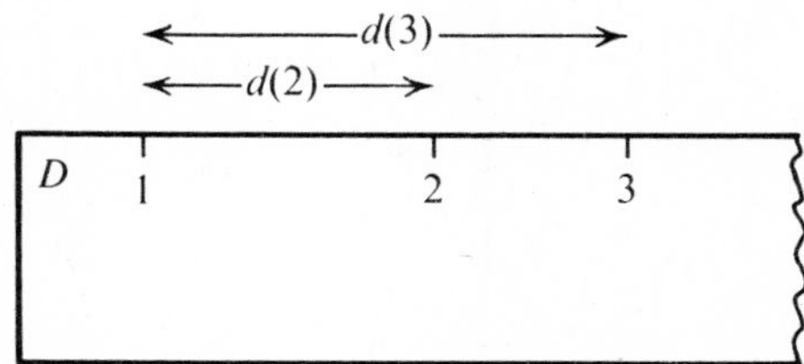

Fig. 1

The function d maps a number on the D scale onto its distance from 1. ($d(2)$ and $d(3)$ are shown.)
By measuring distances on your slide rule, complete the table:

x	1	2	3	4	5	6	7	8	9	10
↓	↓	↓	↓	↓	↓	↓	↓	↓	↓	↓
$d(x)$	0									

What sort of function is d?

2 The logarithm function

(*a*) John Napier, a seventeenth-century Scots mathematician, was the first person to make use of the properties of growth functions and their inverses. He realized that, by using properties (i) and (ii) of Section 1(*e*), multiplication and division could be avoided altogether and replaced by the much simpler processes of addition and subtraction.

(*b*) Napier used the word *logarithm* for a function which is the inverse of a growth function. We shall make use of this name and write the inverse of $x \to 2^x$ as

$$x \to \log_2 x.$$

How will the inverse of $x \to 5^x$ be written?
Figure 2 shows some pairs of the mapping $x \to \log_2 x$.

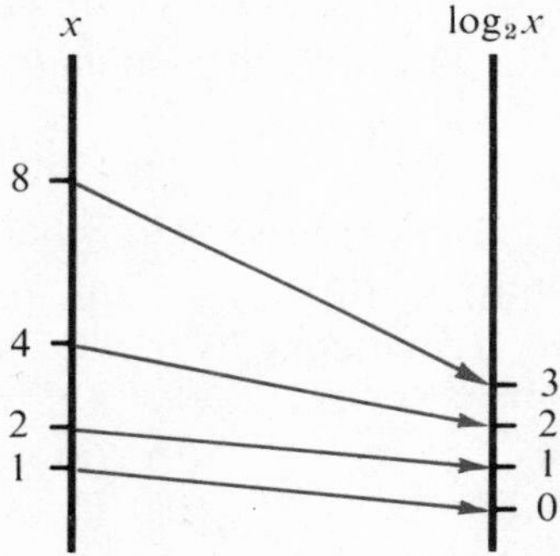

Fig. 2

Since $8 \to 3$, we write $\log_2 8 = 3$ and say 'the logarithm of 8, to the *base* 2, is 3'. Notice that the statements

$$2^3 = 8 \quad \text{and} \quad \log_2 8 = 3$$

are equivalent.

Complete the following:

(i) $2^4 = 16$ and $\log_2 16 =$.
(ii) $2^2 = 4$ and $\log_2 4 =$.
(iii) $2^{-1} = \frac{1}{2}$ and $\log_2 \frac{1}{2} =$.

(*c*) Can you give the value of $\log_2 5$? Why do you find it difficult? For what values of x can you give the value of $\log_2 x$ without difficulty?

In Figure 3(*a*) values of 2^x which correspond to natural number values of x have been plotted and a smooth curve drawn through them. Figure 3(*b*) shows a graph of $x \to \log_2 x$ which has been drawn in a similar way. How are the two graphs related?

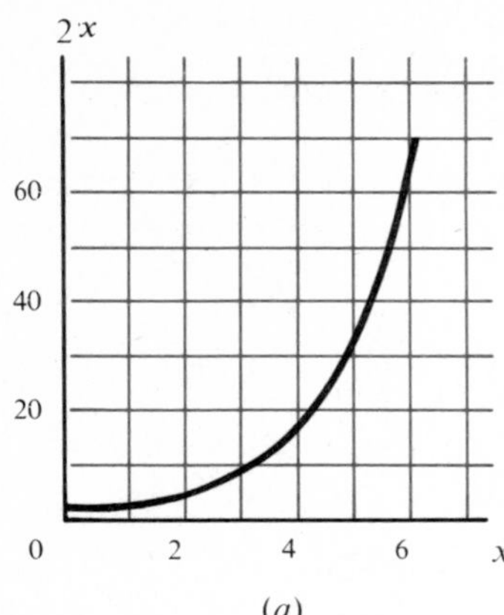

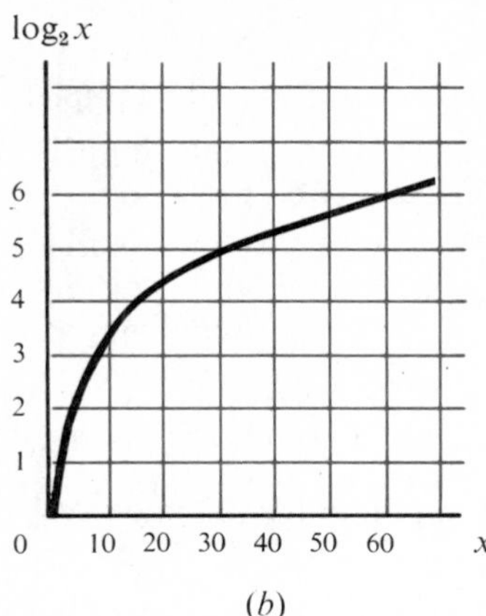

Fig. 3 (*a*) (*b*)

Use the graph in Figure 3(*b*) to find suggested values for $\log_2 5$ and $\log_2 40$. Do your results seem to indicate that

$$\log_2 40 = \log_2 5 + \log_2 8\,?$$

It seems reasonable to use the graph to find an approximate value for $\log_2 x$ where $x \geqslant 1$. We are now in a position to use the logarithm function to avoid multiplication and do addition instead. The table below shows the simple multiplication $8 \times 5 = 40$ carried out using the logarithm function.

x	$\rightarrow$	$\log_2 x$	
8	$\rightarrow$	3	(since $2^3 = 8$)
5	$\rightarrow$	2·3	(obtained from the graph: we may write $2^{2\cdot3} \approx 5$)
40	$\leftarrow$	5·3	($3 + 2\cdot3 = 5\cdot3$ is the only computation we have to do. We see from the graph that $2^{5\cdot3} \approx 40$)
2^x	$\leftarrow$	x	

(*d*) The example used in section (*c*) was simple and we knew the answer beforehand. It should, however, give you some confidence that the logarithm method works. In Example 1 which follows we consider a more realistic calculation and again manage to avoid doing a multiplication.

Example 1

Calculate $2\cdot5 \times 17$.

x	$\rightarrow$	$\log_2 x$		
2·5	$\rightarrow$	1·3	or	$2^{1\cdot3} \approx 2\cdot5$
17	$\rightarrow$	4·1	or	$2^{4\cdot1} \approx 17$
42	$\leftarrow$	5·4	or	$2^{5\cdot4} \approx 42$
2^x	$\leftarrow$	x		

The only piece of direct computation is $1\cdot3 + 4\cdot1 = 5\cdot4$.

Can you explain why 42 is an approximate answer? What is the exact answer for $2\cdot5 \times 17$?

Exercise B

1 Use the first statement to complete the second statement.

(*a*) $2^5 = 32$ and $\log_2 32 =$;
(*b*) $2^9 = 512$ and $\log_2 \quad = 9$;
(*c*) $2^{-2} = \frac{1}{4}$ and $\log_2 \frac{1}{4} =$;
(*d*) $2^{2\cdot 8} \approx 7$ and $\log_2 7 \approx$;
(*e*) $2^{4\cdot 7} \approx 25$ and $\log_2 \quad \approx 4\cdot 7$.

2 What are the values of:

(*a*) $\log_2 128$; (*b*) $\log_2 1$; (*c*) $\log_2 2^6$; (*d*) $\log_2 \frac{1}{8}$?

3 Copy and fill in the missing numbers.

(*a*) $2 \times 4 =$ and $\log_2 2 + \log_2 4 = \log_2$;
(*b*) $16 \times 32 =$ and $\log_2 16 + \log_2 32 = \log_2$;
(*c*) $3 \times 5 =$ and $\log_2 3 + \log_2 5 = \log_2$;
(*d*) $64 \div 4 =$ and $\log_2 64 - \log_2 4 = \log_2$;
(*e*) $36 \div 9 =$ and $\log_2 36 - \log_2 9 = \log_2$;
(*f*) $7 \times 7 =$ and $2 \times \log_2 7 = \log_2$.

4 Remember that $x \rightarrow \log_{10} x$ is the inverse of $x \rightarrow 10^x$ and write down the values of:

(*a*) $\log_{10} 1000$; (*b*) $\log_{10} 10\,000$; (*c*) $\log_{10} \frac{1}{10}$;
(*d*) $\log_{10} 10^6$; (*e*) $\log_{10} 1$; (*f*) $\log_{10} \frac{1}{1000}$.

5 Copy and use the graphs of Figure 3 to complete and obtain a value for $3\cdot 5 \times 14$.

x	$\rightarrow$	$\log_2 x$
$3\cdot 5$	$\rightarrow$	
14	$\rightarrow$	
	$\leftarrow$	
2^x	$\leftarrow$	x

$3\cdot 5 \times 14 \approx$

6 Use the graphs of Figure 3 to work out the following (you are advised to use a layout similar to that in the last question):

(*a*) $4\cdot 5 \times 11$; (*b*) $9 \times 6\cdot 5$; (*c*) $7\cdot 5^2$.

7 What can you say about $\log_n 1$ if n is *any* natural number?

3 Logarithms to base 10

(*a*) From now on we shall use the particular logarithm function $x \rightarrow \log_{10} x$. The reason for this should become clearer as you work through the rest of the chapter. It is important to remember that $x \rightarrow \log_{10} x$ is the inverse of $x \rightarrow 10^x$.

What is (i) $\log_{10} 100$ and (ii) $\log_{10} 10$?

(*b*) In Section 2 we used graphs of the functions $x \to 2^x$ and $x \to \log_2 x$. For convenience and accuracy we shall now use a three-figure table instead of graphs.

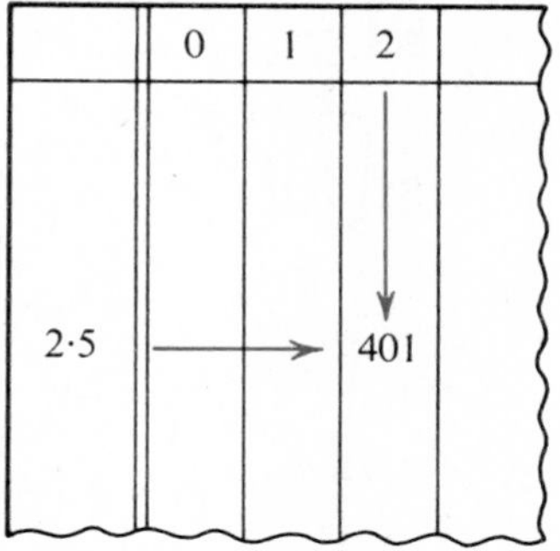

Fig. 4

Figure 4 shows how the pair $2{\cdot}52 \to 0{\cdot}401$ appears in the table. This means that

$$\log_{10} 2{\cdot}52 \approx 0{\cdot}401.$$

The table can also be used for the inverse function $x \to 10^x$.

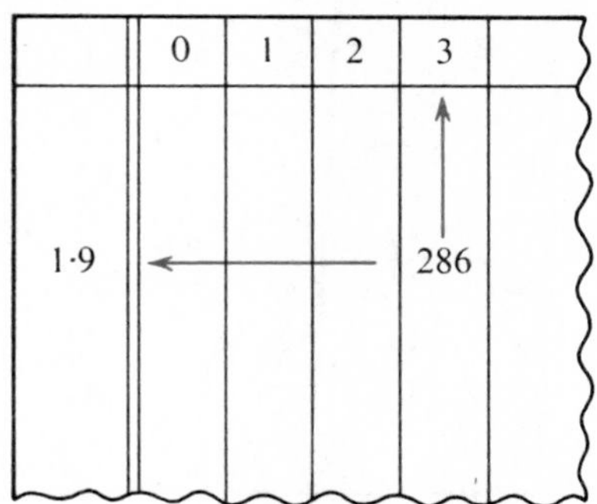

Fig. 5

Figure 5 shows the pair $0{\cdot}286 \to 1{\cdot}93$, that is

$$10^{0{\cdot}286} \approx 1{\cdot}93.$$

(*c*) A more complete extract of the three-figure table is shown in Figure 6.

	0	1	2	3	4	5	6	7	8	9
1·0	0·000	004	009	013	017	021	025	029	033	037
1·1	0·041	045	049	053	057	061	064	068	072	076
1·2	0·079	083	086	090	093	097	100	104	107	111
1·3	0·114	117	121	124	127	130	134	137	140	143
1·4	0·146	149	152	155	158	161	164	167	170	173
1·5	0·176	179	182	185	188	190	193	196	199	201
1·6	0·204	207	210	212	215	217	220	223	225	228
1·7	0·230	233	236	238	241	243	246	248	250	253
1·8	0·255	258	260	262	265	267	270	272	274	276
1·9	0·279	281	283	286	288	290	292	294	297	299
2·0	0·301	303	305	307	310	312	314	316	318	320
2·1	0·322	324	326	328	330	332	334	336	338	340
2·2	0·342	344	346	348	350	352	354	356	358	360
2·3	0·362	364	365	367	369	371	373	375	377	378
2·4	0·380	382	384	386	387	389	391	393	394	396
2·5	0·398	400	401	403	405	407	408	410	412	413
2·6	0·415	417	418	420	422	423	425	427	428	430
2·7	0·431	433	435	436	438	439	441	442	444	446
2·8	0·447	449	450	452	453	455	456	458	459	461
2·9	0·462	464	465	467	468	470	471	473	474	476
3·0	0·477	479	480	481	483	484	486	487	489	490
3·1	0·491	493	494	496	497	498	500	501	502	504
3·2	0·505	507	508	509	511	512	513	515	516	517
3·3	0·519	520	521	522	524	525	526	528	529	530
3·4	0·531	533	534	535	537	538	539	540	542	543
3·5	0·544	545	547	548	549	550	551	553	554	555
3·6	0·556	558	559	560	561	562	563	565	566	567
3·7	0·568	569	571	572	573	574	575	576	577	579
3·8	0·580	581	582	583	584	585	587	588	589	590
3·9	0·591	592	593	594	595	597	598	599	600	601
4·0	0·602	603	604	605	606	607	609	610	611	612
4·1	0·613	614	615	616	617	618	619	620	621	622
4·2	0·623	624	625	626	627	628	629	630	631	632
4·3	0·633	634	635	636	637	638	639	640	641	642
4·4	0·643	644	645	646	647	648	649	650	651	652
4·5	0·653	654	655	656	657	658	659	660	661	662
4·6	0·663	664	665	666	667	667	668	669	670	671
4·7	0·672	673	674	675	676	677	678	679	679	680
4·8	0·681	682	683	684	685	686	687	688	688	689
4·9	0·690	691	692	693	694	695	695	696	697	698
5·0	0·699	700	701	702	702	703	704	705	706	707
5·1	0·708	708	709	710	711	712	713	713	714	715
5·2	0·716	717	718	719	719	720	721	722	723	723
5·3	0·724	725	726	727	728	728	729	730	731	732
5·4	0·732	733	734	735	736	736	737	738	739	740

Fig. 6

Look carefully at the table and confirm that $\log_{10} 2{\cdot}3 \approx 0{\cdot}362$ and that $\log_{10} 1{\cdot}76 \approx 0{\cdot}246$. What are the values of:

(i) $\log_{10} 3{\cdot}5$; (ii) $\log_{10} 2{\cdot}47$?

By using the table as in Figure 5 we can find values of 10^x. For example $10^{0 \cdot 398} \approx 2{\cdot}5$ and $10^{0 \cdot 274} \approx 1{\cdot}88$.

Find, from the table, the values of (iii) $10^{0 \cdot 279}$ and (iv) $10^{0 \cdot 365}$.

(*d*) The method of using a logarithm function to carry out a multiplication has already been shown in detail (Example 1 in Section 2(*d*)) using the function $x \rightarrow \log_2 x$. The next two examples show a multiplication and a division carried out using logarithms to base ten.

Example 2

Calculate $2{\cdot}3 \times 3{\cdot}5$.

x		$\log_{10} x$
2·3	→	0·362
3·5	→	0·544
8·05	←	0·906
10^x	←	x

$10^{0{\cdot}362} \approx 2{\cdot}3$
$10^{0{\cdot}544} \approx 3{\cdot}5$
$10^{0{\cdot}906} \approx 8{\cdot}05$

The answer is 8·05.

Example 3

Calculate $2{\cdot}47 \div 1{\cdot}76$.

x		$\log_{10} x$
2·47	→	0·393
1·76	→	0·246
1·40	←	0·147
10^x	←	x

$10^{0{\cdot}393} \approx 2{\cdot}47$
$10^{0{\cdot}246} \approx 1{\cdot}76$
$10^{0{\cdot}147} \approx 1{\cdot}40$

The answer is 1·40. (There is no entry in the table corresponding to $10^{0{\cdot}147}$. Do you agree that we have taken the nearer of the two values on either side, $10^{0{\cdot}146}$ and $10^{0{\cdot}149}$?)

Exercise C

1 Use a table of logarithms to find:

(*a*) $\log_{10} 2{\cdot}7$, $\log_{10} 5{\cdot}37$, $\log_{10} 9{\cdot}32$, $\log_{10} 7{\cdot}89$, $\log_{10} 6{\cdot}04$;
(*b*) $10^{0{\cdot}380}$, $10^{0{\cdot}555}$, $10^{0{\cdot}486}$, $10^{0{\cdot}817}$, $10^{0{\cdot}954}$;
(*c*) the following numbers as powers of 10:

4·2, 6·18, 1·09, 2·17, 9·82.

2 Use a table of logarithms to calculate:

(*a*) $1{\cdot}5 \times 1{\cdot}4$; (*b*) $1{\cdot}6 \times 3{\cdot}5$; (*c*) $2{\cdot}71 \times 3{\cdot}63$;
(*d*) $3{\cdot}58 \times 2{\cdot}14$; (*e*) $5{\cdot}31 \times 1{\cdot}47$; (*f*) $5{\cdot}99 \times 1{\cdot}09$;
(*g*) $2{\cdot}82^2$; (*h*) $2{\cdot}17 \times 1{\cdot}85 \times 1{\cdot}20$.

3 Use a table of logarithms to calculate:

(*a*) $9{\cdot}8 \div 3{\cdot}5$; (*b*) $8{\cdot}5 \div 3{\cdot}4$; (*c*) $7{\cdot}46 \div 1{\cdot}70$;
(*d*) $9{\cdot}03 \div 8{\cdot}62$; (*e*) $6{\cdot}21 \div 3{\cdot}27$; (*f*) $5{\cdot}82 \div 5{\cdot}31$.

4 Use a table of logarithms to see if the following are correct:

(*a*) $\log_{10} 2{\cdot}5 + \log_{10} 3{\cdot}0 = \log_{10} 7{\cdot}5$;
(*b*) $\log_{10} 4{\cdot}0 + \log_{10} 1{\cdot}7 = \log_{10} 6{\cdot}8$.

Can you explain your findings?

4 Extending the domain to include numbers greater than 10

We can read directly from the table the value of $\log_{10}x$ provided $1 \leqslant x \leqslant 10$ and this has been sufficient for the calculations up to now. For general use, however, we must find a way of overcoming this restriction. The work of this section indicates a simple way of using the properties of the function $x \to \log_{10}x$ to find the logarithm of numbers greater than ten.

For example, what is the value of $\log_{10} 27$? We can write

$$27 = 2{\cdot}7 \times 10 \text{ and say that}$$

$$\begin{aligned}\log_{10} 27 &= \log_{10}(2{\cdot}7 \times 10)\\ &= \log_{10} 2{\cdot}7 + \log_{10} 10\\ &\approx 0{\cdot}431 + 1\\ &= 1{\cdot}431.\end{aligned}$$

This process is shown in the mapping diagram of Figure 7.

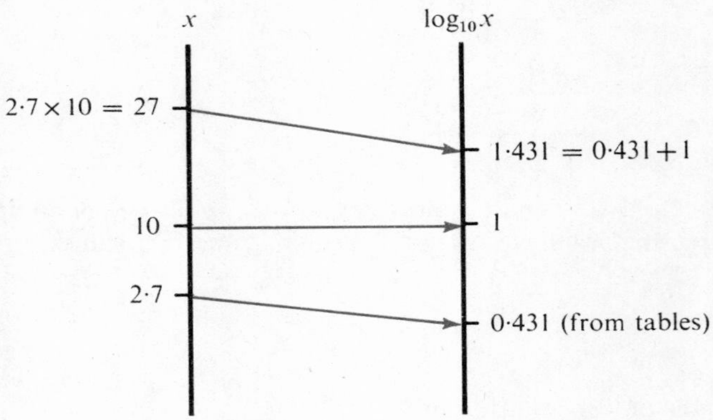

Fig. 7

What is (i) $\log_{10} 36$ and (ii) $\log_{10} 29{\cdot}5$?

To find $\log_{10} 270$, we can write 270 as $2{\cdot}7 \times 100$.

So
$$\begin{aligned}\log_{10} 270 &= \log_{10}(2{\cdot}7 \times 10^2)\\ &= \log_{10} 2{\cdot}7 + \log_{10} 100\\ &\approx 0{\cdot}431 + 2\\ &= 2{\cdot}431.\end{aligned}$$

Notice that we have, in effect, made use of standard index form to find the logarithms of numbers greater than 10.

x	standard index form	$\log_{10} x$
2·7	$2{\cdot}7 \times 10^0$	0·431
27	$2{\cdot}7 \times 10^1$	1·431
270	$2{\cdot}7 \times 10^2$	2·431

Example 4

Find (i) $\log_{10} 57{\cdot}3$; (ii) $\log_{10} 4650$.

(i) $$57{\cdot}3 = 5{\cdot}73 \times 10,$$

so $$\log_{10} 57{\cdot}3 = \log_{10} 5{\cdot}73 + \log_{10} 10 \approx 0{\cdot}758 + 1 = 1{\cdot}758.$$

(ii) $$4650 = 4{\cdot}650 \times 1000,$$

so $$\log_{10} 4650 = \log_{10} 4{\cdot}65 + \log_{10} 1000 \approx 0{\cdot}667 + 3 = 3{\cdot}667.$$

Example 5

Find (i) $10^{2{\cdot}721}$; (ii) x if $\log_{10} x = 1{\cdot}774$.

(i) $10^{2{\cdot}721} = 10^2 \times 10^{0{\cdot}721}$
$\approx 100 \times 5{\cdot}26$ (using the table as in Figure 5)
$= 526.$

(ii) If $\log_{10} x = 1{\cdot}774$

then $$x = 10^{1{\cdot}774} = 10^1 \times 10^{0{\cdot}774} \approx 10 \times 5{\cdot}94 = 59{\cdot}4.$$

We are now in a position to use logarithms for calculations involving numbers greater than 10. Look carefully at Example 6 and remember that only the part of the logarithm to the right of the decimal point comes from the table. The 'standard index form' column should help in the early stages but you may find that, with experience, you can safely omit it.

Example 6

Calculate (i) $46{\cdot}3 \times 162$; (ii) $2370 \div 59{\cdot}3$.

x		standard index form	$\log_{10} x$
46·3	=	$46{\cdot}3 \times 10^1$	→ 1·666
162	=	$1{\cdot}62 \times 10^2$	→ 2·210
7510	=	$7{\cdot}51 \times 10^3$	← 3·876

$$46{\cdot}3 \times 162 \approx 7510$$

(ii)

x		standard index form	$\log_{10} x$
2370	=	$2{\cdot}37 \times 10^3$	→ 3·375
59·3	=	$5{\cdot}93 \times 10^1$	→ 1·773
40	=	$4{\cdot}00 \times 10$	← 1·602

$$2370 \div 59{\cdot}3 \approx 40$$

Exercise D

1 Find:

(*a*) $\log_{10} 43$; (*b*) $\log_{10} 430$; (*c*) $\log_{10} 4300$;
(*d*) $\log_{10} 38{\cdot}2$; (*e*) $\log_{10} 3820$; (*f*) $\log_{10} 87\,600$.

2 Find:

(*a*) $10^{1\cdot568}$; (*b*) $10^{1\cdot752}$; (*c*) $10^{2\cdot752}$; (*d*) $10^{4\cdot909}$.

3 Calculate:

(*a*) $6{\cdot}71 \times 11{\cdot}9$; (*b*) $28{\cdot}7 \times 9{\cdot}2$; (*c*) 303×29;
(*d*) $7{\cdot}62 \times 8310$; (*e*) $53{\cdot}6 \times 8{\cdot}4 \times 32{\cdot}7$; (*f*) $17{\cdot}2^2$.

4 Calculate:

(*a*) $426 \div 58$; (*b*) $3970 \div 84{\cdot}8$; (*c*) $44{\cdot}7 \div 19{\cdot}8$; (*d*) $947 \div 1{\cdot}82$.

5 (*a*) If $\log_{10} x = 2{\cdot}537$, find the value of x.
(*b*) If $\log_{10} 25{\cdot}6 = y$, find the value of y.

6 (*a*) Describe how you found the value of $17{\cdot}2^2$ in 3(*f*).
(*b*) Calculate $17{\cdot}2^3$ using logarithms.
(*c*) Give a general method for finding the power of a number using logarithms.

7 Calculate (using logarithms for multiplication and division):

(*a*) $582 - (43{\cdot}2 \times 5{\cdot}26)$; (*b*) $79{\cdot}3 + \frac{48{\cdot}1}{27{\cdot}6}$; (*c*) $\frac{79{\cdot}3 + 48{\cdot}1}{27{\cdot}6}$;
(*d*) $24{\cdot}7^2 - (7{\cdot}81 \times 3{\cdot}92)$.

8 (*a*) If $10^x = 762$, find the value of x.
(*b*) If $10^{3\cdot861} = y$, find the value of y.

9 (*a*) Make a rough estimate of $\sqrt{650}$.
(*b*) Work out $\sqrt{650}$ using logarithms (it may help to consider how to *square* a number using logarithms).
(*c*) Check that your answers to (*a*) and (*b*) agree.

5 Extending the domain to include numbers between 0 and 1

In this section, we consider $x \to \log_{10} x$ for $0 < x < 1$. Having done this, we shall then be in a position to use logarithms for calculations involving positive numbers of any size. In fact we already know some particular values of $\log_{10} x$ for $0 < x < 1$.

x	10^{-3} or $0{\cdot}001$	10^{-2} or $0{\cdot}01$	10^{-1} or $0{\cdot}1$	1	10	100	1000
↓	↓	↓	↓	↓	↓	↓	↓
$\log_{10} x$	$^-3$	$^-2$	$^-1$	0	1	2	3

What is (i) $\log_{10} 0{\cdot}1$ and (ii) $\log_{10} 0{\cdot}001$?

What is the value of $\log_{10} 0{\cdot}06$? We can find out by expressing 0·06 in standard index form:

$$0{\cdot}06 = 6 \times 10^{-2} = 6 \times 0{\cdot}01.$$

So
$$\begin{aligned}\log_{10} 0{\cdot}06 &= \log_{10}(6 \times 0{\cdot}01)\\ &= \log_{10} 6 + \log_{10} 0{\cdot}01\\ &\approx 0{\cdot}778 + {}^{-}2.\end{aligned}$$

Example 7

Find (i) $\log_{10} 0{\cdot}784$; (ii) $\log_{10} 0{\cdot}0062$.

(i)
$$0{\cdot}784 = 7{\cdot}84 \times 10^{-1} = 7{\cdot}84 \times 0{\cdot}1.$$

So
$$\begin{aligned}\log_{10} 0{\cdot}784 &= \log_{10}(7{\cdot}84 \times 0{\cdot}1)\\ &= \log_{10} 7{\cdot}84 + \log 0{\cdot}1\\ &\approx 0{\cdot}894 + {}^{-}1.\end{aligned}$$

(ii)
$$0{\cdot}0062 = 6{\cdot}2 \times 10^{-3} = 6{\cdot}2 \times 0{\cdot}001.$$

So
$$\begin{aligned}\log_{10} 0{\cdot}0062 &= \log_{10}(6{\cdot}2 \times 0{\cdot}001)\\ &= \log_{10} 6{\cdot}2 + \log_{10} 0{\cdot}001\\ &\approx 0{\cdot}792 + {}^{-}3.\end{aligned}$$

In practice, however, we can avoid the appearance of negative numbers. Example 8 shows how this can be done in a multiplication by using standard index form.

Example 8

Calculate $159 \times 0{\cdot}076$.

$$\begin{aligned}159 \times 0{\cdot}076 &= 159 \times 7{\cdot}6 \times 10^{-2}\\ &\approx 1210 \times 10^{-2}\\ &= 12{\cdot}1\end{aligned}$$

x	$\rightarrow$	$\log_{10} x$
159	$\rightarrow$	2·201
7·6	$\rightarrow$	0·881
1210	$\leftarrow$	3·082
10^x	$\leftarrow$	x

$$159 \times 0{\cdot}076 \approx 12{\cdot}1.$$

Even if the numbers in a division are both greater than one, a difficulty can arise if the answer is less than one.

Example 9 shows how we can avoid negative numbers in this situation.

Example 9

Calculate $2{\cdot}7 \div 5{\cdot}3$.

If $\quad 2{\cdot}7 \div 5{\cdot}3 = x$

then $\quad 27 \div 5{\cdot}3 = 10x$.

So
$$\begin{aligned}2{\cdot}7 \div 5{\cdot}3 &= (27 \div 5{\cdot}3) \div 10\\ &\approx (5{\cdot}09) \div 10\\ &= 0{\cdot}509\end{aligned}$$

x	$\rightarrow$	$\log_{10} x$
27	$\rightarrow$	1·431
5·3	$\rightarrow$	0·724
5·09	$\leftarrow$	0·707
10^x	$\leftarrow$	x

$$2{\cdot}7 \div 5{\cdot}3 \approx 0{\cdot}509.$$

Exercise E

1 Copy and complete:

(*a*) Calculate $222 \times 0{\cdot}417$.

$$222 \times 0{\cdot}417 = 222 \times 4{\cdot}17 \times 10^{-1}$$
$$\approx \quad \times 10^{-1}$$
$$=$$

x	$\to \log_{10} x$
222	$\to$ 2·346
4·17	$\to$ 0·620
	$\leftarrow$ 2·966
10^x	$\leftarrow x$

$$222 \times 0{\cdot}417 \approx$$

(*b*) Calculate $43{\cdot}9 \times 0{\cdot}0171$.

$$43{\cdot}9 \times 0{\cdot}0171 = 43{\cdot}9 \times 1{\cdot}71 \times 10^{-2}$$
$$\approx$$
$$=$$

x	$\to \log_{10} x$
43·9	$\to$
1·71	$\to$
	$\leftarrow$
10^x	$\leftarrow x$

$$43{\cdot}9 \times 0{\cdot}0171 \approx$$

2 Calculate:

(*a*) $342 \times 0{\cdot}714$; (*b*) $0{\cdot}131 \times 69{\cdot}3$;
(*c*) $8{\cdot}51 \times 0{\cdot}37$; (*d*) $0{\cdot}81 \times 0{\cdot}62$;
(*e*) $0{\cdot}00481 \times 72{\cdot}4$; (*f*) $0{\cdot}0641 \times 0{\cdot}491$.

3 Copy and complete:

(*a*) Calculate $4{\cdot}18 \div 13{\cdot}3$.

$$4{\cdot}18 \div 13{\cdot}3 = (41{\cdot}8 \div 13{\cdot}3) \div 10$$
$$\approx \quad \div 10$$
$$=$$

x	$\to \log_{10} x$
41·8	$\to$ 1·621
13·3	$\to$ 1·124
	$\leftarrow$ 0·497
10^x	$\leftarrow x$

$$4{\cdot}18 \div 13{\cdot}3 \approx$$

(*b*) Calculate $24{\cdot}1 \div 321$.

$$24{\cdot}1 \div 321 = (2410 \div 321) \div 10^2$$
$$\approx$$
$$=$$

x	$\to \log_{10} x$
2410	$\to$
321	$\to$
	$\leftarrow$
10^x	$\leftarrow x$

$$24{\cdot}1 \div 321 \approx$$

4 Calculate:

(*a*) $3{\cdot}97 \div 23{\cdot}8$; (*b*) $\dfrac{5{\cdot}97}{0{\cdot}341}$;
(*c*) $\dfrac{240}{0{\cdot}14}$; (*d*) $0{\cdot}073 \div 19$;
(*e*) $0{\cdot}073 \div 0{\cdot}511$; (*f*) $129 \div 0{\cdot}00306$.

Exercise F (Miscellaneous)

1 Find the value of $\dfrac{64 \times 12{\cdot}7}{5{\cdot}9}$ by completing the following:

x		$\log_{10} x$
64	$\rightarrow$	
12·7	$\rightarrow$	
6·4 × 12·7	$\rightarrow$	
5·9	$\rightarrow$	
	$\leftarrow$	
10^x	$\leftarrow$	x

$$\frac{64 \times 12{\cdot}7}{5{\cdot}9} \approx$$

2 Calculate:

(*a*) $\dfrac{247 \times 2{\cdot}38}{59{\cdot}2}$; (*b*) $\dfrac{374}{11{\cdot}2 \times 9{\cdot}81}$.

3 (*a*) Work out $7{\cdot}93^3$ using logarithms.
(*b*) Devise a way of working out 39^4 using logarithms.

4 Calculate:

(*a*) $\dfrac{47{\cdot}3 \times 10{\cdot}7}{15{\cdot}2 \times 28{\cdot}3}$; (*b*) $\dfrac{72{\cdot}8^2}{572}$; (*c*) $\dfrac{4{\cdot}62^2}{29{\cdot}2}$;

(*d*) $\dfrac{7{\cdot}23 - 4{\cdot}89}{13}$; (*e*) $(0{\cdot}48 \times 3{\cdot}21) - 0{\cdot}5$; (*f*) $\pi \times 3{\cdot}49^2$.

5 (*a*) Find the area of a circle which has a radius of 17·2 cm.
(*b*) Find the radius of a circle which has an area of 1470 cm^2.

6 The value given for $\log_{10} 2$ in five-figure tables is 0·30103.
Calculate, giving five decimal places:

(*a*) $\log_{10} 20$; (*b*) $\log_{10} 4$; (*c*) $\log_{10} 32$; (*d*) $\log_{10} 5$.

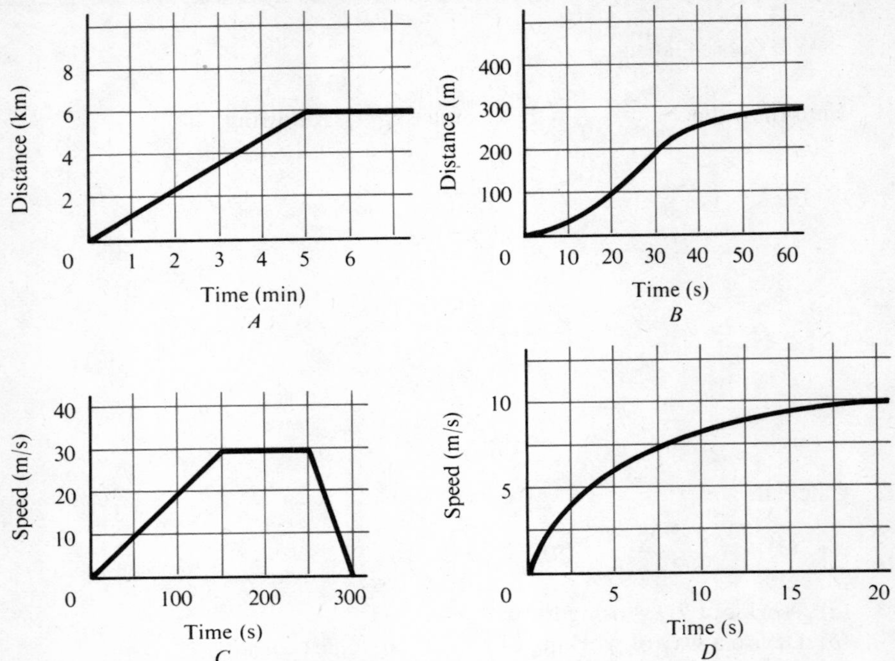

3 Speed and Acceleration

1 Distance–time graphs

The graphs above give information about four different journeys.

(*a*) Distance–time graph *A* describes the journey of a car which travels at a steady speed for 5 minutes and then stops. What are the gradients of the graph

(i) for the time interval 0 to 5,

(ii) for the time interval greater than 5?

Calculate, in metres/second, the car's steady speed.

(*b*) The progress of a cyclist during 1 minute is shown in graph *B*. What is the cyclist's average speed in metres/second over this period of time? Remembering that the gradient of a distance–time graph measures speed at an instant, can you say at what time the cyclist reaches his maximum speed? Explain why graph *B* is more realistic than graph *A*.

Exercise A

1 Sketch a possible distance–time graph of a car journey assuming that the car travels at a constant speed except when it is held up by traffic lights.

2 Draw an accurate distance–time graph given the following details of a journey:

(i) a steady speed of 15 m/s for 3 minutes;
followed by (ii) a $\frac{1}{2}$ minute stop;
followed by (iii) a steady speed of 10 m/s for 4 minutes.

3 The distance travelled at various times on a car journey was recorded as follows:

Time	10.00	10.15	10.30	10.45	11.00	11.15	11.30	11.45	12.00
Distance (km)	0	10	20	33	58	80	98	116	120

Draw a distance–time graph and estimate:

(*a*) the average speed for the whole journey;
(*b*) the average speed between 10.30 and 11.30;
(*c*) the average speed from when the car was 30 km from the start to when it was 90 km from the start.

4 The times at which kilometre posts were passed during a train journey were recorded as follows:

Distance (km)	0	10	20	30	40	45	50	60	70	80
Time (hours)	0	0·17	0·26	0·32	0·42	0·53	0·75	0·86	0·91	1·0

Draw a distance–time graph and estimate:

(*a*) the average speed for the first half-hour of the journey;
(*b*) the speed when the train was 40 km away from the start;
(*c*) the maximum speed.

5

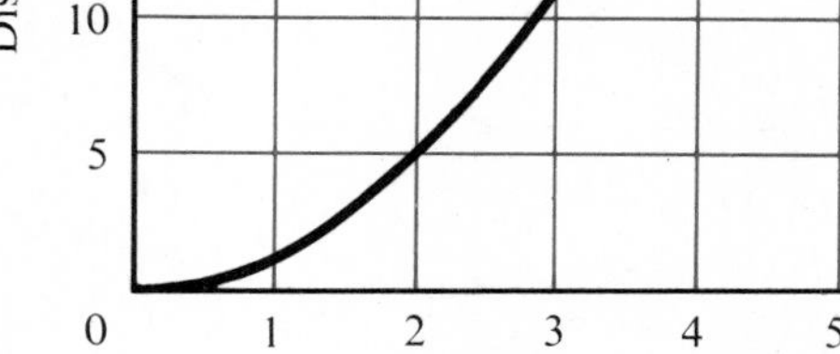

Fig. 1

Make an accurate copy of the graph in Figure 1 and use it to estimate:

(*a*) the speed at 4 seconds;
(*b*) the time when the speed is 5 m/s.

2 Acceleration

(*a*) The speed–time graph *C* at the beginning of this chapter refers to a five-minute train journey between two stations. Do you agree that for the first 150 seconds, the train's speed increases steadily from 0 to 30 m/s? Describe the remainder of the journey in similar terms.

(*b*) The gradient of graph *C* for $0 < t < 150$ is $\frac{30}{150}$ or 0·2. This means that the speed of the train is initially increasing by 0·2 m/s every second. We say that the train has an *acceleration of 0·2 metres per second per second* and write this in short as

$$0{\cdot}2 \text{ m/s}^2 \quad \text{or} \quad 0{\cdot}2 \text{ ms}^{-2}.$$

Acceleration is a measure of rate of change of speed with time. It is derived from the basic quantities length and time and has the dimensions

$$\frac{L}{T^2} \quad \text{or} \quad LT^{-2}.$$

(*c*) For the period $150 < t < 250$, graph *C* is parallel to the time axis. What is the acceleration during this period? Do you agree that the gradient of graph *C* for $250 < t < 300$ is $\frac{-30}{50}$ or $^{-}0{\cdot}6$? During this time interval the train is slowing down steadily at a rate of 0·6 m/s². We describe this as an acceleration of $^{-}0{\cdot}6$ m/s² or a deceleration of 0·6 m/s².

(*d*) Graph *D* is a more realistic speed–time graph. What is the average acceleration over the 20-second interval? At what time is the acceleration greatest?

Exercise B

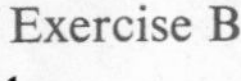

1

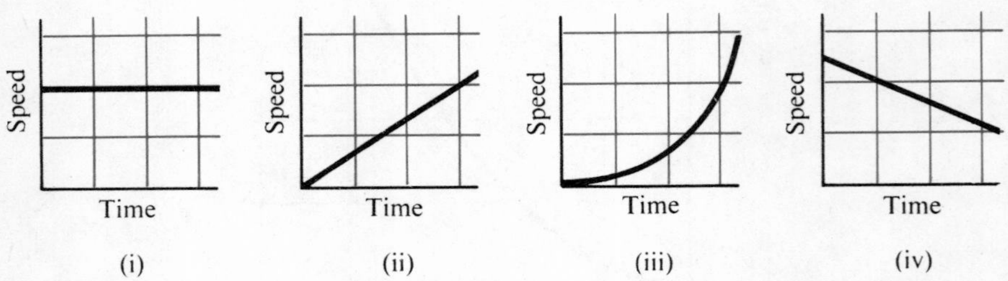

Acceleration

Time

Speed Time (i)

Speed Time (ii)

Speed Time (iii)

Speed Time (iv)

Fig. 2

Figure 2 shows an acceleration–time graph and four speed–time graphs. Which speed–time graph could refer to the same journey as the acceleration–time graph?

2 What is the acceleration of a train which:

(*a*) increases its speed steadily from 0 to 20 m/s in 200 seconds;
(*b*) increases its speed steadily from 15 m/s to 24 m/s in 2 minutes;
(*c*) slows to rest steadily from 40 m/s in a time of 110 seconds?

3 A train starts from rest and travels with constant acceleration until the power is cut off and the brakes applied. It starts to decelerate at a constant rate until it comes to rest again. Sketch a speed-time graph.

4 Draw an accurate acceleration–time graph for the train journey described by graph *C*.

3 Areas under graphs

(*a*) We have already seen in *Book X*, Chapter 12 that the area under a speed–time graph is a measure of distance. Using Figure 3, which is an enlarged version of graph *C*, we can find how far the train travelled.

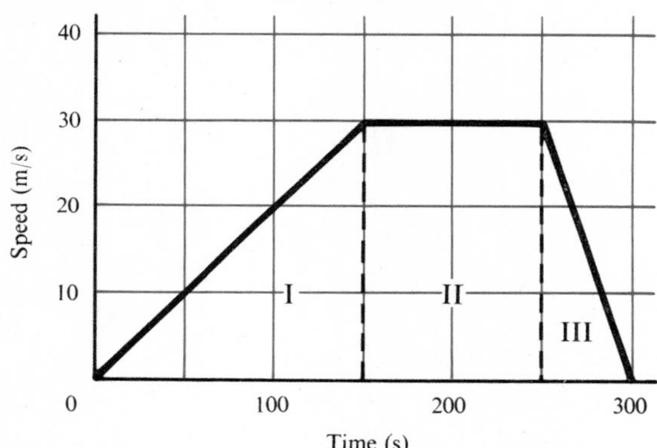

Fig. 3

$$\begin{aligned} \text{Area I} &= \tfrac{1}{2} \times 150 \times 30 = 2250 \\ \text{Area II} &= 100 \times 30 = 3000 \\ \text{Area III} &= \tfrac{1}{2} \times 50 \times 30 = 750 \end{aligned}$$

The area under the graph is 6000 square units, showing that the train travelled 6000 metres or 6 km.

(*b*) Figure 4 is the acceleration–time graph of a missile moving with a constant acceleration of 2 m/s^2 for a period of 10 seconds.

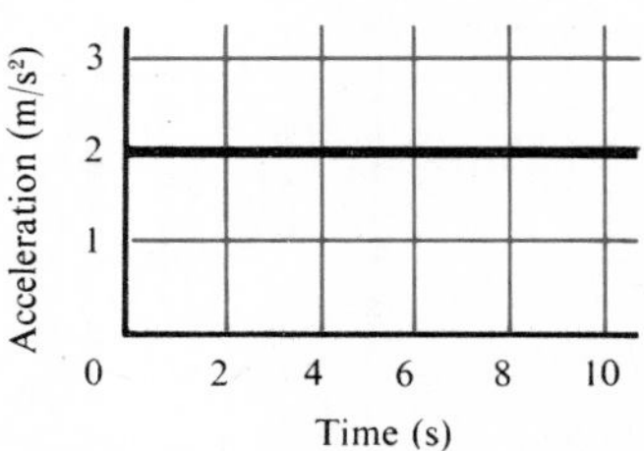

Fig. 4

Do you agree that over the 10-second interval the speed of the missile has increased by 20 m/s ? The area under the graph is 20 square units and this suggests that the area under an acceleration–time graph provides a measure of the increase in speed over the interval.

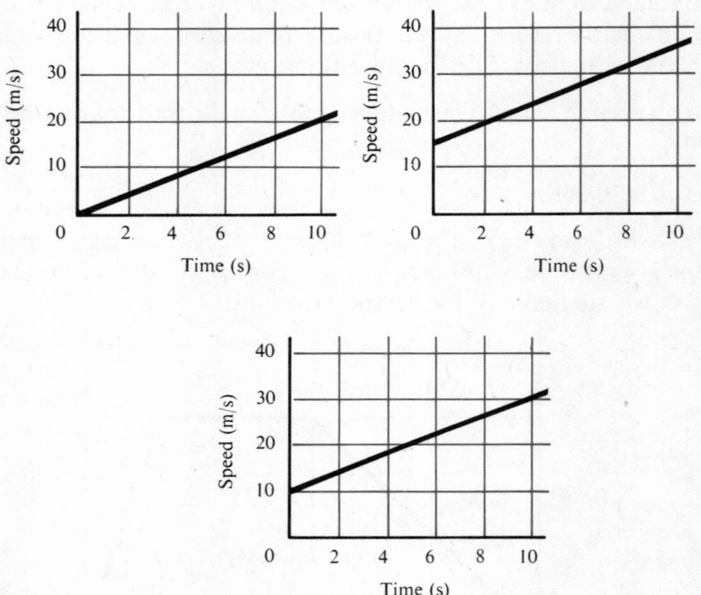

Fig. 5

The three speed–time graphs in Figure 5 all show an increase in speed of 20 m/s gained steadily during a 10-second period. That is, they show an acceleration of 2 m/s². Any one of them could be the speed–time graph of the missile.

Exercise C

1 Sketch three possible speed–time graphs which correspond to each of the graphs in Figure 6.

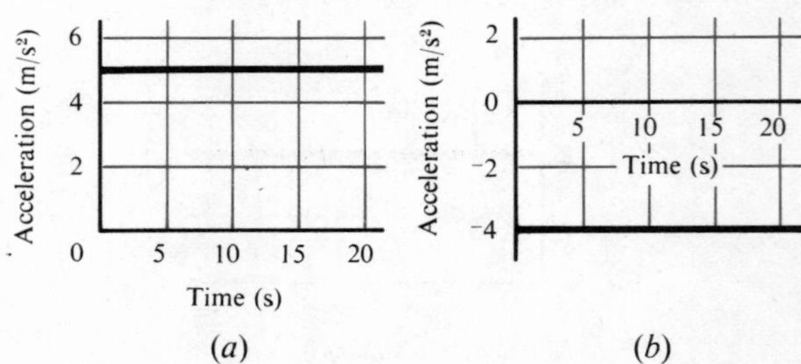

Fig. 6

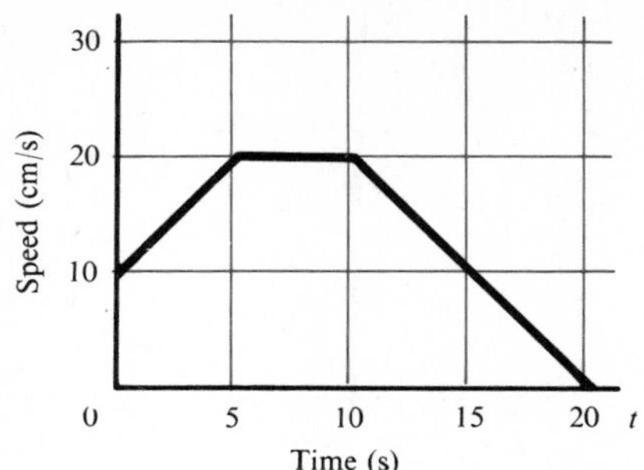

Fig. 7

2 Figure 7 shows the speed–time graph of a particle moving for 20 seconds. Which of the following statements are true?

(*a*) The initial speed is 10 cm/s.
(*b*) The total distance covered is 225 cm.
(*c*) The acceleration for $0 < t < 5$ is 2 cm/s^2.
(*d*) The acceleration for $10 < t < 20$ is 1 cm/s^2.

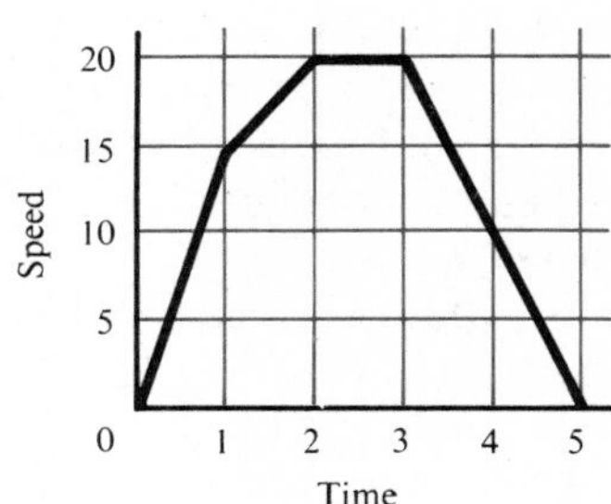

Fig. 8

3 Figure 8 is a speed–time graph; the unit of time is 1 s and the unit of speed is 1 m/s.

(*a*) Give a meaning to the reduction of gradient at $t = 1$ on the graph.
(*b*) Say what is happening from $t = 2$ to $t = 3$.
(*c*) Say what is happening at $t = 4$.
(*d*) Calculate the distance travelled between $t = 3$ and $t = 5$.

4 A train starts from rest and travels with constant acceleration until it reaches a speed of x m/s. It immediately starts to decelerate at a constant rate until it comes to rest. The total distance travelled is 810 metres and the time taken is 120 seconds.

(*a*) Copy Figure 9 and draw a graph of speed against time.

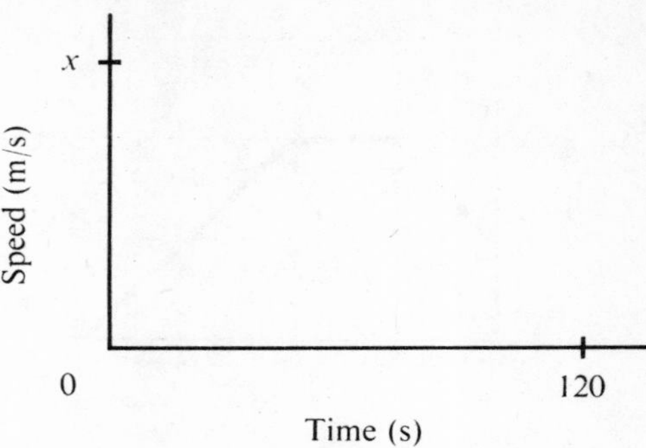

Fig. 9

(*b*) Calculate x.

5 Make an accurate copy of graph *D* and use the trapezium method with four trapeziums to estimate the distance travelled in the 20-second period.

4 Free fall

(*a*) In the late sixteenth century, Galileo conducted a series of experiments at Pisa. By dropping a variety of objects simultaneously from the top of the leaning tower he was able to show that they fell with a constant acceleration. Further, he demonstrated that the value of the acceleration was the same for all the objects dropped since they reached the ground together.

Since then, Galileo's experiment has been repeated many times with the same result. For example, it has been demonstrated that, in a vacuum, a feather and a ball-bearing fall at the same rate.

(*b*) Figure 10(*a*) shows the speed–time graph of a heavy object falling freely from rest.

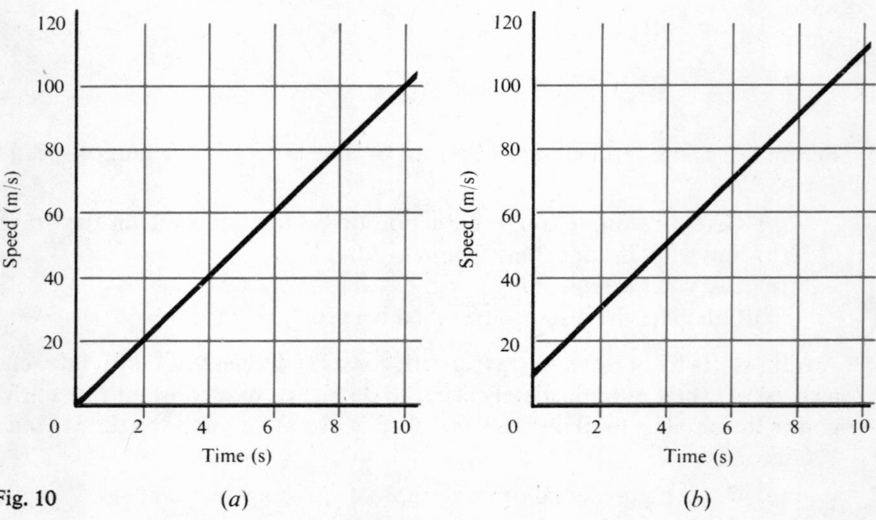

Fig. 10

Use it to find:

(i) the value of the constant acceleration;
(ii) the total distance fallen.

Figure 10(*b*) is the speed–time graph of a freely falling object which is thrown downwards with an initial speed of 10 m/s. In what respect are the two graphs alike?

Exercise D

1 How much further does the object referred to in Figure 10(*b*) travel compared with that in Figure 10(*a*)?

2 A stone dropped down a well travels with a constant acceleration of 9·8 m/s² and is heard to hit the water after 3 seconds.

(*a*) Draw an acceleration–time graph and use it to calculate the speed at which the stone hits the water.
(*b*) Draw a speed–time graph and use it to calculate the distance travelled by the stone before reaching the surface of the water.

3

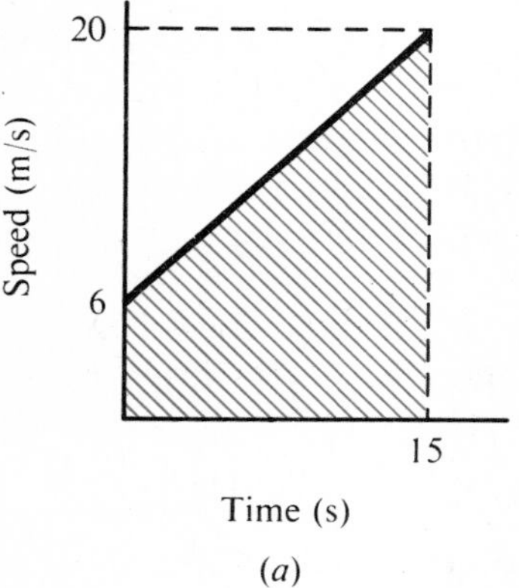

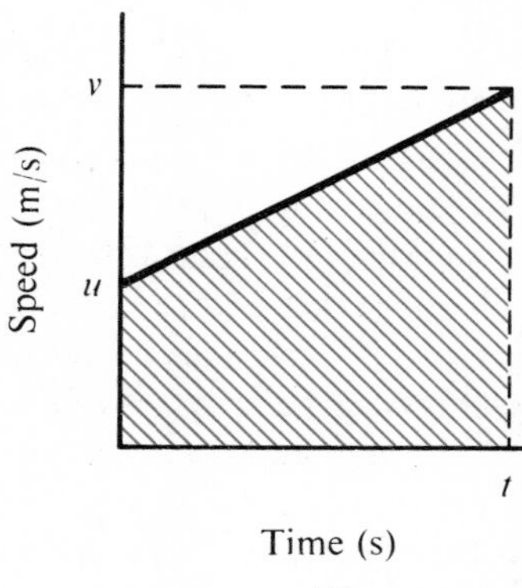

Fig. 11

(*a*) Figure 11(*a*) is the speed–time graph of a car accelerating constantly from a speed of 6 m/s to one of 20 m/s in 15 seconds. What distance has it travelled? (*b*) Figure 11(*b*) is the speed–time graph of a car accelerating constantly from a speed of u m/s to a speed of v m/s in t seconds. If it travels a distance of s metres in this time, find a formula giving s in terms of u, v and t.

4

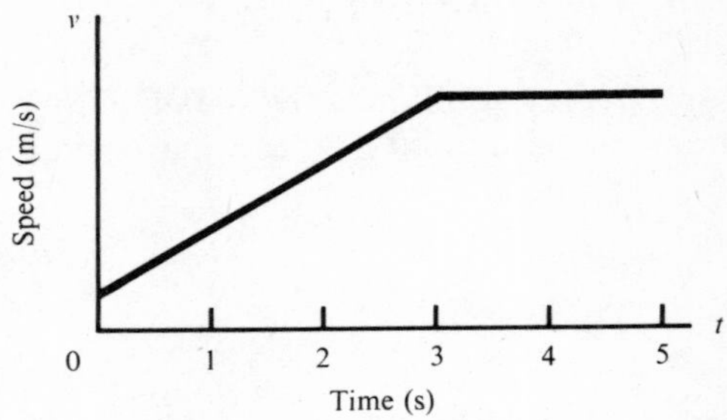

Fig. 12

The graph in Figure 12 shows the motion of a car for 5 seconds. During the first 3 seconds its speed v (m/s) after t seconds is given by the formula $v = 2t + 1$.

It then travels at constant speed.

(*a*) What is this constant speed?
(*b*) What is the acceleration during the first 3 seconds?
(*c*) What is the average speed over the first 2 seconds?
(*d*) How far does the car travel in the 5 seconds?

5 A main road $ABCD$ is crossed by minor roads at A, B, C and D. At each junction there are traffic lights. The lights all change together, remaining red for half a minute and then immediately green for half a minute, and so on. The distances AB, BC and CD are each 1 km.

(*a*) A motorist P passes A just after the lights go green. By drawing a distance–time graph, show that if he travels at 60 km/h he will not have to stop. Take 2 cm to represent both 1 km and 1 minute. State how long it takes before he passes D.
(*b*) On the same axes as your answer to (*a*) draw a distance–time graph showing the progress of a second motorist Q who leaves A at the same time as P and travels at 90 km/h when he can. State how long it takes before Q passes D.
(*c*) In order to pass D before P does it is necessary for Q to exceed a certain speed V km/h on at least one stage of the journey. Find the value of V.

6 A particle moving in a straight line has a speed v metres/second at time t seconds, where $v = 6\sin 15t°$. *Sketch* the speed–time graph for values of t from 0 to 24 and briefly describe the movement of the particle. Explain in particular what is happening at times 6 s and 12 s.

4 Invariance

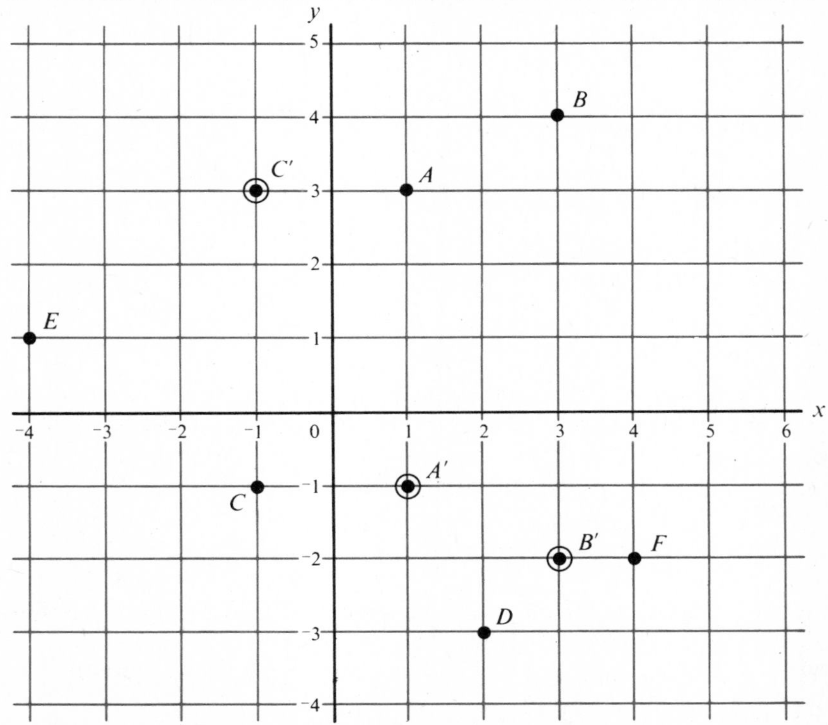

Fig. 1

1 Invariant points and lines of invariant points

(*a*) Figure 1 shows a set of points A, B, C, D, E and F in the (x,y) plane and the images A', B', C' under a certain reflection. Copy the diagram and complete it by drawing in the images of D, E and F.

Which of the points are mapped onto themselves by this transformation? Describe the set of points *of the* (x,y) *plane* which are mapped onto themselves. What is the equation of the line of reflection?

(*b*) Figure 2 shows another set of points and some of their images under an enlargement. Copy the diagram and complete it by marking in the missing images.

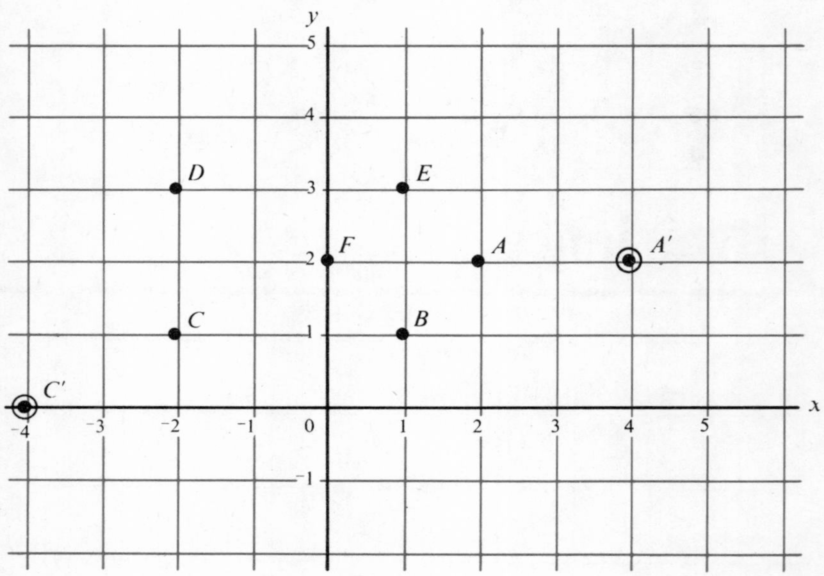

Fig. 2

Which of the points are mapped onto themselves?

Are there any more points of the (x,y) plane other than those labelled in the diagram which are mapped onto themselves by this transformation?

What are the coordinates of the centre of enlargement?

(*c*) Points which are mapped onto themselves by a transformation are said to be INVARIANT under the transformation.

The points on the line $y = 1$ in Figure 1 are invariant, and we can speak of the points on a line being invariant under a reflection. Which other simple transformations leave all points on a line invariant?

The only point which is invariant for an enlargement is the centre of enlargement (F in Figure 2).

Which other simple transformations leave exactly one point invariant?

(*d*) Figure 3 represents the image of the (x,y) plane after the translation $\begin{pmatrix} \frac{1}{2} \\ \frac{1}{2} \end{pmatrix}$. The image plane is shown in red. Which points are invariant under this transformation?

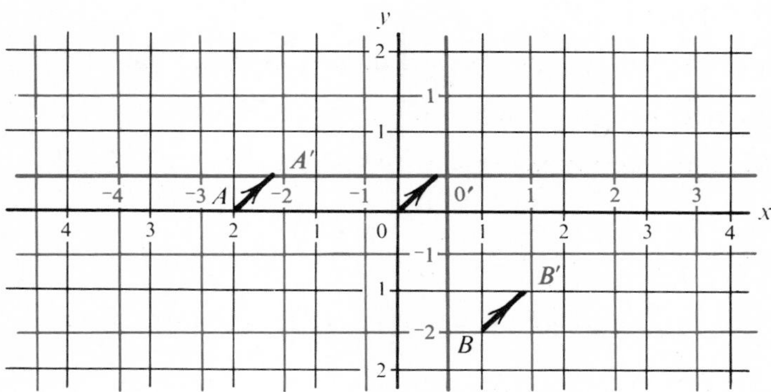

Fig. 3

Which other transformations leave no points invariant?

(*e*) Copy and complete Table 1 showing which transformations leave (i) no points (ii) one point (iii) the points on a line (iv) all points of the (x,y) plane invariant.

Transformation	No points	One point	Points on a line	All points
Reflection			✓	
Rotation				
Translation	✓			
Enlargement		✓		
One-way stretch				
Two-way stretch				
Shear				
Glide reflection	✓			
Identity				

TABLE 1

Summary

1 The invariant points under a transformation are those which are mapped onto themselves.

2 Rotations, enlargements and two-way stretches leave one and only one point invariant.

3 Reflections, one-way stretches and shears leave points on a line invariant.

4 Translations and glide reflections leave no points invariant.

5 The identity transformation leaves all points invariant.

Exercise A

1 Figure 4 shows a triangle T and its image T' after four different transformations.

(*a*) Describe the transformation in each case.
(*b*) Describe the set of invariant points *of the* (x,y) *plane* in each case.

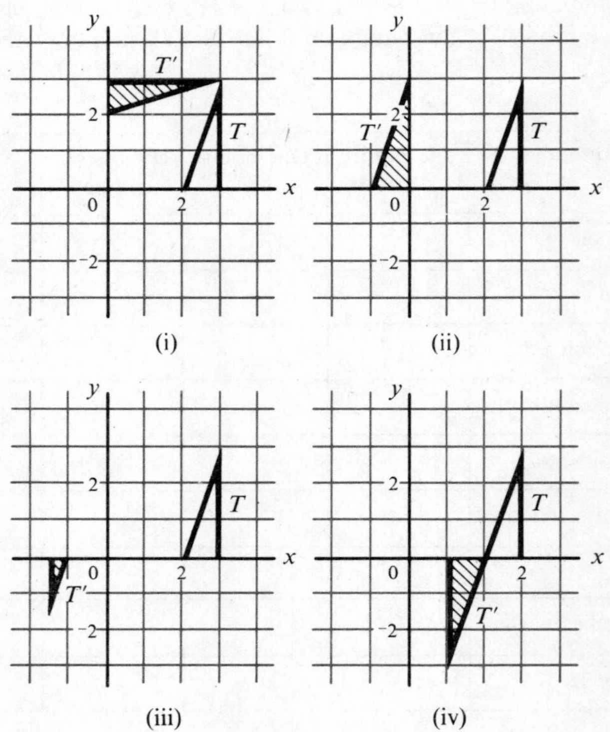

Fig. 4

2 Describe each of the transformations which

(i) leaves $y = 2$ invariant and maps $(1,0) \to (1,4)$;
(ii) leaves $x = 2$ invariant and maps $(3,0) \to (4,0)$;
(iii) leaves only $(1,1)$ invariant and maps $(2,1) \to (4,1)$;
(iv) leaves only $(1,1)$ invariant and maps $(2,1) \to (1,2)$.

3 Multiply $\begin{pmatrix} a & b \\ c & d \end{pmatrix}\begin{pmatrix} 0 \\ 0 \end{pmatrix}$. What does you answer suggest about all transformations which can be described by a matrix of this form?

4 Multiply $\begin{pmatrix} 0 & 1 \\ ^-1 & 2 \end{pmatrix}\begin{pmatrix} 2 \\ 2 \end{pmatrix}$. Interpret your answer in terms of invariance. Find two more points for which the same kind of result is true.

5 By drawing a diagram to show the images of several points of your own choice find the set of invariant points of the transformations which map

(i) $(x,y) \to (x, {}^-y)$;
(ii) $(x,y) \to ({}^-x, {}^-y)$;
(iii) $(x,y) \to (x,0)$;
(iv) $(x,y) \to (2x,2y)$;
(v) $(x,y) \to (x,y+2)$;
(vi) $(x,y) \to (y,x)$;
(vii) $(x,y) \to (x+1, y+1)$;
(viii) $(x,y) \to (x, 2y+2)$;
(ix) $(x,y) \to (2x+1, 1-y)$.

2 Finding invariant points and lines of invariant points algebraically

(*a*) We already know that all transformations which can be represented by a matrix of the form $\begin{pmatrix} a & b \\ c & d \end{pmatrix}$ leave the origin invariant (see Question 3, Exercise A).

We also know that the base vector $\begin{pmatrix} 1 \\ 0 \end{pmatrix}$ is mapped onto $\begin{pmatrix} a \\ c \end{pmatrix}$, $\begin{pmatrix} 0 \\ 1 \end{pmatrix}$ is mapped onto $\begin{pmatrix} b \\ d \end{pmatrix}$ and $\begin{pmatrix} 1 \\ 1 \end{pmatrix}$ is mapped onto $\begin{pmatrix} a+b \\ c+d \end{pmatrix}$.

These facts enable us to draw a diagram which will help us to spot the invariant points under any transformation.

Let us see what effect the transformation with matrix $\begin{pmatrix} 0 & 1 \\ ^-1 & 2 \end{pmatrix}$ has upon the (x, y) plane. (It will help you to refer to Figure 5 as you work through this section.)

Under this transformation $\begin{pmatrix} 1 \\ 0 \end{pmatrix} \to \begin{pmatrix} 0 \\ ^-1 \end{pmatrix}$ and $\begin{pmatrix} 0 \\ 1 \end{pmatrix} \to \begin{pmatrix} 1 \\ 2 \end{pmatrix}$. What is the image of $\begin{pmatrix} 1 \\ 1 \end{pmatrix}$?

Draw a diagram to show the image of the unit square $A(0, 0)$; $B(1, 0)$; $C(1, 1)$; $D(0, 1)$ under this transformation.

Describe the transformation as accurately as you can. Which points of the plane are invariant?

(*b*) Now consider a general point of the (x, y) plane with position vector $\begin{pmatrix} x \\ y \end{pmatrix}$.

First of all check that $\begin{pmatrix} 0 & 1 \\ ^-1 & 2 \end{pmatrix}\begin{pmatrix} x \\ y \end{pmatrix} = \begin{pmatrix} y \\ ^-x + 2y \end{pmatrix}$.

What are the coordinates of the image of the point with position vector $\begin{pmatrix} a \\ c \end{pmatrix}$?

(*c*) Since $\begin{pmatrix} a \\ c \end{pmatrix} \to \begin{pmatrix} c \\ ^-a + 2c \end{pmatrix}$ then, for example, $\begin{pmatrix} 2 \\ 1 \end{pmatrix} \to \begin{pmatrix} 1 \\ 0 \end{pmatrix}$.
Now find the images of (i) $(2, 0)$; (ii) $(^-1, 3)$; (iii) $(^-1, ^-1)$; (iv) $(0, 5)$.
Which of these points are invariant under the transformation?

(*d*) If $\begin{pmatrix} a \\ c \end{pmatrix}$ is the position vector of an invariant point, then $\begin{pmatrix} a \\ c \end{pmatrix}$ and $\begin{pmatrix} c \\ ^-a + 2c \end{pmatrix}$ are position vectors of the same point, i.e.

$$\begin{pmatrix} a \\ c \end{pmatrix} = \begin{pmatrix} c \\ ^-a + 2c \end{pmatrix}.$$

Thus to find all the invariant points of the transformation we need to solve the equations

and
$$\left.\begin{aligned} a &= c \\ c &= {}^-a + 2c \end{aligned}\right\} \Rightarrow \left.\begin{aligned} a &= c \\ c &= a \end{aligned}\right\}.$$

This tells us that $\begin{pmatrix} a \\ c \end{pmatrix}$ and $\begin{pmatrix} c \\ ^-a + 2c \end{pmatrix}$ are position vectors of the same point when $a = c$. $(1, 1)$, $(2, 2)$, $(^-1, ^-1)$, for example, are invariant points.

Obviously, the transformation leaves the points on the line $x = y$ invariant. Did you obtain this answer in (*a*)?

The transformation is, in fact, a shear which maps $\begin{pmatrix} 1 \\ 0 \end{pmatrix} \to \begin{pmatrix} 0 \\ ^-1 \end{pmatrix}$ and $\begin{pmatrix} 0 \\ 1 \end{pmatrix} \to \begin{pmatrix} 1 \\ 2 \end{pmatrix}$, as shown in Figure 5.

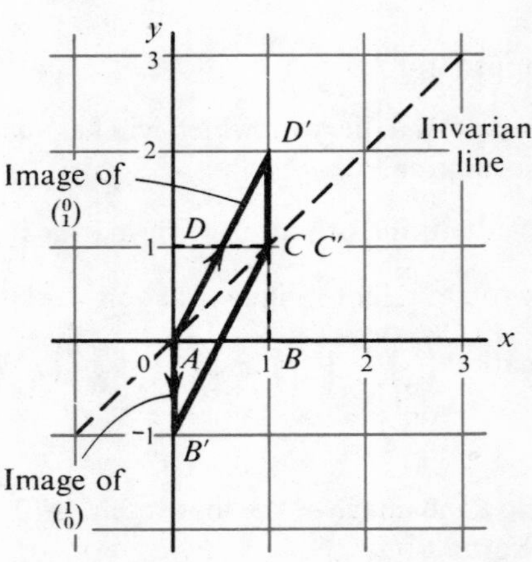

Fig. 5

Now read the following examples before attempting Exercise C.

Example 1

Find the invariant points of the transformation defined by the matrix $\begin{pmatrix} 0 & 3 \\ ^-1 & 4 \end{pmatrix}$.

If $\begin{pmatrix} a \\ b \end{pmatrix}$ is the position vector of an invariant point, then

$$\begin{pmatrix} 0 & 3 \\ ^-1 & 4 \end{pmatrix}\begin{pmatrix} a \\ b \end{pmatrix} = \begin{pmatrix} a \\ b \end{pmatrix}$$

$$\begin{pmatrix} 3b \\ -a+4b \end{pmatrix} = \begin{pmatrix} a \\ b \end{pmatrix}$$

$$\left.\begin{aligned} 3b &= a \\ -a+4b &= b \end{aligned}\right\}$$

$$\left.\begin{aligned} 3b &= a \\ 3b &= a \end{aligned}\right\}.$$

Hence the points on the line $x = 3y$ are invariant.

Example 2

Find the matrix of the transformation which leaves the points on the line $x = 2y$ invariant and maps $(2, 0) \to (3, ^-2)$.

The matrix of the transformation is of the form $\begin{pmatrix} a & b \\ c & d \end{pmatrix}$. (Why ?).

Since $(2,0) \to (3, ^-2)$,

$$\begin{pmatrix} a & b \\ c & d \end{pmatrix}\begin{pmatrix} 2 \\ 0 \end{pmatrix} = \begin{pmatrix} 3 \\ ^-2 \end{pmatrix}$$

$$\begin{pmatrix} 2a \\ 2c \end{pmatrix} = \begin{pmatrix} 3 \\ ^-2 \end{pmatrix}$$

$$\left.\begin{aligned} 2a &= 3 \\ 2c &= {}^-2 \end{aligned}\right\}$$

$$a = \tfrac{3}{2}$$

$$c = {}^-1.$$

We now need to find b and d.

Since all points on the line $x = 2y$ are invariant, the point (2, 1) for example, is invariant. Hence

$$\begin{pmatrix} a & b \\ c & d \end{pmatrix}\begin{pmatrix} 2 \\ 1 \end{pmatrix} = \begin{pmatrix} 2 \\ 1 \end{pmatrix}$$

$$\begin{pmatrix} 2a+b \\ 2c+d \end{pmatrix} = \begin{pmatrix} 2 \\ 1 \end{pmatrix}$$

$$\left.\begin{aligned} 2a+b &= 2 \\ 2c+d &= 1 \end{aligned}\right\}$$

$$\left.\begin{aligned} 3+b &= 2 \\ {}^-2+d &= 1 \end{aligned}\right\}$$

$$\left.\begin{aligned} b &= {}^-1 \\ d &= 3 \end{aligned}\right\}$$

The required matrix is therefore $\begin{pmatrix} \frac{3}{2} & {}^-1 \\ {}^-1 & 3 \end{pmatrix}$.

Now check that all the points on the line $x = 2y$, that is, points with position vectors $\begin{pmatrix} 2k \\ k \end{pmatrix}$, are invariant.

Exercise B

1 By solving two simultaneous equations, find the invariant points of the transformation which maps (x, y) onto

(i) $({}^-x, {}^-y)$; (ii) $(x+1, y+1)$; (iii) (y, x);
(iv) $(0, y)$; (v) $(2x+1, 2y)$; (vi) $(2x, 2y+1)$;
(vii) $(x+y, y)$.

2 Draw diagrams to show the effect of the transformations in Question 1 on the unit square. Describe each transformation accurately. Hence find the invariant points of each transformation and check your answers to Question 1.

3 Find algebraically the invariant points of the transformations defined by the matrices:

(i) $\begin{pmatrix} 2 & 0 \\ 1 & 1 \end{pmatrix}$; (ii) $\begin{pmatrix} 0 & 1 \\ 1 & 0 \end{pmatrix}$; (iii) $\begin{pmatrix} 2 & 2 \\ 3 & 3 \end{pmatrix}$;

(iv) $\begin{pmatrix} 2 & 1 \\ 1 & 2 \end{pmatrix}$; (v) $\begin{pmatrix} 0 & {}^-1 \\ {}^-1 & 0 \end{pmatrix}$; (vi) $\begin{pmatrix} 1 & 0 \\ \frac{1}{2} & {}^-1 \end{pmatrix}$.

4 A transformation leaves the points on the line $x+y=0$ invariant and maps $(1, 0)$ onto $(3, 0)$. Use the method of Example 2 to find the matrix of the transformation.

5 Find the matrix of the shear which leaves the points on the line $x = 3y$ invariant and maps $(0, 1) \to (3, 2)$.

6 The transformation **S** given by

$$\begin{pmatrix} x \\ y \end{pmatrix} \to \begin{pmatrix} 5 & 8 \\ {}^-2 & {}^-3 \end{pmatrix} \begin{pmatrix} x \\ y \end{pmatrix}$$

describes a shear. Find the equation of the line of invariant points.

7 A one-way stretch leaves points on the line $x = y$ invariant and has scale factor 2. Use the method of Example 2 to find the matrix of the transformation.

8 Discuss whether or not lines perpendicular to an axis of reflection are invariant under the reflection.

3 Invariant lines and lines of invariant points

(*a*) The lines $x = c$ are perpendicular to the line $y = 0$ (see Figure 6). Under a reflection in $y = 0$ these lines are therefore mapped onto themselves. Are the *points* on the lines $x = c$ invariant?

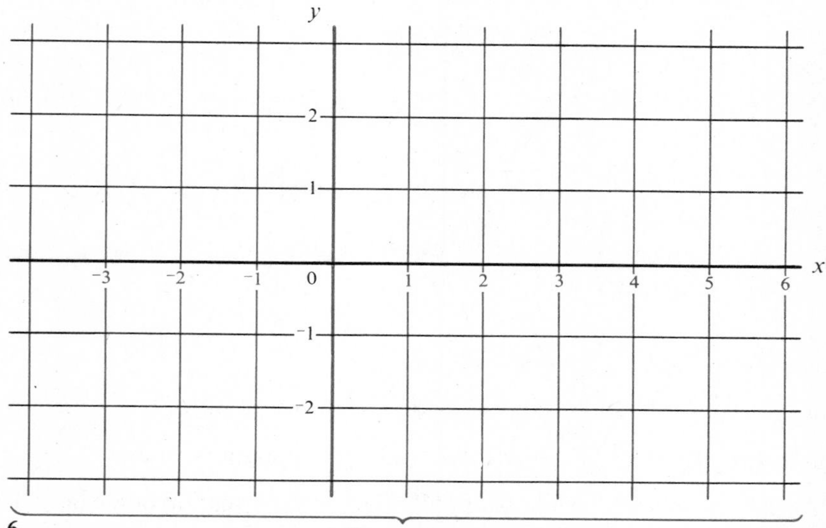

Fig. 6

(*b*) Although the points on the lines $x = c$ are not invariant (other than the points $(c, 0)$), the lines themselves may be termed 'invariant lines'. It is important that we distinguish between 'invariant lines' and 'lines of invariant points' ($y = 0$ is a 'line of invariant points'). Which lines are invariant under (i) a reflection in $x = 0$; (ii) a reflection in $y = x$? Which lines are 'lines of invariant points' under each transformation?

(*c*) Other transformations leave lines invariant. For example, under a translation of the plane defined by $\begin{pmatrix} 2 \\ 0 \end{pmatrix}$, there are no lines of invariant points, but the lines $y = c$ parallel to the direction of translation are invariant. Which lines are invariant under a shear which has $x = 0$ as a line of invariant points?

Exercise C

1 State which lines are invariant and which are lines of invariant points under each of the following transformations:

(i) reflection in $x = 1$; (ii) reflection in $x + y = 0$;

(iii) translation $\begin{pmatrix} 0 \\ 1 \end{pmatrix}$; (iv) translation $\begin{pmatrix} 1 \\ 1 \end{pmatrix}$;

(v) an enlargement centre $(0, 0)$ scale factor 2;

(vi) shear leaving $(0, 0)$ and $(5, 0)$ invariant;

(vii) shear leaving $(0, 0)$ and $(2, 1)$ invariant;

(viii) a glide reflection with glide line $y = 2x$;
(ix) a half turn, centre (0, 0).

2 Are (*a*) any points; (*b*) any lines; (*c*) any lines of points invariant under each of the following combined transformations?

(i) A reflection in $y = 0$ followed by a reflection in $x = 0$;
(ii) a translation $\begin{pmatrix}0\\1\end{pmatrix}$ followed by a translation $\begin{pmatrix}-1\\0\end{pmatrix}$;
(iii) an enlargement scale factor 2, centre (0, 0) followed by a translation $\begin{pmatrix}2\\0\end{pmatrix}$;
(iv) a one-way stretch scale factor 2, with $x = 0$ invariant, followed by a one-way stretch scale factor 2 with $y = 0$ invariant;
(v) a reflection in $y = 0$ followed by a shear with $y = 0$ invariant mapping $(1, 0) \to (2, 1)$.

4 Invariants under the general transformation

(*a*) The transformations we discussed in Section 2 could all be represented by a matrix of the form $\begin{pmatrix}a & b\\c & d\end{pmatrix}$, and each one left the origin invariant.

(*b*) We have already seen in *Book Y* that any transformation of the (x, y) plane can be represented by an expression of the form

$$\begin{pmatrix}x\\y\end{pmatrix} \to \begin{pmatrix}a & b\\c & d\end{pmatrix}\begin{pmatrix}x\\y\end{pmatrix} + \begin{pmatrix}h\\k\end{pmatrix}$$

i.e. as the combination of a transformation which leaves (0, 0) invariant, followed by a translation.

(*c*) Figure 7 represents a quarter turn about (1, 1) which maps T onto T'. Figure 8 shows that this is equivalent to a quarter turn about (0, 0) followed by a translation $\begin{pmatrix}2\\0\end{pmatrix}$.

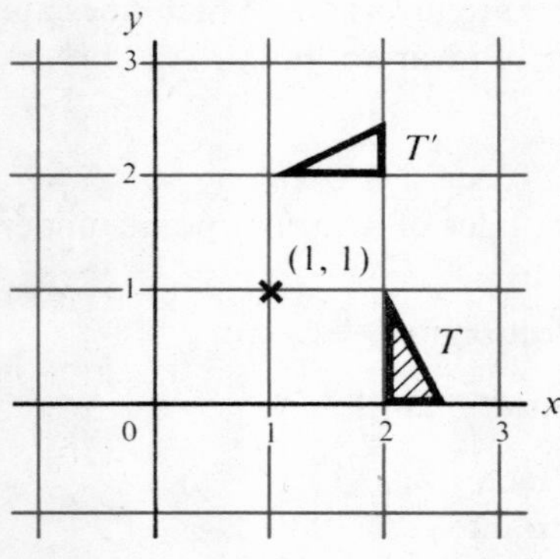

Fig. 7

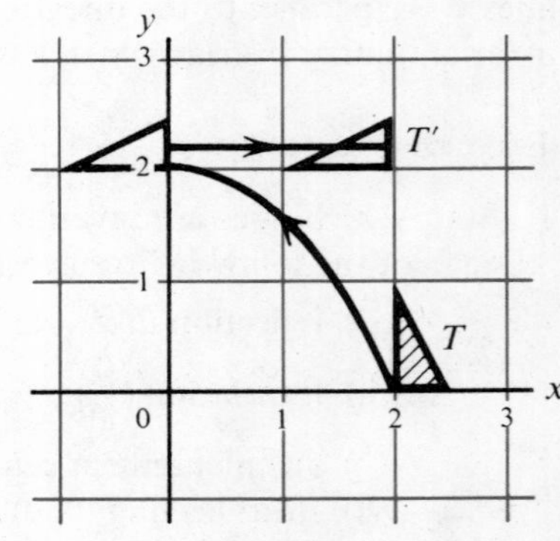

Fig. 8

Write down the expression which represents this transformation.

(*d*) Using equivalent methods to those in Section 2 we can again find expressions representing general transformations.

Example 3

Find the expression which represents a half-turn about (2, 2).

Our experience tells us that this is equivalent to a half-turn about the origin followed by a certain translation. The matrix of the half-turn about the origin is $\begin{pmatrix} ^-1 & 0 \\ 0 & ^-1 \end{pmatrix}$. Hence the required expression is of the form $\begin{pmatrix} ^-1 & 0 \\ 0 & ^-1 \end{pmatrix}\begin{pmatrix} x \\ y \end{pmatrix} + \begin{pmatrix} h \\ k \end{pmatrix}$.

To find $\begin{pmatrix} h \\ k \end{pmatrix}$:

since (2, 2) is invariant,
$$\begin{pmatrix} ^-1 & 0 \\ 0 & ^-1 \end{pmatrix}\begin{pmatrix} 2 \\ 2 \end{pmatrix} + \begin{pmatrix} h \\ k \end{pmatrix} = \begin{pmatrix} 2 \\ 2 \end{pmatrix}$$
$$\begin{pmatrix} ^-2 \\ ^-2 \end{pmatrix} + \begin{pmatrix} h \\ k \end{pmatrix} = \begin{pmatrix} 2 \\ 2 \end{pmatrix}$$
$$\left.\begin{aligned} ^-2 + h &= 2 \\ ^-2 + k &= 2 \end{aligned}\right\}$$
$$\left.\begin{aligned} h &= 4 \\ k &= 4 \end{aligned}\right\}$$

and the required expression is

$$\begin{pmatrix} ^-1 & 0 \\ 0 & ^-1 \end{pmatrix}\begin{pmatrix} x \\ y \end{pmatrix} + \begin{pmatrix} 4 \\ 4 \end{pmatrix}.$$

Example 4

Which single transformation is equivalent to a reflection in $y = 0$ followed by a translation $\begin{pmatrix} 0 \\ 2 \end{pmatrix}$?

The expression for the combined transformation is

$$\begin{pmatrix} x \\ y \end{pmatrix} \to \begin{pmatrix} 1 & 0 \\ 0 & ^-1 \end{pmatrix}\begin{pmatrix} x \\ y \end{pmatrix} + \begin{pmatrix} 0 \\ 2 \end{pmatrix}$$

so
$$\begin{pmatrix} x \\ y \end{pmatrix} \to \begin{pmatrix} x \\ ^-y + 2 \end{pmatrix}.$$

The invariant points of the transformation are given by solving the equation

$$\begin{pmatrix} x \\ y \end{pmatrix} = \begin{pmatrix} x \\ ^-y + 2 \end{pmatrix}$$

that is,
$$\left.\begin{aligned} x &= x \\ y &= {}^-y + 2 \end{aligned}\right\}$$
$$\left.\begin{aligned} x &= x \\ y &= 1 \end{aligned}\right\}.$$

This tells us that the x coordinate of each point remains constant and that the points on the line $y = 1$ are invariant. The combined transformation is therefore equivalent to a reflection in $y = 1$.

Exercise D

1 Find the invariant points of the combined transformation which maps (x, y) onto (i) $(1 - x, 1 - y)$; (ii) $(2x + 1, 2y + 1)$; (iii) $(4 - x, y)$; (iv) $(x + 3, 2 - y)$. (In each case one of the transformations is a translation.)
By finding the images of several points of your own choice describe each combined transformation accurately.

2 Find the invariant points (if any) of the transformations:

(i) $\begin{pmatrix} x \\ y \end{pmatrix} \to \begin{pmatrix} 2 & 0 \\ 0 & 2 \end{pmatrix}\begin{pmatrix} x \\ y \end{pmatrix} + \begin{pmatrix} 3 \\ 1 \end{pmatrix}$; (ii) $\begin{pmatrix} x \\ y \end{pmatrix} \to \begin{pmatrix} 1 & 0 \\ 1 & 1 \end{pmatrix}\begin{pmatrix} x \\ y \end{pmatrix} + \begin{pmatrix} 2 \\ 2 \end{pmatrix}$;

(iii) $\begin{pmatrix} x \\ y \end{pmatrix} \to \begin{pmatrix} ^-1 & 0 \\ 0 & ^-1 \end{pmatrix}\begin{pmatrix} x \\ y \end{pmatrix} + \begin{pmatrix} 2 \\ 0 \end{pmatrix}$; (iv) $\begin{pmatrix} x \\ y \end{pmatrix} \to \begin{pmatrix} 1 & 0 \\ 0 & ^-1 \end{pmatrix}\begin{pmatrix} x \\ y \end{pmatrix} + \begin{pmatrix} 1 \\ 0 \end{pmatrix}$.

Describe each transformation accurately.

3 Find the translation which, when combined with a half-turn about (0, 0) leaves ($^-1$, 0) invariant.

4 Find the image of (x, y) after a reflection in $x = y$ followed by a translation $\begin{pmatrix} 3 \\ 3 \end{pmatrix}$. Hence show that the transformation leaves no points invariant. To which single transformation is this equivalent?

5 Find the expressions which represent

(i) a half-turn about (0, $^-1$);
(ii) an enlargement centre (2, 2) scale factor 3;
(iii) a reflection in $x = 4$;
(iv) a shear which leaves $y = 1$ invariant and maps $(1, 0) \to (2, 0)$.

6 Find the expressions which represent the transformations of Question 1.

7 When combined, each of the following pairs of transformations is equivalent to a reflection. In each case express the transformations algebraically and hence find the equation of the mirror line.

(*a*) A reflection in $y = 0$ followed by a translation $\begin{pmatrix} 0 \\ 6 \end{pmatrix}$;

(*b*) a reflection in $x = 0$ followed by a translation $\begin{pmatrix} ^-5 \\ 0 \end{pmatrix}$;

(*c*) a reflection in $y = x$ followed by a translation $\begin{pmatrix} 3 \\ ^-3 \end{pmatrix}$.

What happens if the pairs of transformations are combined in the reverse order?

8 Find the expression for the glide reflection which maps

(*a*) $(x, y) \rightarrow (x+2, {}^{-}y)$; (*b*) $(x, y) \rightarrow ({}^{-}x, y-4)$.

5 Invariant properties – area and sense

(*a*) Figure 9 shows a rectangle $ABCD$ and its image after a shear. What is the area of $ABCD$? What is the area of $ABC'D'$?

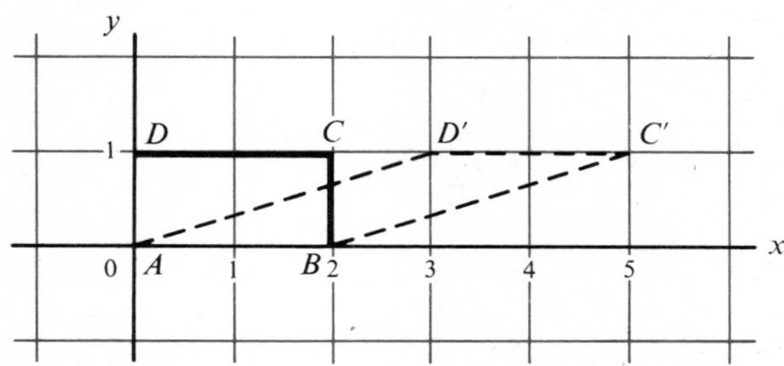

Fig. 9

Area is an invariant property under the shearing transformation. Which other transformations preserve area?

(*b*) The matrix of the shear represented by Figure 9 is $\begin{pmatrix} 1 & 3 \\ 0 & 1 \end{pmatrix}$.

What is the value of the determinant of this matrix?

What significance has this value?

(*c*) The value of the determinant of a matrix ($ad-bc$ for the matrix $\begin{pmatrix} a & b \\ c & d \end{pmatrix}$) gives us the area scale factor of the transformation (*Book Y*, Chapter 7).

For the shear with matrix $\begin{pmatrix} 1 & 3 \\ 0 & 1 \end{pmatrix}$, the determinant equals 1. This tells us immediately that $ABC'D'$ in Figure 9 has the same area as $ABCD$.

Which of the transformations defined by the following matrices leave area invariant?

(i) $\begin{pmatrix} 2 & 0 \\ 0 & 2 \end{pmatrix}$; (ii) $\begin{pmatrix} 1 & 0 \\ 1 & 1 \end{pmatrix}$;

(iii) $\begin{pmatrix} 2 & 2 \\ 1 & 0 \end{pmatrix}$; (iv) $\begin{pmatrix} 2 & 3 \\ 1 & 1 \end{pmatrix}$.

(*d*) The determinant of (ii) is 1 and the determinant of (iv) is ${}^{-}1$. Each of the transformations preserves area. The sign of the determinant, however, provides us with additional information about the transformation. Can you say what this is?

(*e*) Write down the matrices of (i) a reflection in $x = y$; (ii) a quarter turn about the origin; (iii) a shear with $x = 0$ invariant taking (1, 1) to (1, 4); (iv) an enlargement scale factor $^-1$, centre (0, 0). Write down the value of the determinant of each of these matrices. Which of the determinants are negative?

Now write down the matrices of some combined transformations formed from those above. Which of these have negative determinants?

(*f*) Your answers to (*c*) may have suggested to you that whenever a combined or single transformation involves a reflection then the determinant is negative.

For example, consider the matrix $\begin{pmatrix} 2 & 3 \\ 1 & 1 \end{pmatrix}$ in Section (*c*) above.

Check that

$$\begin{pmatrix} 2 & 3 \\ 1 & 1 \end{pmatrix} = \begin{pmatrix} 1 & 2 \\ 0 & 1 \end{pmatrix} \begin{pmatrix} 1 & 0 \\ 1 & 1 \end{pmatrix} \begin{pmatrix} 0 & 1 \\ 1 & 0 \end{pmatrix}.$$

The transformation is equivalent to a reflection in $x = y$ followed by two shears.

The determinant $= 1 \times 1 \times {}^-1 = {}^-1$.

(*g*) Geometrically, this means that whenever the determinant has a negative value then the image of the base vector $\begin{pmatrix} 1 \\ 0 \end{pmatrix}$ makes a greater angle with the positive x-axis than does the image of the base vector $\begin{pmatrix} 0 \\ 1 \end{pmatrix}$.

This is shown by Figure 10 for the transformation with matrix $\begin{pmatrix} 2 & 3 \\ 1 & 1 \end{pmatrix}$.

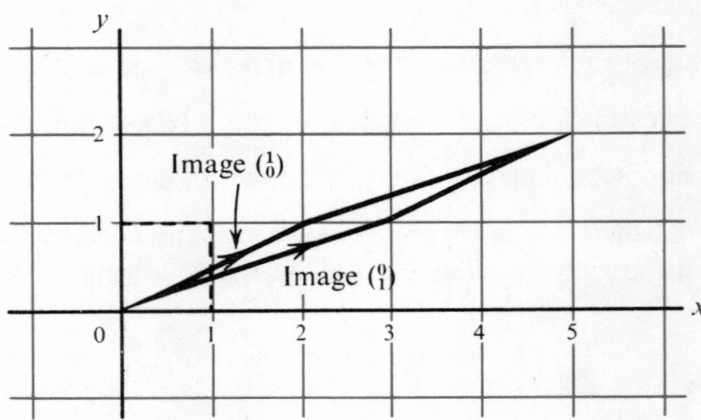

Fig. 10

In such cases we say that there has been a change in SENSE in the figure we are considering. When the determinant is positive there is no change in sense.

Exercise E

1 By finding the value of the determinant of each matrix, state which of the transformations preserve (*a*) area; (*b*) sense.

(i) $\begin{pmatrix} ^-1 & 1 \\ 1 & 2 \end{pmatrix}$; (ii) $\begin{pmatrix} ^-1 & ^-1 \\ 3 & 5 \end{pmatrix}$; (iii) $\begin{pmatrix} 2 & 1 \\ 1 & 2 \end{pmatrix}$;

(iv) $\begin{pmatrix} 1 & 0 \\ 0 & ^-1 \end{pmatrix}$; (v) $\begin{pmatrix} 1 & ^-1 \\ ^-1 & ^-1 \end{pmatrix}$.

2 Study Figure 11. Which of the transformations of the triangle *T* have involved a change in sense?

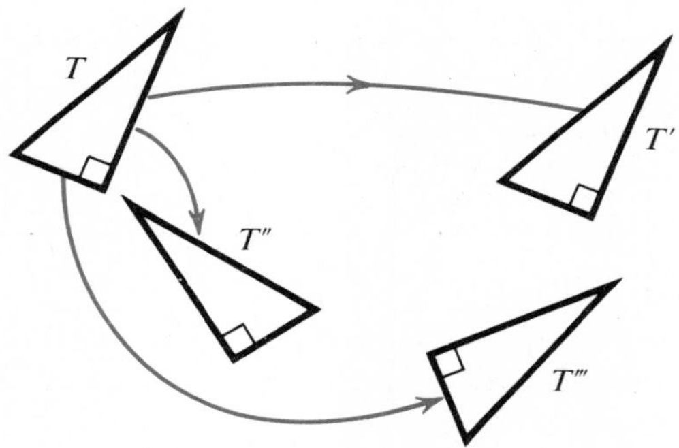

Fig. 11

3 Write down the images of the base vectors $\begin{pmatrix} 1 \\ 0 \end{pmatrix}$ and $\begin{pmatrix} 0 \\ 1 \end{pmatrix}$ under each of the following transformations. By drawing a diagram to show these images decide which of the transformations preserve sense.

(i) $\begin{pmatrix} 1 & 0 \\ 2 & ^-1 \end{pmatrix}$; (ii) $\begin{pmatrix} ^-1 & 1 \\ 0 & 2 \end{pmatrix}$; (iii) $\begin{pmatrix} 4 & 5 \\ ^-1 & ^-1 \end{pmatrix}$; (iv) $\begin{pmatrix} \frac{1}{2} & \frac{1}{2} \\ 1 & 0 \end{pmatrix}$.

4 Show that $\begin{pmatrix} 2 & 2 \\ 0 & ^-2 \end{pmatrix} = \begin{pmatrix} 1 & 0 \\ 0 & ^-1 \end{pmatrix}\begin{pmatrix} 1 & 1 \\ 0 & 1 \end{pmatrix}\begin{pmatrix} 2 & 0 \\ 0 & 2 \end{pmatrix}$. Describe the transformations defined by each of the three matrices on the right-hand side of this equation. Which of the three transformations does not preserve sense? Does the combined transformation preserve sense? What is the connection between the determinant of the matrix for the combined transformation and the determinants of the matrices of the single transformations?

5 Show that if two transformations involve a change in sense, then the combined transformation formed from these does not.

6 Without drawing a diagram describe the effect of the transformation defined by the matrix $\begin{pmatrix} ^-1 & ^-1 \\ 0 & 1 \end{pmatrix}$.

6 Classifying transformations – length, angle and parallelism as invariants

(*a*) Figure 12 represents an enlargement scale factor 2, centre (0, 0). Are the corresponding angles of the object and image figures equal?

The lines through AD and BC are parallel. Are the images of these lines parallel?

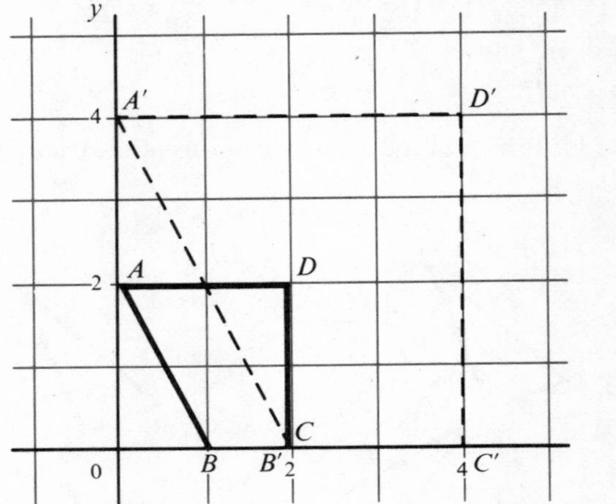

Fig. 12

(*b*) An enlargement preserves angle and parallelism. That is, the angles of the image figure are equal to the corresponding angles of the object figure, and parallel lines are mapped onto parallel lines. Is this true of any enlargement? Are lengths of line segments invariant under an enlargement?

(*c*) Use your knowledge of simple transformations to help you complete the table of invariant properties (Table 2).

Transformation	Invariant properties		
	Length	Angle	Parallelism
Reflection	✓	✓	✓
Rotation			
Translation			
Enlargement	×	✓	✓
One-way stretch			
Two-way stretch			
Shear			
Glide reflection			
Identity			

TABLE 2

(*d*) Which one property is invariant under each transformation in the table?

Transformations which preserve parallelism are called AFFINE transformations. All the transformations we have discussed in this chapter are members of this class.

(*e*) Transformations which preserve angle are called the SIMILARITIES.

(*f*) Length-preserving transformations are called ISOMETRIES. If a transformation preserves angle does it also preserve length? Is the converse true?

(*g*) The relationship between these three classes of transformations is represented by the Venn diagram in Figure 13. Notice that all isometries are similarities, but the converse is not true.

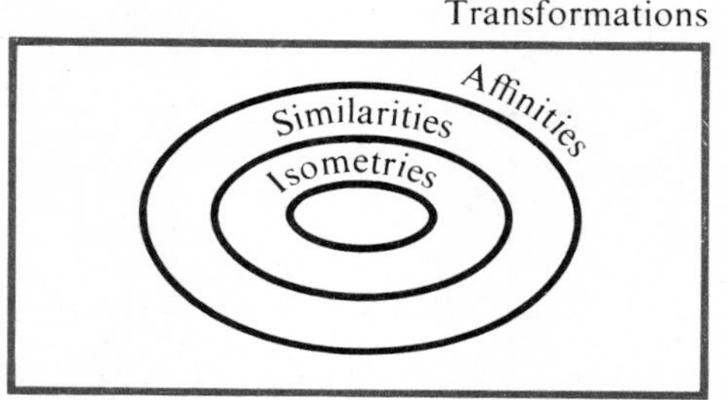

Fig. 13

Name a similarity which is also an isometry. Name a similarity which is not an isometry. Name an affine transformation which is not a similarity.

(*h*) You will notice in Figure 13 that a Universal set of transformations has been suggested. Can you think of a set of transformations of which the affine transformations are only a sub-set?

(*i*) In Figure 14 the network (*a*) has been transformed to (*b*) by a TOPOLOGICAL transformation. Which properties are invariant under such a transformation? Are all the transformations we have discussed above topological transformations?

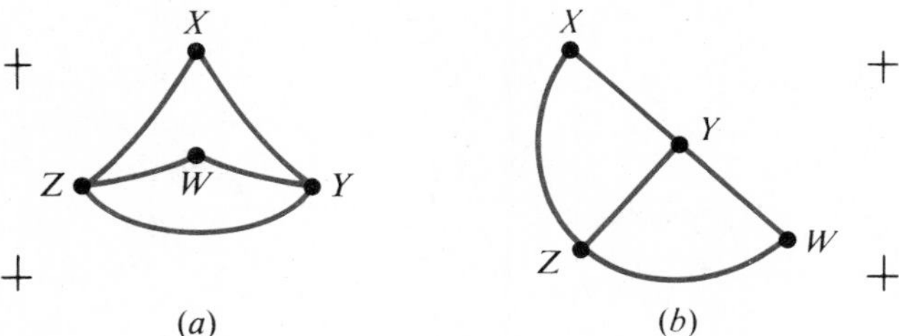

Fig. 14

Summary

TOPOLOGIES preserve relative positions of points on a line and order of nodes (all transformations discussed in this chapter)

AFFINITIES preserve parallelism (shears, one-way stretches, two-way stretches, the similarities and the isometries)

SIMILARITIES preserve angle (enlargements and the isometries)

ISOMETRIES preserve length (reflections, rotations, translations, glide reflections)

IDENTITY preserves all properties.

Exercise F

1 Study the transformations of the unit square represented in Figure 15 (assume D is the origin). Which of the following properties are invariant under each transformation?

(i) Length; (ii) angle; (iii) parallelism; (iv) order of nodes; (v) shape; (vi) sense; (vii) area.

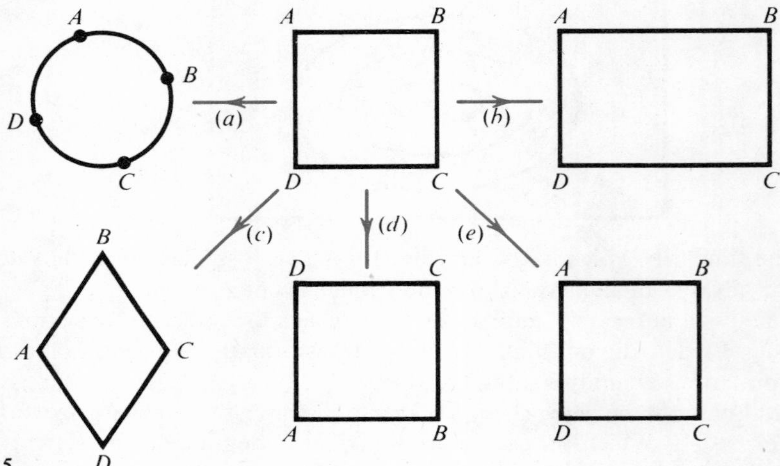

Fig. 15

To which set does each transformation probably belong?

2 The shaded square $ABCD$ in Figure 16 can be mapped onto itself by eight different isometries. Describe five of them and draw sketches to show the images of A, B, C and D in each case. Which of these transformations leave each of the following invariant?

Area; length; direction; angles between lines; sense.

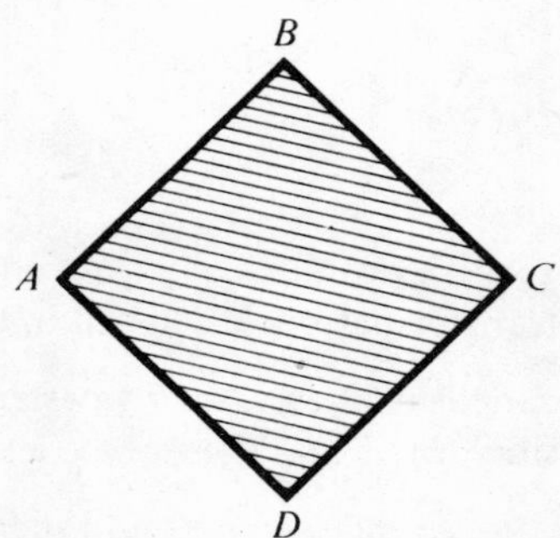

Fig. 16

3 The value of the determinant of a 2×2 matrix is 1. To which class of transformations does the transformation with this matrix *necessarily* belong?

4 A transformation **T** is defined between the points of a plane π and the points of a plane π' as follows (see Figure 17).

From every point P in π, a line p is drawn perpendicular to the plane π'. P', the image of P is defined as $p \cap \pi'$.

$ABCD$ is a square and $A'B'C'D$ is the image of $ABCD$ in π'.

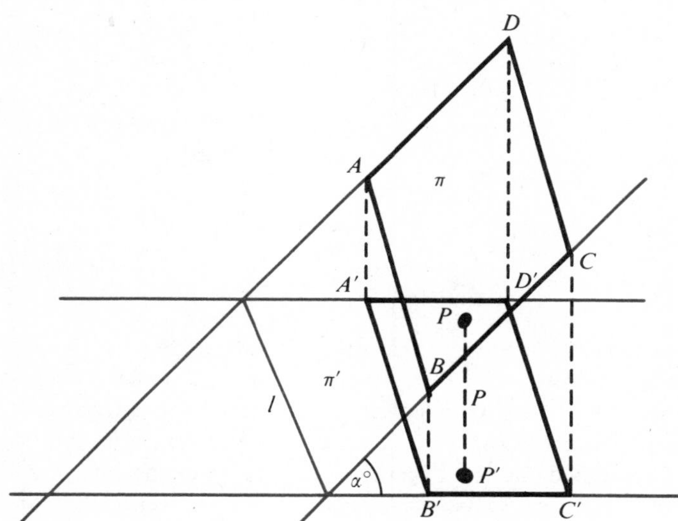

Fig. 17

(*a*) Does **T** map straight lines into straight lines?
(*b*) Does **T** preserve the sense of a figure?
(*c*) Does **T** map any line segments onto line segments of equal length?
(*d*) Are parallel lines mapped onto parallel lines?
(*e*) What shape will $A'B'C'D'$ be?
(*f*) What is the area scale factor of the transformation?
(*g*) The angle between AC and BD is a right-angle. Is the angle between $A'C'$ and $B'D'$ also a right-angle?
(*h*) What would be the image of a circle touching the sides of $ABCD$?

5 **A** is a reflection, **B** is an enlargement, **C** is a two-way stretch, **S** is a shear. To which class of transformations do the following combined transformations belong:

(i) **AB**; (ii) **CS**; (iii) **BC**; (iv) **AS**;
(v) $\mathbf{A}^2$; (vi) $\mathbf{AB}^{-1}$; (vii) $\mathbf{C}^2$?

7 Congruence

(*a*) The triangle P in Figure 18 can be mapped onto P', P'' and P''' by three different isometries. Which particular kind of transformation is involved in each case?

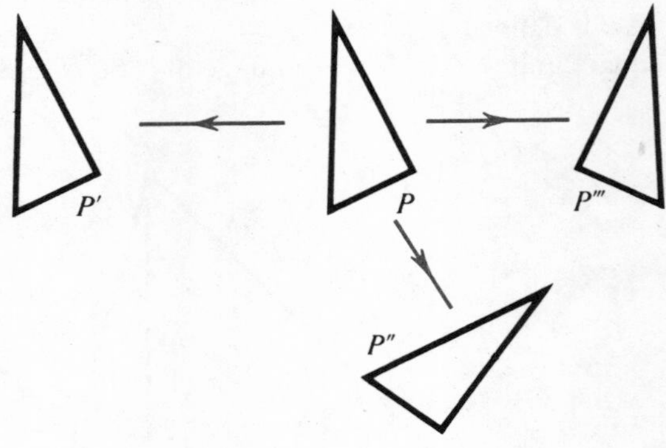

Fig. 18

Which transformation changes the sense of the triangle?

(*b*) P can be mapped onto P' by a translation, and onto P'' by a rotation. Neither transformation involves a change in sense. The three triangles P, P' and P'' are said to be *directly* congruent. Rotations are *direct* isometries.

(*c*) P can be mapped onto P''' by a reflection. Here a change in sense is involved and P and P''' are said to be *oppositely* congruent. Reflections are *opposite* isometries.

(*d*) Explain how P' can be mapped onto P'''. Are P' and P''' directly or oppositely congruent?

Explain how P'' can be mapped onto P'''. Is P'' directly or oppositely congruent to P'''?

(*e*) Make a tracing of triangle P and fit the tracing first over P' then P'' and P'''. Notice that the tracing paper had to be turned over to fit P onto P'''. This fact will help you to decide whether two figures are directly or oppositely congruent. Use your tracing paper to check that P'' and P''' are oppositely congruent.

Summary

1 Figures which can be mapped onto each other by an isometry are said to be congruent.

2 If a change in sense is involved in mapping one congruent figure onto another then the figures are oppositely congruent and the transformation is an opposite isometry. If no change in sense is involved the figures are directly congruent and the transformation is a direct isometry.

3 Translations and rotations are direct isometries. Reflections are opposite isometries.

Exercise G

1 Which of the pennants in Figure 19 are

(i) directly congruent to P; (ii) oppositely congruent to P?

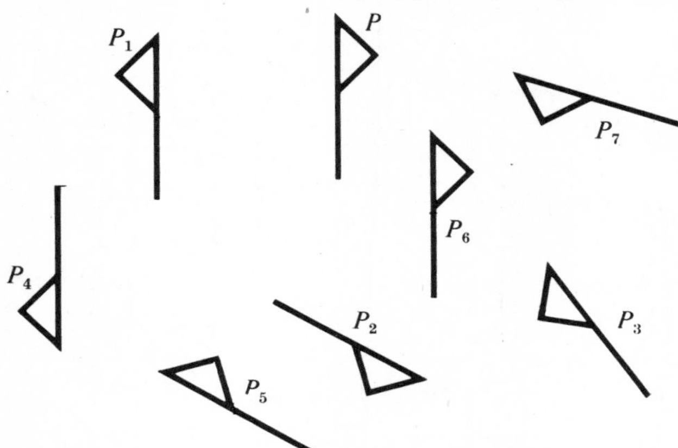

Fig. 19

2 **G** is a glide reflection. If P is any figure, are P and **G**(P) directly or oppositely congruent? Draw a diagram to explain your answer.

3 (*a*) Draw diagrams to show the image of triangle T in Figure 20 under each of the combined transformations:

(i) a reflection in $x = 0$ followed by a half turn about (0, 0);

(ii) a reflection in $x = 1$ followed by a translation $\begin{pmatrix} 2 \\ 1 \end{pmatrix}$;

(iii) a translation $\begin{pmatrix} 2 \\ 1 \end{pmatrix}$ followed by a half turn about (0, 0).

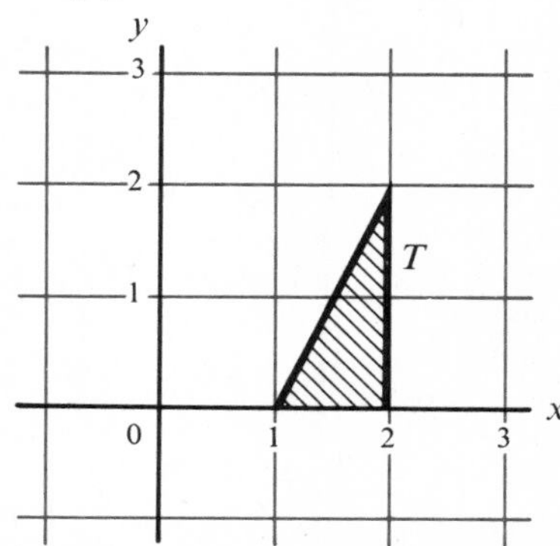

Fig. 20

(*b*) State whether the object and image triangles are oppositely or directly congruent in each case.

(*c*) If **A** is a direct isometry and **B** and **C** are opposite isometries, which of the following are direct isometries?

(i) **AB**; (ii) **AC**; (iii) **CA**; (iv) **BC**;
(v) **ABC**; (vi) **CAB**; (vii) $\mathbf{A}^2$; (viii) $\mathbf{C}^2$.

4 (*a*) In Figure 21, P can be mapped onto P' by a glide reflection in each case. Copy the diagram and mark in the mirror line for each glide reflection. Express the translation 'part' of the glide reflection in the form $\begin{pmatrix} a \\ b \end{pmatrix}$.

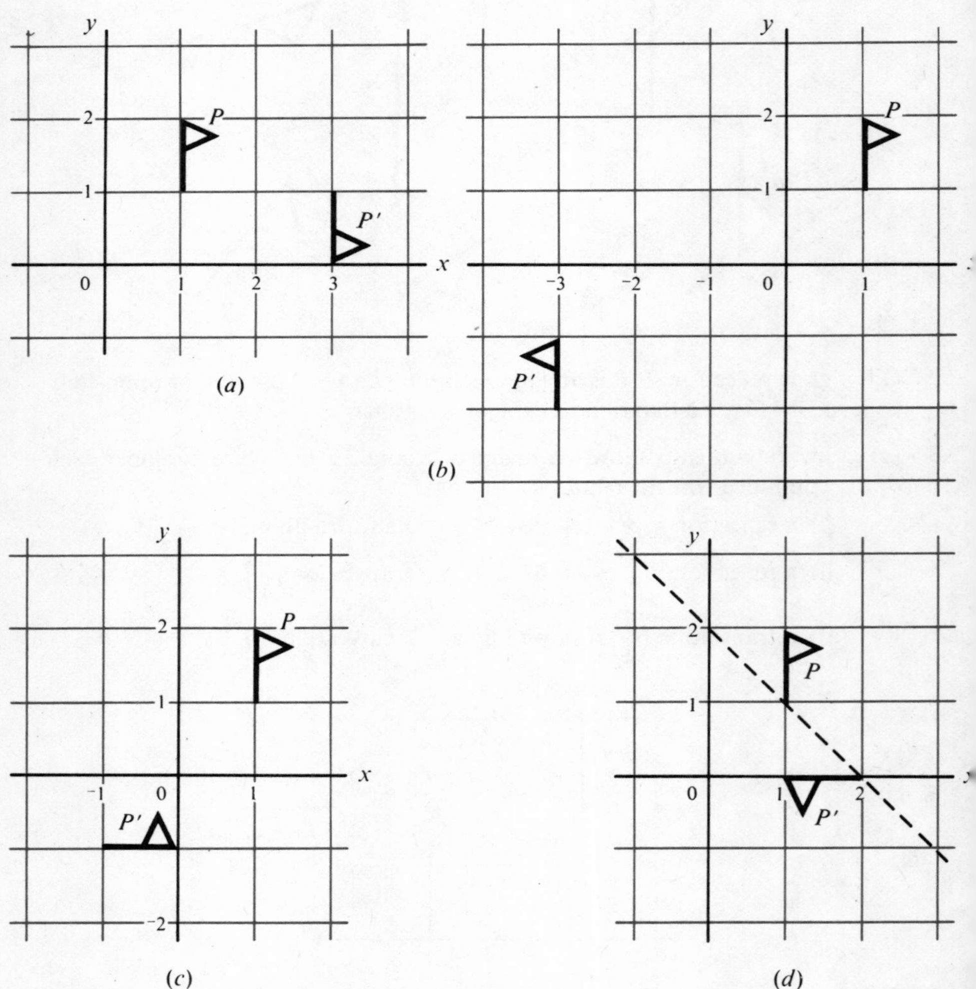

Fig. 21

(*b*) Copy Figure 22 and, using tracing paper if need be, show how P can be mapped onto P' by a glide reflection. Draw in the mirror line and the image of P at the intermediate stage (after the reflection or translation 'part' of the glide reflection).

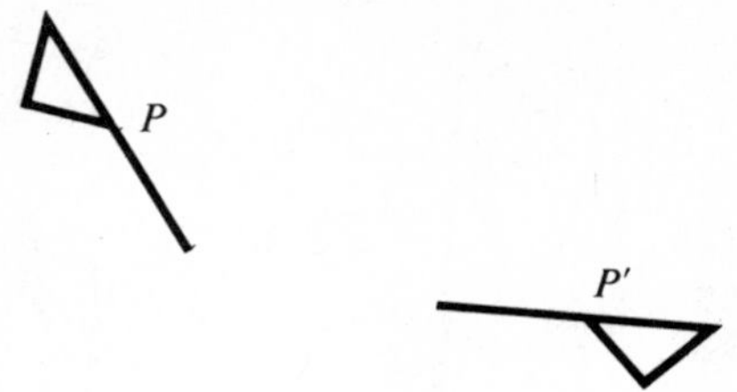

Fig. 22

5 P and P' are two directly congruent pennants. If P cannot be mapped onto P' by a translation, can P always be mapped onto P' by a rotation?

6 (*a*) Triangles ABC, DEF have corresponding sides of equal length. Are ABC, DEF necessarily congruent?
(*b*) $ABCD$, $EFGH$ have corresponding sides of equal length. Are $ABCD$, $EFGH$ necessarily congruent?
(*c*) ABC, DEF have two corresponding sides equal and one corresponding angle equal. Are ABC, DEF necessarily congruent?

Summary

1 Directly congruent figures can always be mapped onto each other by either (i) a rotation or (ii) a translation.

2 Oppositely congruent figures can always be mapped onto each other by either (i) a reflection or (ii) a glide reflection.

5 Using Combination Tables

1 Finite arithmetics

(*a*) We can classify the counting numbers

$$N = \{1, 2, 3, \ldots\}$$

according to whether they are even or odd.

The arrow diagram in Figure 1 shows the mapping from the set N onto $\{E, O\}$, where

$$E = \{2, 4, 6, \ldots\}$$
$$O = \{1, 3, 5, \ldots\}.$$

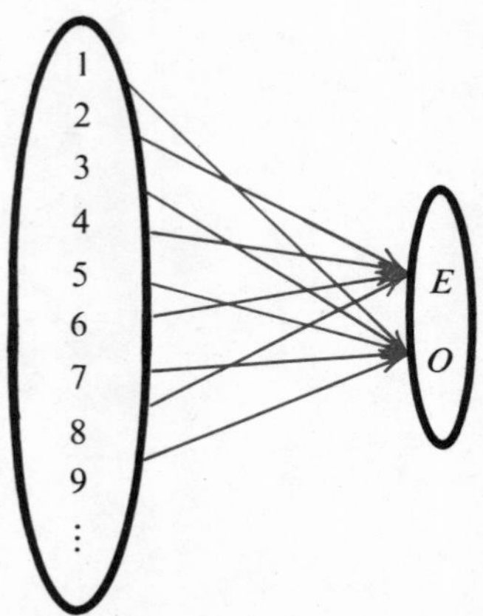

Fig. 1

Choose two numbers, one from the class E and one from the class O, and add them together. In which class is your answer? Is it always in the same class no matter which numbers you choose?

We say that $E+O=O$.

Copy the following combination tables and investigate whether it is possible to complete them:

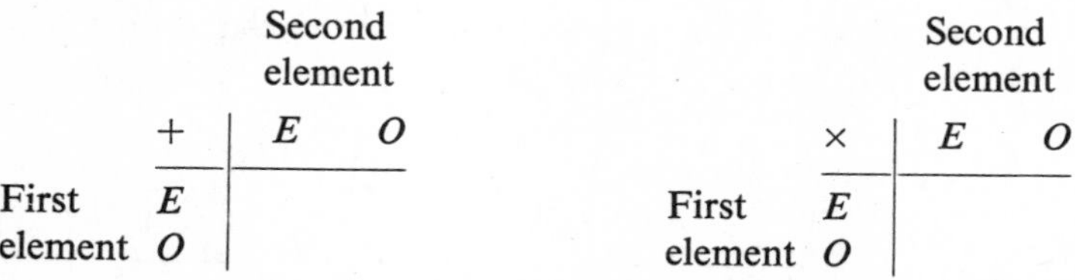

	+	Second element E	Second element O
First element	E		
	O		

	$\times$	Second element E	Second element O
First element	E		
	O		

(*b*) We can also classify the counting numbers according to whether they are squares or not.

If
$$S=\{1, 4, 9, 16, 25, 36, \ldots\}$$
$$S'=\{2, 3, 5, 6, 7, 8, 10, \ldots\},$$

try to complete these tables:

	+	Second element S	Second element S'
First element	S		
	S'		

	$\times$	Second element S	Second element S'
First element	S		
	S'		

What happens?

We see that the result of combining two numbers, one from each of two classes, does not always belong to some definite class.

(*c*) Counting numbers are classified as even or odd according to their relationship with the number 2.

A number belongs to the class E if it is exactly divisible by 2, that is, if it leaves a remainder of 0 when divided by 2.

A number belongs to the class O if it leaves a remainder of 1 when divided by 2.

We can also classify the counting numbers by the way in which they are related to numbers other than 2. For example, Figure 2 shows a way of writing them down so that they are classified according to their relationship with the number 4.

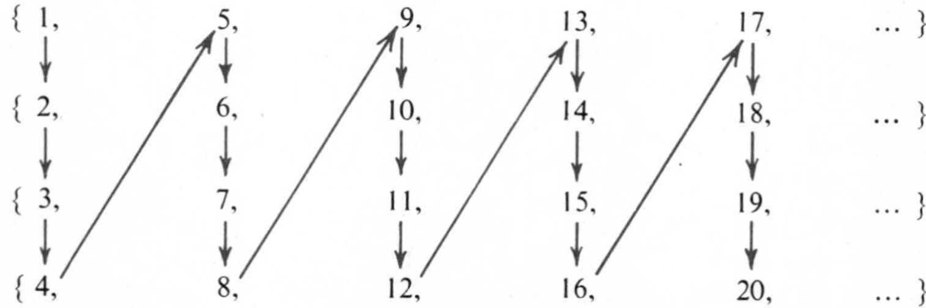

Fig. 2

There are no special names such as 'even' or 'odd' for these sets, so we shall use **0**, **1**, **2**, **3** to denote the sets of numbers which when divided by 4 leave remainders of 0, 1, 2, 3 respectively. Thus:

$\mathbf{0} = \{4, 8, 12, 16, \ldots\}$, $\mathbf{1} = \{1, 5, 9, 13, \ldots\}$,
$\mathbf{2} = \{2, 6, 10, 14, \ldots\}$, $\mathbf{3} = \{3, 7, 11, 15, \ldots\}$.

Now copy the following combination tables for the classes **0**, **1**, **2**, **3** under addition and multiplication and investigate whether it is possible to complete them.

	+	Second element **0**	**1**	**2**	**3**
First element	**0**		**1**		
	1				**0**
	2	**2**			
	3			**1**	

	×	Second element **0**	**1**	**2**	**3**
First element	**0**	**0**			
	1		**1**		
	2			**0**	
	3				**1**

The arithmetic of the finite set of classes {**0**, **1**, **2**, **3**} is called arithmetic modulo 4 or, more briefly, arithmetic mod 4.

(*d*) Check your combination tables for arithmetic modulo 4 by comparing them with those on page 63.

Exercise A

1 Classify the counting numbers according to their remainders when divided by 3. Label the classes **0**, **1**, **2** and write out addition and multiplication tables for arithmetic modulo 3.

2 Write out the addition table for arithmetic modulo 5. Use your table to find out which class **x** must be if:

(*a*) $\mathbf{x}+\mathbf{2}=\mathbf{1}$; (*b*) $\mathbf{4}+\mathbf{x}=\mathbf{2}$; (*c*) $\mathbf{x}+\mathbf{x}=\mathbf{1}$.

3 (*a*) Write out the multiplication table for arithmetic modulo 5. Use your table to help you to list all the pairs of classes which multiply to give **0**. What do you notice?
(*b*) Now write out the multiplication table for arithmetic modulo 6 and again list all the pairs of classes which multiply to give **0**. What happens this time?
(*c*) Investigate which pairs of classes multiply to give **0** in other finite arithmetics.

4 Figure 3 shows a magic square. The total in each row, each column and each diagonal is 15.
Replace each number by the class to which it belongs in arithmetic modulo 4. Do you now have a magic square for addition modulo 4?
If you replaced each number by the class to which it belongs in arithmetic modulo 3, what would be the total (mod 3) in each row, column and diagonal?

2	9	4
7	5	3
6	1	8

Fig. 3

5 Classify the counting numbers according to whether they are prime numbers or not. Label the classes P, P' and investigate whether it is possible to write out addition and multiplication tables for these classes.

2 Four properties

(*a*) The combination tables for addition and multiplication in arithmetic modulo 4 are:

	+ mod 4	Second element 0	1	2	3
First element	**0**	**0**	**1**	**2**	**3**
	1	**1**	**2**	**3**	**0**
	2	**2**	**3**	**0**	**1**
	3	**3**	**0**	**1**	**2**

	× mod 4	Second element 0	1	2	3
First element	**0**	**0**	**0**	**0**	**0**
	1	**0**	**1**	**2**	**3**
	2	**0**	**2**	**0**	**2**
	3	**0**	**3**	**2**	**1**

The addition table shows that the set {**0**, **1**, **2**, **3**} is closed under addition, since the body of the table can be completed using only the elements in the master row and column.

Is {**0**, **1**, **2**, **3**} closed under multiplication?

What is (i) the additive identity, (ii) the multiplicative identity for {**0**, **1**, **2**, **3**}?

Notice that an identity element can easily be found from a combination table since the row and column opposite the identity are a repeat of the master row and column, thus:

	× mod 4	Second element 0	1	2	3
First element	**0**		**0**		
	1	**0**	**1**	**2**	**3**
	2		**2**		
	3		**3**		

Every element of {**0**, **1**, **2**, **3**} has an additive inverse: **0** and **2** are self-inverse and **1** and **3** form an inverse pair.

Does every element have a multiplicative inverse? For {**0**, **1**, **2**, **3**} under multiplication modulo 4, state which elements, if any, (i) are self-inverse, (ii) form an inverse pair, (iii) have no inverse.

Addition modulo 4 is associative. We can use the addition table to verify this for any particular triple. For example,

$$(3+1)+2 = 0+2 = 2,$$
$$3+(1+2) = 3+3 = 2,$$

and so
$$(3+1)+2 = 3+(1+2).$$

Do you think that multiplication modulo 4 is associative?

(*b*) Transformations which map a shape onto itself are called symmetry transformations. The effects of the six symmetry transformations of an equilateral triangle are shown in Figure 4.

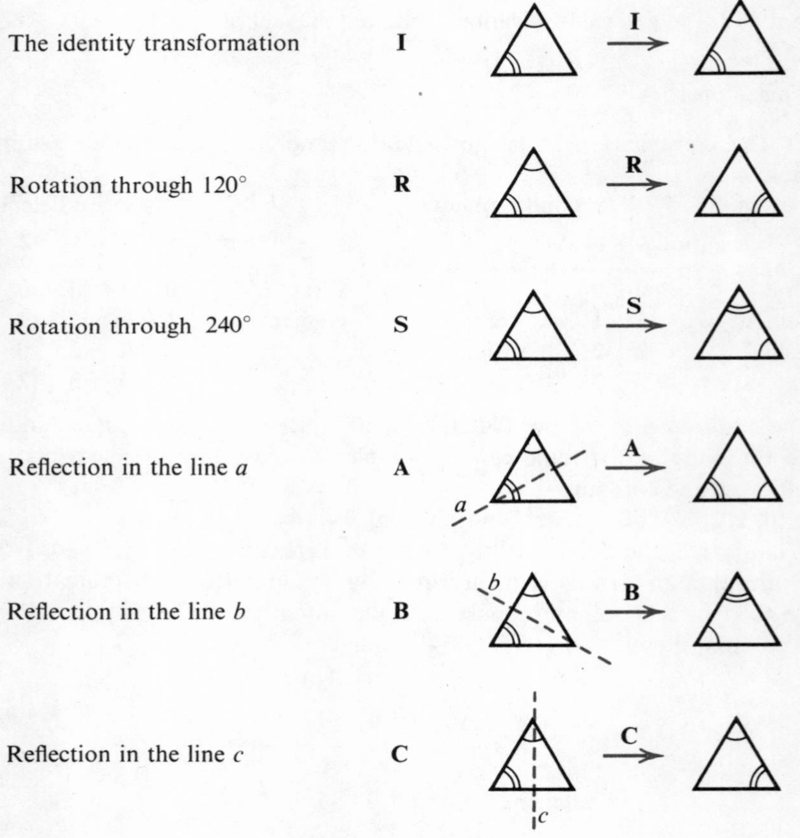

Fig. 4

These symmetry transformations can be combined in the usual way. For example, Figure 5 shows that **SC** has the same effect as **A** and we can write

$$\mathbf{SC} = \mathbf{A}.$$

Remember that **SC** is the combined transformation 'first **C** and then **S**'.

Fig. 5

Copy and complete the combination table for these transformations:

	follows	I	R	S	A	B	C
		Second element (First transformation)					
	I						**C**
	R				**C**	**A**	**B**
First element (Second transformation)	**S**						**A**
	A						
	B						
	C						

The combination of transformations is always associative. Verify from the table that **A**(**RC**) = (**AR**)**C**.

Is {**I**, **R**, **S**, **A**, **B**, **C**} closed under the operation of combination of transformations?

Is there an identity element?

Does every element have an inverse?

(*c*) You will find that many sets on which an operation is defined possess all four of the properties of closure, identity, inverse and associativity. Because of this a set of elements with an operation has been given the name of a *group* if it possesses all four properties.

The symmetry transformations of an equilateral triangle form a group.

Does {**0**, **1**, **2**, **3**} under addition modulo 4 form a group?

Why is {**0**, **1**, **2**, **3**} under multiplication modulo 4 not a group?

Summary

1 A set S is said to be *closed* under an operation $*$ if

$$a * b \in S \text{ whenever } a, b \in S.$$

2 If a set S with operation $*$ contains an element e such that

$$e * a = a = a * e \text{ whenever } a \in S,$$

then e is an *identity* element for S under $*$.

3 If a set S with operation $*$ contains an identity element e and $a, b \in S$ such that

$$a * b = e = b * a,$$

then b is an *inverse* of a.

If b is different from a, then a and b form an *inverse pair*.
If a and b are the same element, then a is *self-inverse*.

4 An operation $*$ defined on a set S is associative if, for any $a, b, c, \in S$,

$$(a * b) * c = a * (b * c).$$

Exercise B

1 (*a*) Compile combination tables for the following sets under multiplication modulo 12.

(i) {**4, 8**}; (ii) {**2, 4, 8**}; (iii) {**1, 2, 4, 8**}; {**1, 5, 7, 11**}.

(*b*) Which of the tables show closure?
(*c*) For each set, find the identity element if there is one; if not, say so.
(*d*) For which of the tables does each element have an inverse?
(*e*) Assuming that multiplication modulo 12 is associative, state which of the tables are group tables.

2 (*a*) Compile combination tables for:

(i) {**1, 2**} under multiplication modulo 3;
(ii) {**1, 2, 3**} under multiplication modulo 4;
(iii) {**1, 2, 3, 4**} under multiplication modulo 5;
(iv) {**1, 2, 3, 4, 5**} under multiplication modulo 6;
(v) {**1, 2, 3, 4, 5, 6**} under mutiplication modulo 7;
(vi) {**1, 2, 3, 4, 5, 6, 7**} under multiplication modulo 8;
(vii) {**1, 2, 3, 4, 5, 6, 7, 8**} under multiplication modulo 9.

[Keep your combination tables; you will need some of them again.]
(*b*) In which cases do some elements have no inverse? Try to state a general rule for deciding whether or not all elements will have inverses.

3 Two pennies are placed side by side on a table. They can be 'moved' in four ways:

(i) leave them as they are, **I**;
(ii) turn the right one over, **R**;
(iii) turn the left one over, **L**;
(iv) turn them both over, **B**.

(*a*) What single move is equivalent to **R** followed by **L**?
(*b*) Compile the combination table for {**I, R, L, B**} under 'followed by'. Is it a group table? [You may assume that 'followed by' is associative.]

4 Look carefully at each of the following tables. State whether or not they are group tables, giving your reasons fully and clearly.

(i)

$*$	a	b	c
a	a	b	c
b	b	d	a
c	c	a	b

(ii)

$\circ$	a	b	c
a	c	b	a
b	b	a	c
c	a	c	b

(iii)

$\bullet$	a	b	c
a	c	a	a
b	a	b	c
c	a	c	b

5 (*a*) Compile the combination table for {**2, 4, 6, 8**} under multiplication modulo 10.

(*b*) Does the table show closure?
(*c*) State the identity element.
(*d*) Which elements, if any (i) are self-inverse, (ii) form an inverse pair, (iii) have no inverse?

6 Compile combination tables for:

(i) {**1, 2, 3, 4**} under multiplication modulo 5;
(ii) {**1, 3, 7, 9**} under multiplication modulo 10;
(iii) {**0, 1, 2, 3**} under addition modulo 4.

In each case state which elements are self-inverse. Now rearrange the third table with the elements in the order **0, 1, 3, 2** and compare the pattern of this table with that in the other two.

Investigate whether or not it is possible to rearrange the tables for {**2, 4, 6, 8**} under multiplication modulo 10 and for {**1, 5, 7, 11**} under multiplication modulo 12 so that they also show the same pattern.

7 The four functions i, j, k, l are defined by:

$$i: x \to x, \qquad j: x \to {}^{-}x, \qquad k: x \to \frac{1}{x}, \qquad l: x \to -\frac{1}{x}.$$

(*a*) Verify that the composite function kj is equal to l.
(*b*) Compile a combination table for $\{i, j, k, l\}$ under composition of functions and comment on the table which you obtain.

3 Solving equations from combination tables

(*a*) Look again at the combination tables for arithmetic modulo 4:

(*a*) Second element across; first element down.

+	0	1	2	3
0	0	1	2	3
1	1	2	3	0
2	2	3	0	1
3	3	0	1	2

(*b*) Second element across; first element down.

	0	1	2	3
0	0	0	0	0
1	0	1	2	3
2	0	2	0	2
3	0	3	2	1

Fig. 6 (*a*) (*b*)

The flow diagram in Figure 6(*a*) shows that the solution of the equation

$$\mathbf{x + 2 = 1}$$

is

$$\mathbf{x = 3}.$$

Follow the flow diagram in Figure 6(*b*) and hence solve the equation

$$\mathbf{3 \times x = 2}.$$

Use the tables in Figure 6 to solve:

(i) $\mathbf{x}+\mathbf{1}=\mathbf{0}$; (ii) $\mathbf{3}+\mathbf{x}=\mathbf{2}$;
(iii) $\mathbf{x}\times\mathbf{3}=\mathbf{1}$; (iv) $\mathbf{1}\times\mathbf{x}=\mathbf{0}$.

Can you find any equations of the form $\mathbf{a}+\mathbf{x}=\mathbf{b}$ or $\mathbf{x}+\mathbf{a}=\mathbf{b}$ which have (i) no solution, (ii) more than one solution?

Can you find any equations of the form $\mathbf{a}\times\mathbf{x}=\mathbf{b}$ or $\mathbf{x}\times\mathbf{a}=\mathbf{b}$ which have (i) no solution, (ii) more than one solution?

(*b*) Use the combination table for the symmetry transformations of an equilateral triangle (see Figure 7) to solve the equations:

(i) $\mathbf{SX}=\mathbf{C}$; (ii) $\mathbf{XS}=\mathbf{C}$; (iii) $\mathbf{BX}=\mathbf{I}$;
(iv) $\mathbf{XR}=\mathbf{I}$; (v) $\mathbf{SX}=\mathbf{R}$; (vi) $\mathbf{XA}=\mathbf{B}$.

Can you find any equations of the form $\mathbf{PX}=\mathbf{Q}$ or $\mathbf{XP}=\mathbf{Q}$ which have (i) no solution, (ii) more than one solution?

Second element (first transformation) across the top; First element (second transformation) down the side.

	I	R	S	A	B	C
I	I	R	S	A	B	C
R	R	S	I	C	A	B
S	S	I	R	B	C	A
A	A	B	C	I	R	S
B	B	C	A	S	I	R
C	C	A	B	R	S	I

Fig. 7

(*c*) Let $\varnothing$ represent the empty set, A represent $\{a\}$, B represent $\{b\}$ and $\mathscr{E}$ represent $\{a, b\}$.

Write out the combination table for $\{\varnothing, A, B, \mathscr{E}\}$ under intersection.

Use this table to solve the equations:

(i) $X \cap A=\varnothing$; (ii) $X \cap \varnothing = A$;
(iii) $X \cap B = B$; (iv) $\mathscr{E} \cap X = B$.

Does $\{\varnothing, A, B, \mathscr{E}\}$ form a group under intersection?

Compile combination tables for the sets in (*d*) to (*i*) under the given operations. In each case, state:

(i) whether the set is closed under the given operation or not;
(ii) whether the set contains an identity or not;
(iii) whether each element of the set has an inverse or not;
(iv) whether the operation is associative or not;
(v) whether the set with its operation forms a group or not;
(vi) whether or not every equation of the form $a * x = b$ or $x * a = b$ has one and only one solution.

Enter your results in a copy of the table shown in Figure 8 and comment on them.

	Set	Operation	Closure	Identity	Every element has an inverse	Associativity	Group	All equations have unique solutions
(*a*)	{**0**, **1**, **2**, **3**}	Addition mod 4	Yes	Yes, **0**	Yes	Yes	Yes	Yes
	{**0**, **1**, **2**, **3**}	Multiplication mod 4	Yes	Yes, **1**	No	Yes	No	No
(*b*)	{**I**, **R**, **S**, **A**, **B**, **C**}	Combination of transformations	Yes	Yes, **I**	Yes	Yes	Yes	Yes
(*c*)	{ø, *A*, *B*, $\mathscr{E}$}	Intersection	Yes	Yes, $\mathscr{E}$	No	Yes	No	No
(*d*)	{**0**, **2**, **4**, **6**, **8**}	Addition mod 10	Yes	Yes				
(*e*)	{**1**, **2**, **3**, **4**, **5**, **6**}	Multiplication mod 7	Yes					
(*f*)								

Fig. 8

(*d*) {**0, 2, 4, 6, 8**} under addition modulo 10.
(*e*) {**1, 2, 3, 4, 5, 6**} under multiplication modulo 7.
(*f*) {**0, 2, 3, 4**} under multiplication modulo 6.
(*g*) {**I, J, K**} under matrix multiplication where

$$\mathbf{I}=\begin{pmatrix}1 & 0\\0 & 1\end{pmatrix},\quad \mathbf{J}=\begin{pmatrix}1 & 0\\0 & ^-1\end{pmatrix},\quad \mathbf{K}=\begin{pmatrix}^-1 & 0\\0 & 1\end{pmatrix}.$$

(*h*) {**H, I, J, K**} under matrix multiplication where

$$\mathbf{H}=\begin{pmatrix}^-1 & 0\\0 & ^-1\end{pmatrix},\quad \mathbf{I}=\begin{pmatrix}1 & 0\\0 & 1\end{pmatrix},\quad \mathbf{J}=\begin{pmatrix}1 & 0\\0 & ^-1\end{pmatrix},\quad \mathbf{K}=\begin{pmatrix}^-1 & 0\\0 & 1\end{pmatrix}.$$

(*i*) {0, 1, 2} under ~ (read 'twiddles') where ~ represents the operation 'find the positive difference between'. [For example, 1 ~ 2 = 1.]

4 The solution of equations with one operation

You should have discovered that, whenever a set $S=\{a, b, c, \ldots\}$ with an operation $*$ forms a group, then every equation which is of the form $a * x = b$ or $x * a = b$ has a unique solution, but that, if $(S, *)$ does not form a group, then an equation may have one solution, more than one solution or even no solution at all.

(*a*) Consider the equation

$$\mathbf{3}+\mathbf{x}=\mathbf{2}$$

in arithmetic modulo 4.

We know that {**0, 1, 2, 3**} under addition modulo 4 forms a group, so each element has an inverse. The inverse of **3** is **1**, therefore adding **1** on the left of both sides:

$$\mathbf{1}+(\mathbf{3}+\mathbf{x})=\mathbf{1}+\mathbf{2}.$$

Addition modulo 4 is associative, so

$$(\mathbf{1}+\mathbf{3})+\mathbf{x}=\mathbf{1}+\mathbf{2}.$$

Each element combines with its inverse to give the identity, so

$$\mathbf{0}+\mathbf{x}=\mathbf{1}+\mathbf{2}.$$

Elements remain unchanged when combined with the identity, so

$$\mathbf{x}=\mathbf{1}+\mathbf{2}.$$

{**0, 1, 2, 3**} is closed under addition modulo 4, so an element exists which is equal to **1** + **2**. This element is **3**, so the equation has the unique solution:

$$\mathbf{x}=\mathbf{3}.$$

Check from the table in Figure 6(*a*) that **3** is the correct solution.

(*b*) Consider the equation

$$\mathbf{XC}=\mathbf{S}$$

in the group of symmetry transformations on an equilateral triangle.

C is a reflection and is therefore self-inverse, so operating with **C** on the right,

$$\begin{aligned}(\mathbf{XC})\mathbf{C} &= \mathbf{SC}.\\ \mathbf{X}(\mathbf{CC}) &= \mathbf{SC}. \quad \text{Why?}\\ \mathbf{XI} &= \mathbf{SC}. \quad \text{Why?}\\ \mathbf{X} &= \mathbf{SC}. \quad \text{Why?}\\ \mathbf{X} &= \mathbf{A}, \text{ from Figure 5.}\end{aligned}$$

Check from the table in Figure 7 that **A** is the correct solution.

(*c*) Consider the equation $\mathbf{2} \times \mathbf{x} = \mathbf{5}$
in the system {**1, 2, 3, 4, 5, 6**} under multiplication modulo 7.

Complete a copy of the following solution by finding suitable elements to substitute for the question marks.

$$\begin{aligned}\mathbf{2} \times \mathbf{x} &= \mathbf{5}\\ \mathbf{?} \times (\mathbf{2} \times \mathbf{x}) &= \mathbf{?} \times \mathbf{5}\\ (\mathbf{?} \times \mathbf{2}) \times \mathbf{x} &= \mathbf{?} \times \mathbf{5}\\ \mathbf{1} \times \mathbf{x} &= \mathbf{?} \times \mathbf{5}\\ \mathbf{x} &= \mathbf{?} \times \mathbf{5}\\ \mathbf{x} &= \mathbf{?}\end{aligned}$$

Check from the table which you compiled in Section 3(*e*) that your answer is correct.

(*d*) Use the above method to solve:

(i) $\mathbf{x} + \mathbf{5} = \mathbf{2}$ in the system {**0, 1, 2, 3, 4, 5**} under addition modulo 6;
(ii) $\mathbf{4} \times \mathbf{x} = \mathbf{3}$ in the system {**1, 2, 3, 4**} under multiplication modulo 5;
(iii) $2 + x = {}^{-}6$ in the system {integers} under addition;
(iv) $\begin{pmatrix} 3 & 1 \\ 5 & 2 \end{pmatrix} \mathbf{X} = \begin{pmatrix} ^{-}1 & 2 \\ 7 & ^{-}3 \end{pmatrix}$ in the system {2 by 2 matrices with inverses} under matrix multiplication.

The importance of the method is its great generality, since it can be used to solve equations in any system $(S, *)$ for which the four group properties have been established.

(*e*) Now try using the same method to solve the following equations. In each case, write out the solution as far as you are able and explain clearly why the method breaks down.

(i) $\mathbf{JX} = \mathbf{K}$ in the system defined in Section 3(*g*);
(ii) $X \cap B = \mathscr{E}$ in the system defined in Section 3(*c*);
(iii) $\mathbf{4} \times \mathbf{x} = \mathbf{2}$ in the system defined in Section 3(*f*);
(iv) $x \sim 1 = 1$ in the system defined in Section 3(*i*).

Use the tables you compiled in Section 3 to find the solution set for each of these equations.

Exercise C

1 (*a*) Write out the addition and multiplication tables in arithmetics modulo 3, modulo 6, modulo 7 and modulo 9. [You can save time by using the multiplication tables which you compiled for Exercise B, Question 2 (a).]

(*b*) Has every equation of the form $\mathbf{x} + \mathbf{a} = \mathbf{b}$ a unique solution in these arithmetics?

(*c*) Has every equation of the form $\mathbf{x} \times \mathbf{c} = \mathbf{d}$, where $\mathbf{c} \neq \mathbf{0}$, a unique solution in these arithmetics?

(*d*) If every equation of the form $\mathbf{x} \times \mathbf{c} = \mathbf{d}$, where $\mathbf{c} \neq \mathbf{0}$, has a unique solution in arithmetic modulo p, what can you say about p?

(*e*) Do the same results hold for equations of the types: $\mathbf{a} + \mathbf{x} = \mathbf{b}$ and $\mathbf{c} \times \mathbf{x} = \mathbf{d}$, where $\mathbf{c} \neq \mathbf{0}$?

2 Compile a multiplication table for $\{\mathbf{1}, \mathbf{3}, \mathbf{4}, \mathbf{5}, \mathbf{9}\}$ under multiplication modulo 11.

Has every equation of the form $\mathbf{x} \times \mathbf{c} = \mathbf{d}$ a unique solution? How can you tell just by looking at the table?

3 Solve the equations:

(*a*) $\mathbf{x} + \mathbf{4} = \mathbf{3}$ in arithmetic modulo 6;
(*b*) $\mathbf{3x} = \mathbf{6}$ in arithmetic modulo 9;
(*c*) $\mathbf{4x} = \mathbf{5}$ in arithmetic modulo 7.

4 Answer the following questions for the system {2 × 2 matrices} under matrix addition.

(*a*) Is there an identity element? If so, state what it is.
(*b*) Does every element have an inverse?
(*c*) Does every equation of the form $\mathbf{A} * \mathbf{X} = \mathbf{B}$ have a unique solution?

5 Repeat Question 4 for the system {2 × 2 matrices} under matrix multiplication.

6 Repeat Question 4 for the system {sets} under union.

7 Use the method of Section 4 to solve the following equations in the system {rational numbers} under addition:

(*a*) $2 + x = 7$; (*b*) $x + 13 = 1$; (*c*) $x + \frac{1}{4} = {}^{-}2$.

Which of these equations, if any, could be solved by the method of Section 4 in (i) {integers} under addition, (ii) {counting numbers} under addition?

8 (*a*) If P and Q are two sets, $P \triangle Q$ is defined as the set of elements which belong to P or Q *but not to both*. Thus the shaded region in Figure 9 represents $P \triangle Q$.

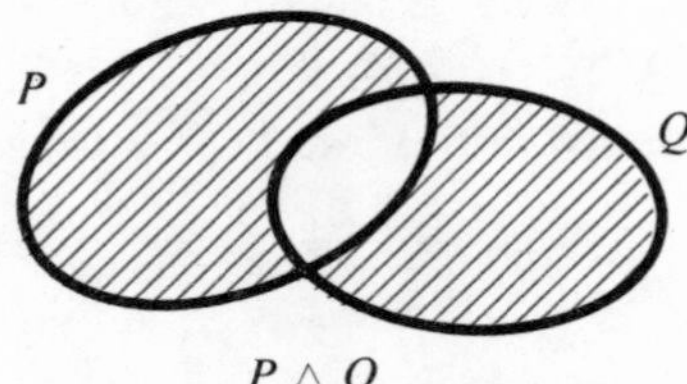

Fig. 9 $P \triangle Q$

Draw diagrams to illustrate (i) $(P \triangle Q) \triangle R$, (ii) $P \triangle (Q \triangle R)$. Is $\triangle$ associative?

(*b*) Compile a combination table for {$\varnothing$, *A*, *B*, $\mathscr{E}$} under $\triangle$ where $\varnothing$, *A*, *B* and $\mathscr{E}$ represent the empty set, {*a*}, {*b*} and {*a*, *b*} respectively.

What can you say about inverses in this system?

(*c*) Is it possible to solve every equation of the form

$$P \triangle X = Q$$

in (i) the system {$\varnothing$, *A*, *B*, $\mathscr{E}$} under $\triangle$, (ii) the system {sets} under $\triangle$? Give reasons for your answers.

9 The operations $\circ$ and $*$ are defined by the tables:

	$\circ$	Second element *a*	*b*	*c*	*d*
	a	*a*	*c*	*d*	*a*
First element	*b*	*b*	*a*	*c*	*d*
	c	*d*	*b*	*a*	*c*
	d	*c*	*d*	*b*	*a*

	$*$	Second element *a*	*b*	*c*	*d*
	a	*b*	*c*	*d*	*a*
First element	*b*	*c*	*d*	*a*	*b*
	c	*d*	*a*	*b*	*c*
	d	*a*	*b*	*c*	*d*

Give the complete solution set for *x* in each of the following:

(*a*) $b \circ x = c$; (*b*) $(a \circ x) * c = a$;
(*c*) $(a \circ x) * b = c$; (*d*) $[a * (x \circ b)] \circ c = b$.

6 Latitude and Longitude

1 The sphere

(*a*) The pictures above show (i) a donkey tethered to a stake; (ii) a bird tethered to a swivel on the top of a pole.

What region of grass can the donkey eat if it can reach a distance of 8 m from the foot of the stake?

What region of space can the bird fly through if it can reach a distance 2 m from the pole?

(*b*) The donkey can eat grass within a circle of radius 8 m with the base of the pole as its centre. In set notation this region can be represented by

$$\{P: OP \leqslant 8 \text{ m}\},$$

where O is fixed and P moves in *two dimensions* (see Figure 1(*a*)).

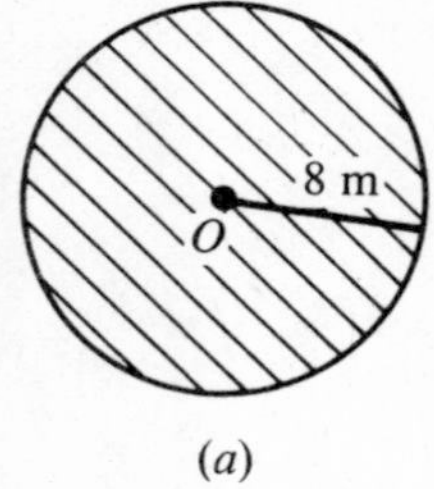

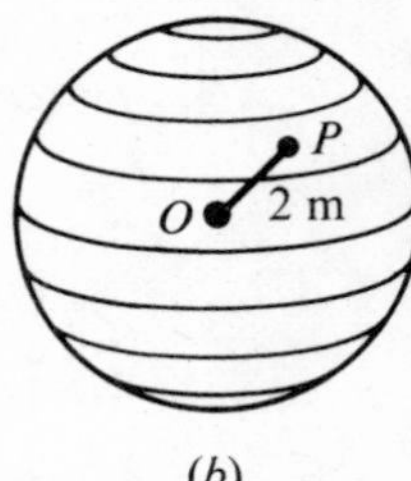

Fig. 1 (*a*) (*b*)

Describe (i) in words; (ii) in set notation, the region of grass the donkey *cannot eat*.

(*c*) The 'flying zone' for the bird is the set of points inside a *sphere* of radius 2 m with the swivel as its centre. In set notation such a region can be described by:

$$\{P: OP \leqslant 2 \text{ m}\},$$

where O is fixed and P moves in *three dimensions* (in space) – see Figure 1(*b*).

Describe (i) in words; (ii) in set notation, the 'non-flying zone' for the bird.

(*d*) In mathematics, we define a sphere of radius k as the set of points:

$$\{P: OP = k \text{ units}\}$$

where O is fixed and P moves in three dimensions. This is a similar definition to that of a circle in two dimensions:

$$\{P: OP = k \text{ units}\}$$

where O is fixed.

The points $\{P: OP \leqslant k \text{ units}\}$ in three dimensions form a *solid sphere* of radius k. Similarly, in two dimensions, the points $\{P: OP \leqslant k \text{ units}\}$ form a *disc* of radius k.

Exercise A

1 Figure 2(*a*) shows a cone (with a circular base). If you cut the cone parallel to its base, the cut will reveal a circular cross-section (see Figure 2(*b*)). Draw diagrams to show the other cross-sectional shapes that could be produced by different cuts.

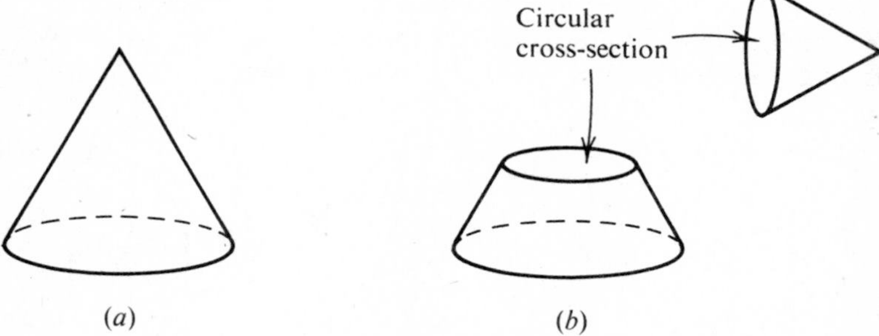

Fig. 2

2 (*a*) Explain how a circular cross-section can be produced by cutting (i) a cylinder (Figure 3(*a*)); (ii) a tyre (Figure 3(*b*)); (iii) a sphere.
(*b*) Draw diagrams to show the other cross-sectional shapes that can be produced by other cuts.

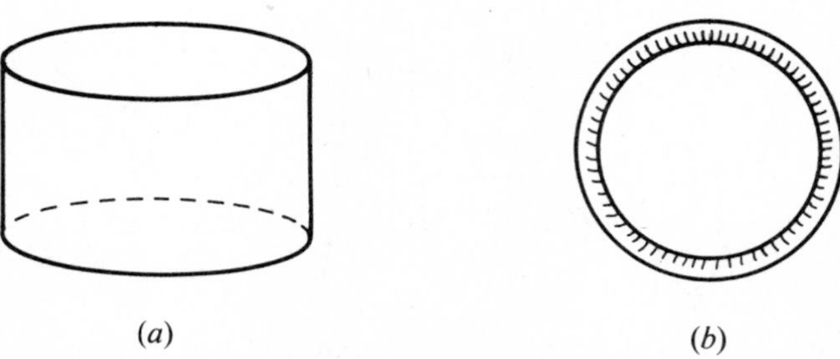

Fig. 3

2 Shortest distance along the surface of a sphere

(*a*) Use compasses and scissors to make three discs of radii 2 cm, 3 cm and 5 cm from thin card. Mark two points A and B in your book at a distance apart of less than 4 cm. Position the smallest disc as shown in Figure 4 and draw in the red (minor) arc.

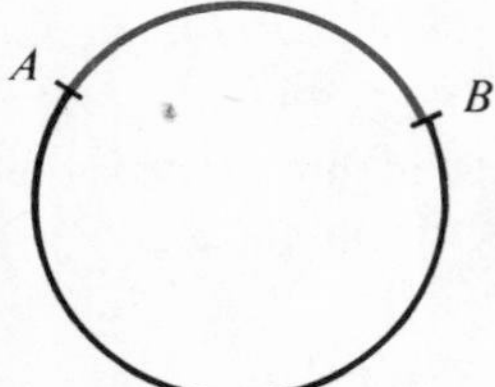

Fig. 4

Repeat the experiment with the remaining discs. Which of the arcs you have drawn is the shortest?

(*b*) You should have found in (*a*) that as the radius of the disc increases, the arc distance AB decreases.

Figure 5 shows three ways in which a sphere can be cut into two parts by a plane passing through two fixed points A and B (O is the centre of the sphere). Notice that the larger the radius of the cutting circle, the smaller the (minor) arc distance AB. You might like to experiment yourself by cutting sections from an apple.

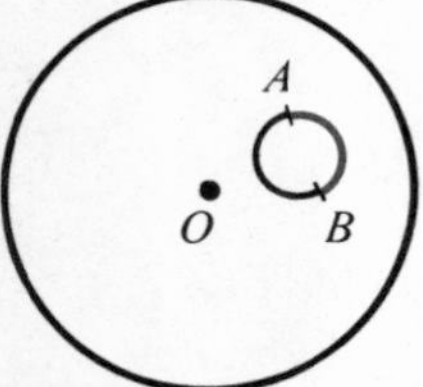

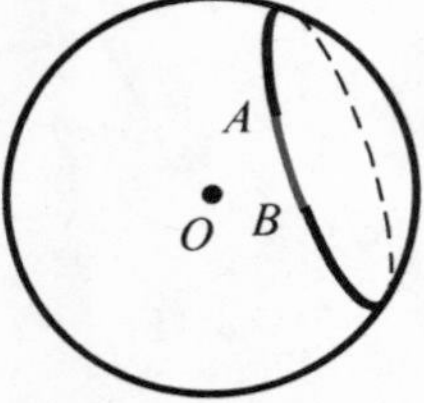

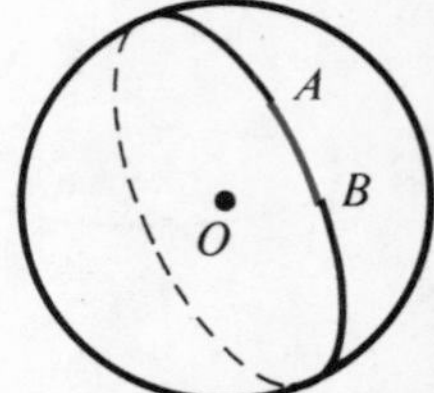

Fig. 5

If you cut a sphere of radius 6 cm as in Figure 5, what would be the radius of the largest circular section you could produce? Do you agree that (i) the cut must pass through O, the centre of the sphere; (ii) in this case the sphere is cut into two congruent parts; and (iii) the minor arc distance AB is then the shortest distance between A and B measured along the *surface of the sphere*? (See Figure 6.)

A great circle cuts a sphere into two hemispheres

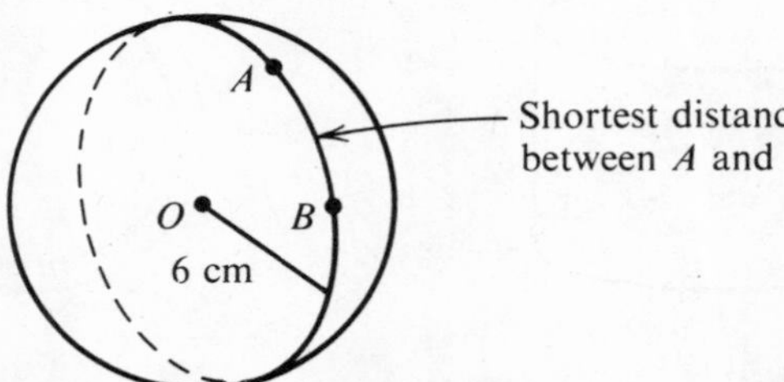

Fig. 6

O is the centre of the sphere, and also the centre of the circle through A and B.

(*c*) Circles on the surface of a sphere which have the centre of the sphere as their own centre are called GREAT CIRCLES. Three great circles are shown in Figure 7(*a*). All other circles, like those shown in Figure 7(*b*), are called SMALL CIRCLES.

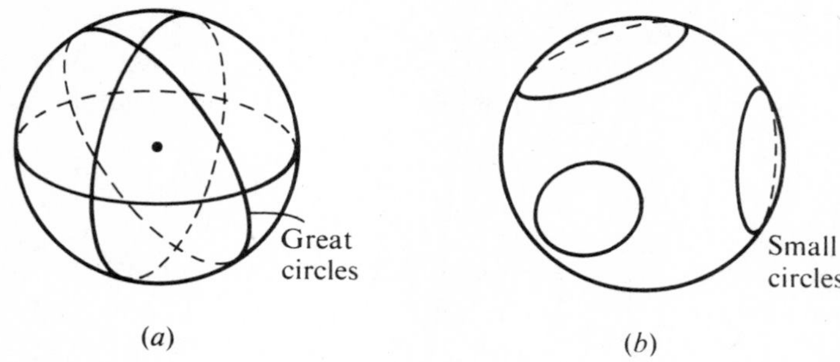

Fig. 7

(*d*) The earth is often regarded as a sphere. In fact, the distance from the earth's centre to the equator is about 21 km greater than the distance to the poles.

Many yachtsmen in the trans-Atlantic single-handed yacht race choose the 'Great circle' route between Plymouth and Newport, Rhode Island, USA because this is the shortest route. In the 1972 race *British Steel* (Brian Cooke) followed the Great circle route and was placed fourth in the race.

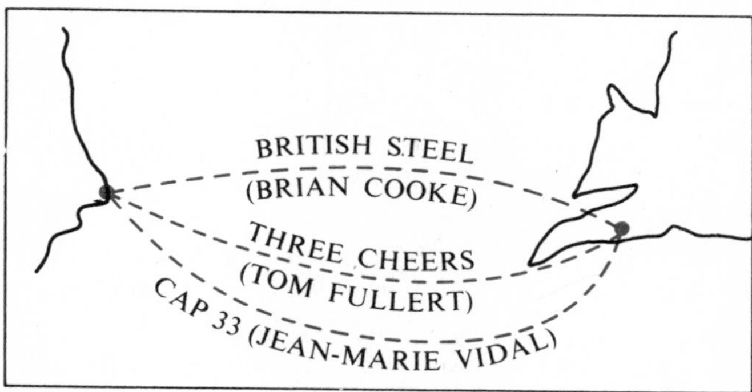

Fig. 8

Investigation 1

Obtain a table-tennis ball or other spherical object. Mark two points *A* and *B* on its surface. Cut circular holes of various sizes in a sheet of card. Rest the table-tennis ball in each hole and draw the circle of intersection of the ball with the 'plane' in each case, so that this circle passes through *A* and *B*.

How many circles of the same radius can you draw passing through *A* and *B*?

Mark the shortest route between *A* and *B* on the ball.

Investigation 2

On a plastic football, or a spherical balloon, mark the North and South poles, the equator and the positions of Plymouth and Rhode Island. (You will need an atlas and a globe to help you.) Using the method of Investigation 1 draw the Great circle route between Plymouth and Rhode Island. Use a piece of string to compare the distances travelled by the yachts in Figure 8. Mark the positions of some towns other than Plymouth and Newport, Rhode Island, on the same Great circle route.

Exercise B

1 (*a*) How many Great circles can be drawn to pass through one particular point on a sphere?
(*b*) How many Great circles can be drawn to pass through *two* particular points on a sphere?

2 The radius of a Great circle of a sphere is 3 cm. What is (i) the diameter of the sphere; (ii) the length of longest 'circular route' round the sphere?

3 A hemi-spherical goldfish bowl of radius 10 cm contains water whose surface is a disc of radius 8 cm. What is the depth of water?

4 Figure 9 represents a circular lid of a powder compact making an angle of 90° with its circular base. There is a small hinge at A. If the radius of each circle is 3 cm, calculate the distance O_1O_2, where O_1, O_2 are the centres of the base and the lid.
What is the greatest possible distance between a point of the lid and a point of the base when the compact is open at 90°?
Find the greatest distance when it is open at 150°.

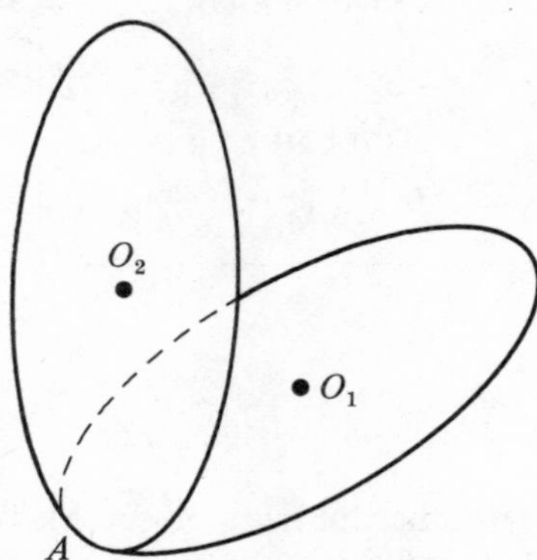

Fig. 9

5 Discuss the symmetries of a sphere.

6 A small circle of radius 4 cm is drawn on a sphere of radius 6 cm. How far from the centre of the sphere is the centre of the small circle?

7 To what radius would you set a pair of compasses to draw a small circle of radius 5 cm on a sphere of radius 10 cm? Could you draw a Great circle on the same sphere with a pair of compasses?

3 The earth

In the following work we will assume that the earth is spherical.

(*a*) *A skeleton model for the earth.* Cut three discs of radius 5 cm from thin card, and cut slots in each one as shown in Figure 10. The short cuts are each of length 2·7 cm, and the long ones of length 5 cm.

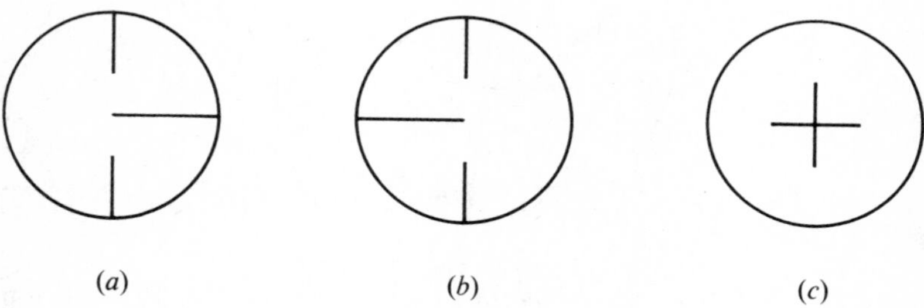

Fig. 10

We can assemble these to make a 'skeleton earth' as follows:

(i) slot (*a*) and (*b*) together and fold each semi-circular projecting part back onto itself as shown in Figure 11;

(ii) push this assembly through the cross-shaped slot in (*c*) and straighten the semi-circles again to form the model shown in Figure 12.

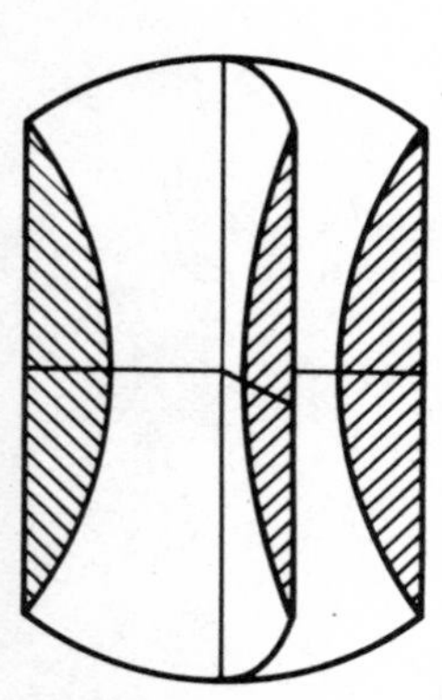

Fig. 11

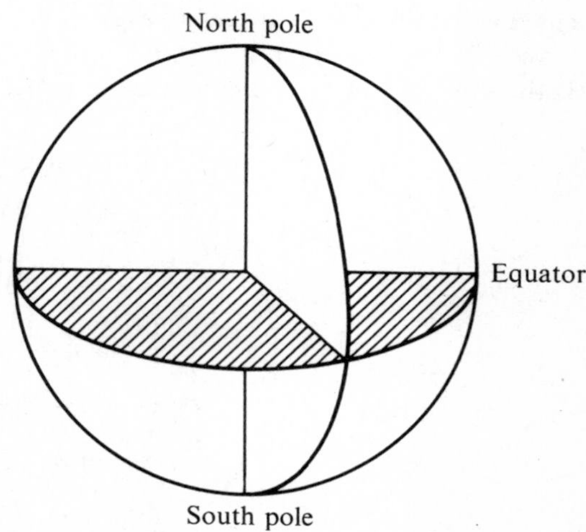

Fig. 12

(*b*) Look at Figure 13. The North and South poles are the points where the axis of rotation of the earth cuts its surface. Mark the North and South poles with a dot on your model. The equator is the Great circle lying in the plane which meets the polar axis in a right angle. The *half* of a Great circle which passes through the poles is called a MERIDIAN. The particular meridian which passes through Greenwich, London, has a special significance and is referred to as the *Greenwich meridian.*

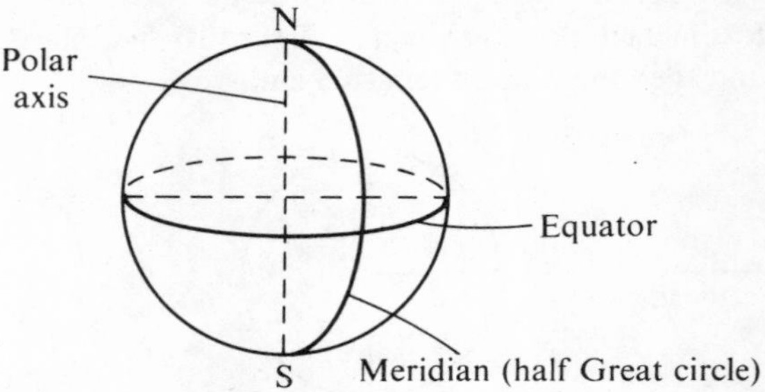

Fig. 13

(*c*) *Representing points on the earth's surface.* To describe the position of a point in a *plane* we:

(i) define an origin;
(ii) draw a grid of x and y lines;
(iii) represent points by ordered pairs (x, y).

Let us now look at the problem of describing the position of a point on the earth's surface. Our 'plane' is now the surface of a sphere.

(*d*) Suppose we wish to describe the position of Glasgow (G). At first sight, it might appear a good idea to produce a grid as shown in Figure 14, choosing the origin O as the intersection of the two Great circles marked in red, and the x and y lines as small circles 'parallel' to these.

According to our grid G is the point (1, 1). Why is this method not satisfactory? (Is there more than one point with coordinates (1, 1)?)

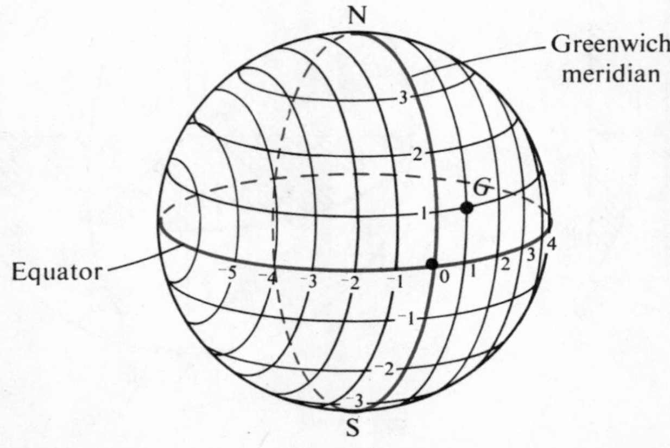

Fig. 14

(*e*) In Figure 15 Great circles have been drawn passing through the poles together with small circles parallel to the equatorial plane. Again we can try defining G as (1, 1). Is this satisfactory?

Now try using *half* Great circles (meridians) as 'y lines'. Is there more than one point with coordinates (1, 1) now?

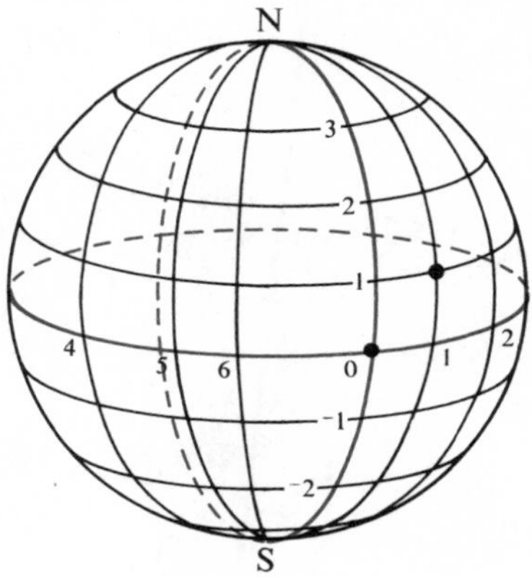

Fig. 15

(*f*) You should have found that *meridians* as 'y' lines and small circles parallel to the equatorial plane as 'x lines' give us a satisfactory coordinate system. The problem now is how to number these lines on a globe. It is particularly important that a common system be used for navigation purposes.

The method we use is described below, but before you read on, discuss the meaning of the distress signal 'Mayday, mayday. This is S.S. *Jodan*. We are sinking. Our position is 40° N, 30° E'.

(*g*) Colour the edge of one of the meridians on your model to represent the Greenwich meridian. Colour also the 'half-disc' (shaded in Figure 16) of which this meridian is a partial boundary.

Suppose this half-disc is rotated to the East with the earth's axis fixed. As it rotates, it defines our '*y* lines'. In Figure 16 the plane has turned through 30°. The angle between such a plane and the plane of the Greenwich meridian is called the angle of LONGITUDE of the plane. The plane of the Greenwich meridian has longitude 0°. The point *P* in Figure 16 lies on the *line of longitude* 30° E. (Why east?)

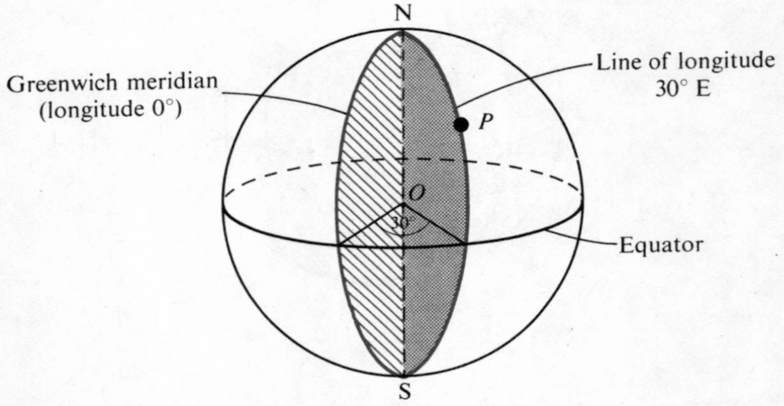

Fig. 16

(*h*) To describe the position of *P* exactly we need a second coordinate. Consider Figure 17. Suppose the radius *OH* rotates to the north with *O* fixed and *H* moving along the line of longitude 30° E. When *OH* has turned through 40°, *H* coincides with *P*. The small circle which passes through *P* and is parallel to the equatorial plane is called the parallel of LATITUDE 40° N. Points on this circle have *latitude* 40° N. (Why north?)

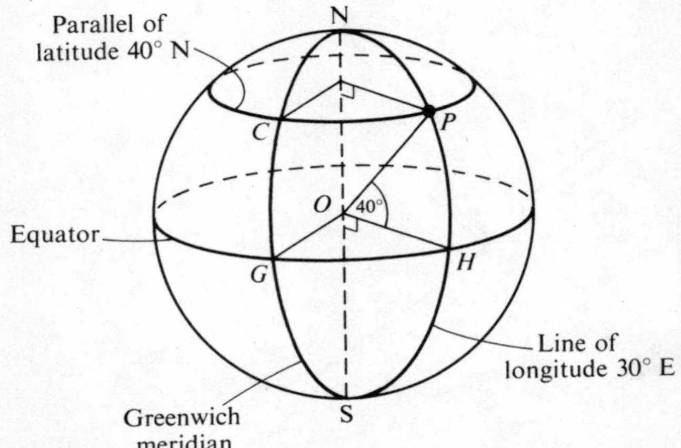

Fig. 17

We can now describe P as the point (40° N, 30° E). Notice that we give latitude first, longitude second.

Write down the positions of (*a*) C; (*b*) H; (*c*) G, in Figure 17.

Use a coloured pencil to mark the positions of (i) A(60° N, 0°); (ii) B(60° S, 90° W); (iii) C(30° S, 180° E) on your model.

Exercise C

1 You are standing at the point (30° S, 60° E). If you dig down through the centre of the earth, at what point do you eventually emerge?

Answer the same question if you are originally at the point (i) (40° N, 70° E); (ii) (80° N, 60° W).

2 Which two points on the earth's surface lie on each and every line of longitude? How would you describe the position of these two points using latitude and longitude?

3 P is the point (30° S, 27° W). Write down the new position of P after (i) a rotation 60° W followed by a rotation 70° N; (ii) a rotation 50° E followed by a rotation 50° S; (iii) a reflection in the Equatorial plane; (iv) a reflection in the plane of the Great circle passing through the poles and containing the Greenwich meridian; (v) a reflection in the plane containing the line of longitude 20° E.

4 You depart from Karachi (50° E, 20° N). What position do you reach if you fly (i) 40° E; (ii) 70° W; (iii) 30° N?

5 Each meridian has another one 'opposite' to it called its ANTIMERIDIAN. The antimeridian of the Greenwich meridian has longitude 180° E (or 180° W). We measure longitude from 0° to 180° E and from 0° to 180° W. What is the antimeridian of

(i) 50° W; (ii) 115° E; (iii) 90° E?

6 Two men start at P(60° N, 20° W) and travel by hovercraft on different Great circle routes. If they travel at the same speed, where will they meet again?

4 Nautical miles

(*a*) Let us take a trip in a hovercraft around a Great circle of the earth (see Figure 18). When we arrive back at our starting point the line segment OH will have turned through 360°. One degree can be divided into 60 smaller parts called *minutes* ($'$). OH therefore turns through $360 \times 60' = 21\,600'$.

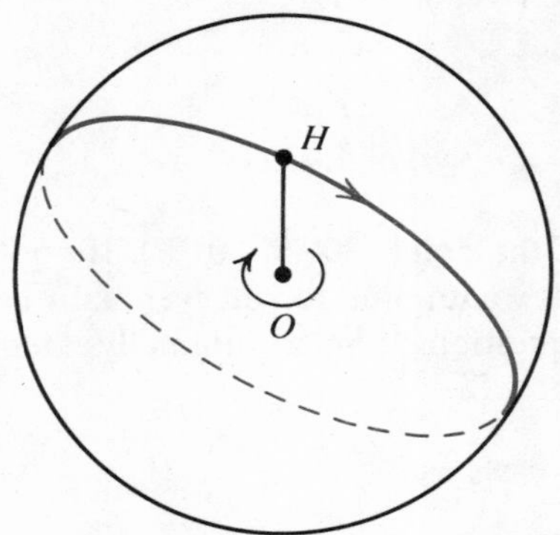

Fig. 18

How far have we travelled? We could measure the distance in kilometres and we would find it to be approximately 40 000 km.

Navigators, however, use another unit called a *nautical mile*. In travelling around the world the distance travelled is 21 600 nautical miles (n.m.). Can you see how a nautical mile is defined?

(*b*) If you travelled in a hovercraft from H to H' (Figure 19) so that OH turns through $1'$, you will have travelled a distance of 1 nautical mile. In the same way, if you travelled from H to H'', where $\angle HOH'' = 10°$, you will have travelled $10 \times 60 = 600$ n.m. A nautical mile is therefore the length of arc of a Great circle which subtends an angle of $1'$ at the earth's centre.

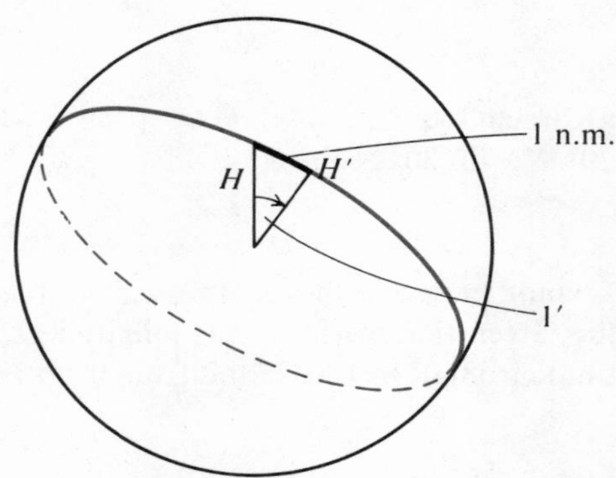

Fig. 19

What distance (in n.m.) would you travel from (i) (20° N, 10° E) to (30° N, 10° E); (ii) (20° N, 30° W) to (30° S, 30° W)?

Is it true that the distance from P(60° N, 30° E) to Q(60° N, 40° E) is $10 \times 60 =$ 600 n.m.? Explain your answer.

Exercise D

1 Calculate the shortest distance, measured along the surface of the earth, between the North and South poles. Give your answer in nautical miles.

2 Calculate the shortest distance in nautical miles between the following pairs of points on the earth's surface:

(i) (50° N, 20° E) and (30° N, 20° E);
(ii) (60° N, 30° W) and (20° S, 30° W);
(iii) (57° 24′ N, 127° 49′ E) and (9° 50′ N, 127° 49′ E);
(iv) (15° 15′ S, 4° 5′ E) and (12° 45′ N, 4° 5′ E);
(v) (0°, 153° E) and (0°, 178° W).

3 The circumference of the earth is 40 000 km. Copy and complete the statement: '1 n.m. is equivalent to . . . km'.

4 Taking π to be 3·14, calculate the radius of the earth in nautical miles.

5 The shortest distance between P and Q on the earth's surface is 1720 n.m. If O is the centre of the earth, calculate $\angle POQ$.

6 Two cities lie on the same meridian and are 3360 n.m. apart, measured along this meridian. One city is on the circle of latitude 25° 54′ N. On which circle of latitude does the other city lie? Is there more than one possible answer?

5 Distance along a parallel of latitude

(*a*) $S(60°$ N, 30° E) and $Q(60°$ N, 50° E) are two towns on the parallel of latitude 60° N (Figure 20). Would it be true to say that the distance SQ measured along this parallel of latitude is 20 × 60 n.m. = 1200 n.m.?

(*b*) Consider Figure 20. We can see that $\angle SXQ = \angle MOL = 20°$. Figure 21 shows the segment SXQ superimposed upon the segment MOL.

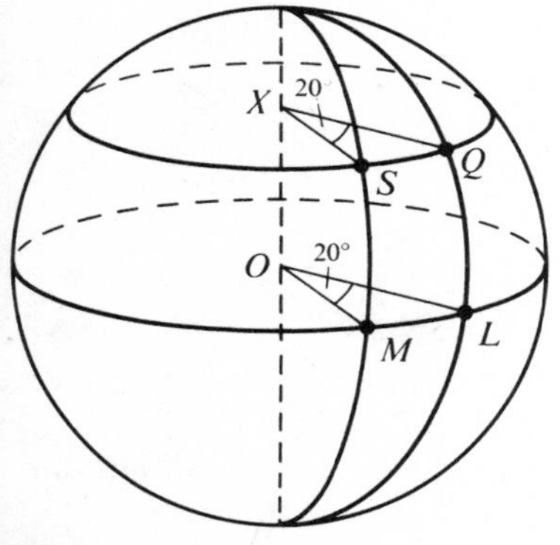

Fig. 20

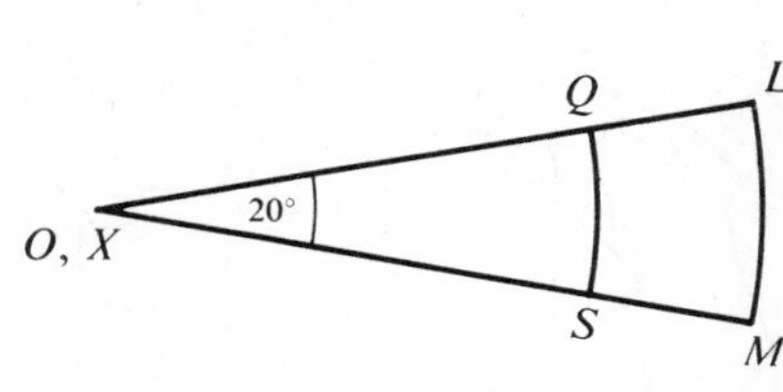

Fig. 21

It is obvious that the segments are similar, and that $QS = LM \times$ scale factor. But $ML = 60 \times 20$ n.m. = 1200 n.m. Hence, if we can find the scale factor, we can calculate QS.

(*c*) The scale factor we are looking for is $\dfrac{QS}{LM} = \dfrac{QX}{LO}$.

Look at triangle OQX in Figure 22.
Since $QO = LO$, the scale factor we require is

$$\frac{QS}{LM} = \frac{QX}{LO} = \frac{QX}{QO} = \cos 60°.$$

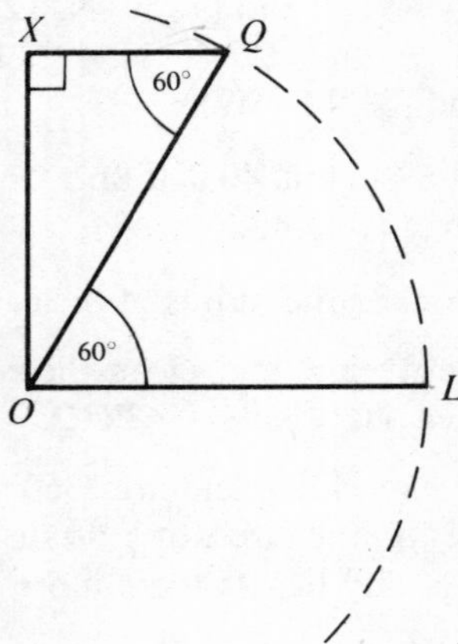

Fig. 22

Hence $QS = LM \times \cos 60°$
$= LM \times \frac{1}{2} = 1200 \times \frac{1}{2}$
$= 600.$

The length of QS is 600 n.m.

Generally, if S and Q lie on the parallel of latitude $\theta°$ N, then the distance SQ is given by

$$SQ = ML \times \cos\theta°.$$

Notice also that the radius of this circle of latitude is given by

$$XQ = OL \times \cos\theta°$$
$$= \text{radius of the earth} \times \cos\theta°.$$

Exercise E

1 Find the distance, in nautical miles, measured along the parallel of latitude between:

(i) (70° N, 30° E) and (70° N, 40° E);
(ii) (30° N, 35° E) and (30° N, 45° E);
(iii) (30° S, 0° E) and (30° S, 20° E);
(iv) (80° S, 100° W) and (80° S, 100° E).

2 (*a*) Stuttgart is at (49° N, 9° E), St Helier (49° N, 2° W).
Find the distance apart of these two towns measured along the parallel of latitude 49° N.
(*b*) Answer the same question for Milan (45° N, 9° E) and Bordeaux (45° N, 1° E).

3 Calculate the circumference of the circle of latitude 80° N, giving your answer in nautical miles.

4 Through what angle does the earth rotate in 1 hour? An aircraft flies along the equator. At what speed and in which direction will it have to fly for the sun to appear stationary to it? Give your answer in nautical miles per hour (knots). Would the aircraft have to fly faster, slower, or at the same speed if it flew along a parallel of latitude from London (52° N)? If not at the same speed, calculate what it should be.

5 The distance between two towns measured along the parallel of latitude 45° S is 1250 n.m. One town is at the point (45° S, 20° E). Where is the other town?

6 Find the new longitude of a hovercraft which travels 460 n.m. due west from (*a*) (30° N, 135° W); (*b*) (30° S, 135° E); (*c*) (50° S, 135° W); (*d*) (50° S, 40° 10′ E).

6 Volume and surface area of a sphere

Two important formulas, which will be proved to you at a later stage, are:

(i) surface area of a sphere of radius $r = 4\pi r^2$ units^2;
(ii) volume of a sphere of radius $r = \frac{4}{3}\pi r^3$ units^3.

Exercise F gives you some practice in using these two formulas and Questions 5 and 6 have been devised to give you an idea of how the formulas are derived.

Example 1

Calculate the volume and surface area of a sphere of radius 2 cm. (Take $\pi = 3{\cdot}14$.)

$$\text{Volume} = \tfrac{4}{3}\pi r^3 = \tfrac{4}{3} \times 3{\cdot}14 \times 8 = 33{\cdot}5 \text{ cm}^3 \text{ (3 S.F.)}.$$
$$\text{Surface area} = 4\pi r^2 = 4 \times 3{\cdot}14 \times 4 = 50{\cdot}2 \text{ cm}^2 \text{ (3 S.F.)}.$$

Exercise F

1 Find the volume and surface area of a sphere of radius 6 cm.

2 Find the radius of a sphere of area 5670 m^2.

3 A sphere passes through the eight vertices of a cube of side 2 cm. Calculate (i) the radius of the sphere; (ii) the volume of the sphere.

4 A ball has volume 9280 cm^3. Calculate its radius.

5 Look at Figure 23. Archimedes discovered that a pair of parallel planes perpendicular to the axis of the cylinder in Figure 5 would cut off the same area of the sphere as of the cylinder. What does this tell you about the surface area of the sphere and the surface area of the cylinder?
If the radius of the cylinder is r, what is its surface area?
Deduce the surface area of the sphere.

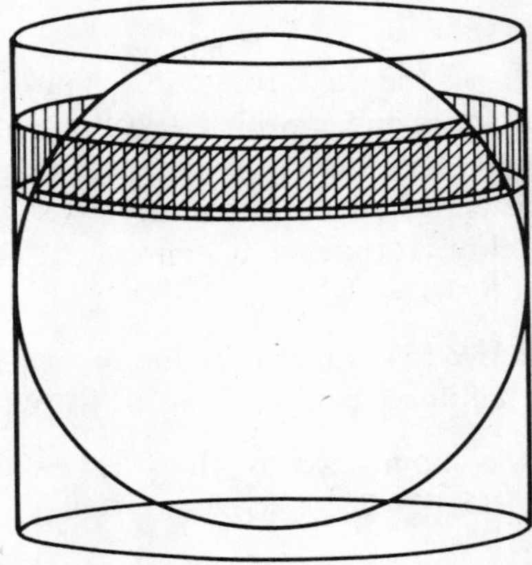

Fig. 23

6 Imagine a sphere filled with small pyramids, each with its vertex at the centre of the sphere and its base composed of a small portion of the sphere. Four of them are shown in Figure 24.

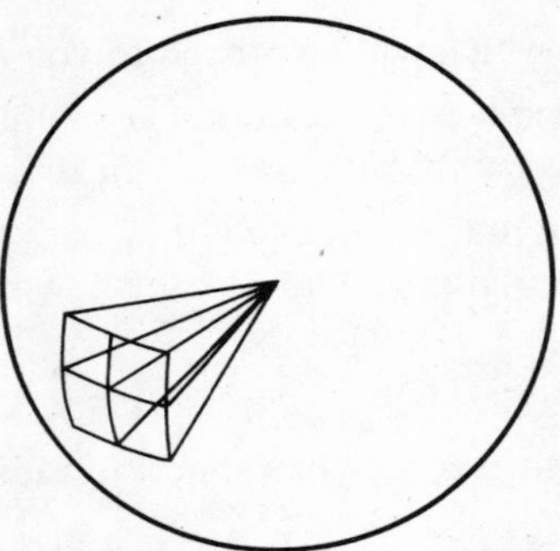

Fig. 24

The height of each pyramid is r. If the pyramid is small enough its base can be taken as being flat. The volume of each pyramid is therefore $\frac{1}{3}$(base area) $\times r$.

Deduce that the volume of the sphere is $\frac{4}{3}\pi r^3$, by using the fact that its surface area is $4\pi r^2$.

7 Find the surface area of the earth in (n.m.)^2.

8 Four spherical balls of radius 2 cm rest on a horizontal table with their centres at the corners of a square of side 4 cm. A fifth ball of radius 1 cm rests symmetrically on them. What is the height of its centre above the table?

9 What is the surface area of a cylindrical hole of radius 4 cm bored centrally through a sphere of radius 6 cm?

Proof

One of the important things you learn to do in mathematics is to 'argue things out'. You have to do that in other activities too. For example, a detective 'deduces' from the evidence that *X* committed a crime, whilst the defending lawyer tries to 'prove' to the jury that he didn't. Both of them have to convince other people by arguments which they are prepared to accept 'beyond all reasonable doubt'. Scientists argue things out too, by careful experiments and reasoned arguments leading to conclusions. We use the words *proof* or *demonstration* to describe reasoned arguments in mathematics. The study of the reasoned arguments is called *logic*.

Although there is some similarity between legal proofs, those in experimental science and mathematical proofs, there is one important difference. New evidence may change our legal or scientific conclusions, but in mathematics the 'evidence' consists of numbers or geometrical facts which we all agree to accept. So the evidence is fixed once we have agreed that we are talking about numbers or a particular kind of space.

If you think about it, you will see that we use logical (reasoned) proofs in the following way. There are three ingredients.

1 We want to convince someone (ourselves or others) that something is true.

2 We need an agreed starting point; some facts that the people concerned are prepared to accept.

3 We use a reasoned argument, which has to be correct (good logic) and convince the people concerned.

A new look at an old proof

Let us see how this works by examining a different proof of the fact that the angle in a semi-circle is a right angle. (*Book Y*, p. 134, Property 3).

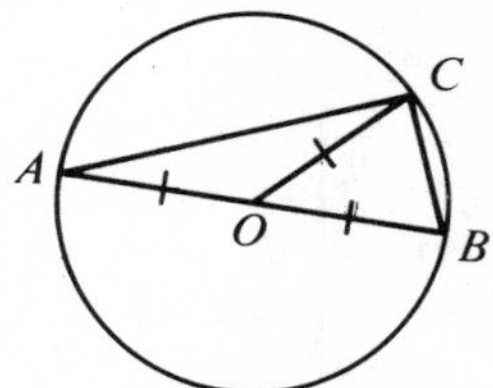

Fig. 1

Mathematical discoveries are usually made by guessing that something is true and then working backwards to an agreed starting point. When you have found a proof (an argument which you find convincing), then you write it out, beginning with an agreed starting point and ending with your discovery (in order to convince others). You will see a proof written in this way on p. 91.

We call the angles at A, B, C of the triangle ABC, $a°$, $b°$, $c°$, respectively. We want to prove that $c = 90$. (See Figure 1.)

What do you know about the angles of a triangle? If $a + b = 90$, what can you say about c?

What do we know about the triangle, as illustrated in Figure 1? AB is a diameter of a circle, O is the centre; so $OA = OB = OC$, because they are all radii of the circle.

Triangles AOC, BOC are isosceles and we know that the angles opposite equal sides are equal. So $\angle OCB = b°$, $\angle ACO = a°$.

Now $\angle ACO + \angle OCB = a° + b°$. But $\angle ACO + \angle OCB$ makes up $\angle ACB = c°$. Can you now convince yourself that $a + b = 90$?

Let us make a list of things we have assumed.

1 The angles of a triangle add up to 180°.

2 The angles opposite the equal sides of an isosceles triangle are equal.

Do you think these are good agreed starting points, or would you prefer to have them proved? (See the exercises at the end of this Interlude.)

About theorems

Mathematical properties or facts are often called *theorems*, which comes from a Greek word meaning 'look at this'. New theorems are often found by combining old ones and trying to get something extra out of looking at them together. Look at Properties 3 and 6 of *Book Y*, Some properties of the circle, and illustrate them in the same picture (Figure 2(a)).

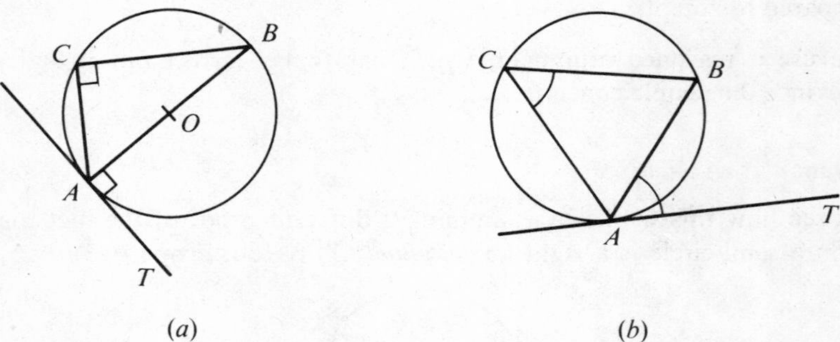

Fig. 2 (a) (b)

O is the centre of a circle, AT the tangent at A. The angle $\angle ACB$ is 90° (Property 3) and $\angle BAT$ is 90° (Property 6). So $\angle ACB = \angle BAT$. So we might ask: is it always true that $\angle ACB = \angle BAT$ when AT is a tangent, even if AB is not a diameter (Figure 2(b))?

Clearly we cannot expect to measure all possible angles; so we must find some other connection between the two angles. What do we know already from Figure 2(a)? We know that the answer to our question is 'yes' in the special case when AB is a diameter. So, as an experiment, draw in a diameter through A, as in Figure 3. Call it AD.

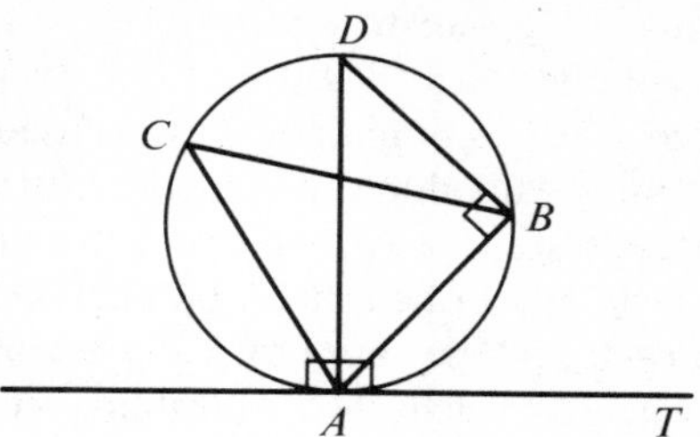

Fig. 3

Draw your own diagram and then answer these questions. Why is $\angle TAD = \angle DBA = 90°$? Does anything about Figure 3 remind you of Property 2, Some properties of the circle, *Book Y*? What can you say about $\angle ACB$ and $\angle ADB$? Why?

Now look at $\angle ADB$ and $\angle DAB$. What do they add up to? So what can you say about $\angle BAT$ and $\angle ADB$? Now what can you say about $\angle BAT$ and $\angle ACB$?

Let us make a list of our assumptions (agreed starting points).

1 The angle between the diameter and the tangent drawn at the end of the diameter is a right angle.

2 The angle in a semi-circle is a right angle.

3 Angles in the same segment are equal.

4 The angles of a triangle add up to 2 right angles.

Can you prove those theorems?

Writing out a proof

When a scientist or mathematician wants to publish his discoveries or write a book, he usually writes out his proofs in a polished way, beginning at the agreed starting points and giving clear reasons all the way through. (At any rate he tries to!) He has to begin by explaining clearly all the technical words he is going to use; that is, by making clear *definitions*. For example, the angle ACB in Figure 2(*b*) is called the angle in the *alternate segment* to $\angle BAT$. So, in a polished way, we should write:

Theorem (Look at this!) The angle between the tangent and the chord drawn through the point of contact is equal to the angle in the alternate segment.

Proof. Through the point of contact A, draw the diameter AD and join DB (see Figure 3). Then

$\angle ABD = 90°$ (angle in a semi-circle),

$\angle DAT = 90°$ (AT is a tangent, AD a diameter).

Now $\angle ACB = \angle ADB$ (angles in the same segment).

But $\angle DAB + \angle ADB = 90°$ (angles of a triangle).

Also $\angle DAB + \angle BAT = \angle DAT = 90°$ (proved above).

We have $\angle DAB + \angle BAT = \angle DAB + \angle ADB$.

Hence $\angle ACB = \angle ADB = \angle BAT$. Q.E.D.

In old-fashioned books they used to write Q.E.D. to show that the proof was completed. (It stands for the Latin 'quod erat demonstrandum', which means 'which was to be proved'. As you might guess, there used to be a corny joke that the letters stood for 'quite easily done'.) You will see that we have written out the argument so that each step has a line to itself and a brief explanation in brackets. That way of writing is not essential, but it is clear. The words 'now' and 'but' which introduce each sentence are not really essential either; they just get the sentences started. The word 'hence' in the last line means that that step comes easily from the preceding one.

When you are reading mathematics books, it is usually best actually to write out the proof for yourself, rather than just reading it as though it were a novel. And, of course, you should practise making proofs for yourself. That is why there are exercises at the end of this interlude.

A proof from arithmetic

Let us have a rest from geometry and do some arithmetic. You know that a prime number is one which is bigger than 1 and has no factors other than itself and 1. You have made lists of prime numbers; do you think such lists go on for ever or is there some number at which they stop? To answer that question it is no use going on calculating, you must produce a proof. What do you think the answer is?

What are our agreed starting points going to be? Obviously, the rules of arithmetic. How are we going to tackle the proof? We want to show that the primes go on for ever. That is, for any given prime number there must always be a bigger one. (So the list must go on for ever.)

Arrange the primes in ascending order of magnitude and write p_n to stand for the nth prime number. (Thus $p_1 = 2, p_2 = 3, p_3 = 5, p_4 = 7, p_5 = 11$, etc.) Consider the number $p_1p_2p_3 \ldots p_n + 1$. It is not divisible by p_1 or $p_2 \ldots$ or p_n, because each time there is a remainder 1 when you divide. So its least prime divisor is different from the first n prime numbers and must therefore be a prime number bigger than p_n. We can always repeat that argument and so the list never comes to an end.

Another sort of proof

'If Lancashire don't win the Gillette Cup, I'll eat my hat.' The speaker has no intention of being so absurd as to eat his hat, he is just expressing his confidence in Lancashire cricket. There is a sort of proof, which works rather like that. 'If that theorem isn't true', you say, 'then something absurd will happen.' (And since mathematical absurdities are not allowed, the theorem must be true.) The method of proof is called *reductio ad absurdum* (you don't need to know Latin to translate that) or *indirect proof*.

Here is an example of indirect proof in arithmetic.

The question is: is it possible to express $\sqrt{2}$ as a fraction? That is, are there whole numbers a, b such that

$$2 = \frac{a^2}{b^2}?$$

Suppose there are two such numbers and that any common factor has been cancelled. Then

$$2b^2 = a^2.$$

Now think about the units digits of a and of b and complete Table 1.

TABLE 1

	a	b	a^2	b^2	$2b^2$
	0	0	0	0	0
	1	1	1	1	2
	2				
Units digits	3				
	4				
	5				
	6				
	7	7	9	9	8
	8	8	4	4	8
	9				

What do you notice about the entries in column 3 and column 5? When are they equal? What are the corresponding entries in columns 1 and 2? Have a and b any factors in common? Why? Have you found an absurdity (a *contradiction*)?

Indirect proof in geometry

We saw in Some properties of the circle, *Book Y*, Property 6, that if you draw a diameter of a circle and a tangent at one end of it, then the angle between them is 90°. Can we turn that round and ask: if you draw a line at 90° at the end of a diameter, must it be a tangent? (Turning the theorem round gives a new statement called the *converse*. We shall meet more converse theorems in the next section.)

First of all, let us agree that a tangent meets a circle in exactly one point, the point where they touch. So our question is this: is it possible to draw a diameter and a line at right angles to it at the point where it meets the circumference, which is not a tangent?

We begin by supposing it is possible to draw such a line and then we shall argue our way to a contradiction. If a correct argument produces a contradiction, our starting point must be wrong. So we shall be able to deduce that our line must be a tangent after all.

If the line is not a tangent, it will meet the circle again. Let AB be the given diameter, O the centre of the circle and C the point where the line meets the circle again (see Figure 4). Since the line AC is given to be at right angles to AB, we know that $\angle CAB = 90°$. (Even though it doesn't look like that.)

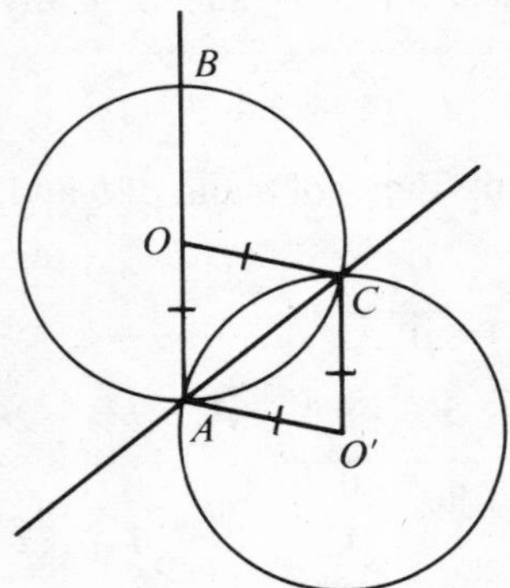

Fig. 4

Imagine that the circle is reflected in the line AC. We know that reflection preserves distances and angles (more agreed starting points! – see the Geometry Review in this book); so OA must be reflected to $O'A$, where O' is the centre of a circle exactly like the given one and $\angle O'AC = 90°$.

Now $\angle OAO' = \angle OAC + \angle CAO' = 180°$; so OAO' is a straight line (in spite of appearances). This means that the figure $OAO'C$ is really a triangle. Now $OC = O'C$, because they are radii of equal circles. But then $OO' = OA + AO' = OC + CO'$. This is impossible, because the lengths of two sides of a triangle added together must be greater than the third. (Do you agree to take that as a starting point?)

So we have our contradiction and we can get out of it only by admitting that the point C is really the point A; that is, that the line meets the circle exactly once and so is a tangent.

Now write that out as a theorem with a polished proof.

Converse theorems

Look again at the two theorems about circles and tangents which we have just been dealing with. One said: 'If you draw a tangent at the end of a diameter, then the angle is a right angle.' The other said: 'If you draw a line at the end of a diameter such that the angle they make is a right angle, then the line is a tangent.' In other words, theorems usually read something like this: 'If THIS fact is given, then THAT fact follows.' THIS is called the *hypothesis*, THAT the *conclusion*. (In practice, both THIS and THAT may be several facts.)

Can you say which was the hypothesis and which the conclusion in each of the two theorems about tangents and diameters? In the second one the conclusion of the first has become the hypothesis and vice versa. If two theorems are related like that, then one is said to be the *converse* of the other.

It is important to realize that both a theorem and its converse need proofs; you cannot be sure that a converse is true just because the theorem is. (You may remember what the Hatter said at the Mad Tea-Party in *Alice in Wonderland.* 'Not the same thing a bit!' said the Hatter. 'Why, you might just as well say that "I see what I eat" is the same thing as "I eat what I see"!' In this case the 'theorem' is: 'if I eat it, I see it' and the converse is 'if I see it, I eat it'.) Here is an example of a mathematical theorem which has a false converse. *If two triangles have equal sides, then they have equal angles.* That is true. The converse is: *If two triangles have equal angles, then they have equal sides.* That is false, one could be an enlargement of the other.

You will find some more examples at the end of this Interlude.

Why bother?

Why bother with proofs at all? Well, if you were an astronaut about to be launched to Mars you would be mildly interested to know whether or not the mathematicians had argued things out correctly when they advised the scientists and engineers about the right orbit. There are a great many situations where mathematics is being applied and where we need to know that the arguments are correct.

Another reason for learning mathematics is that it is interesting to try to find out facts about numbers and space and then try to prove them. Of course you have to take some things for granted. In mathematics part of the fun lies in taking as little for granted as possible. That means we have to prove more, but it also means there is less likelihood of a mistake. It is also more fascinating, in the same way as the story of Robinson Crusoe fascinates, just because he was able to do so much with so little.

The things you decide to take for granted are called *axioms* or *postulates*. The things you manage to prove are called *theorems* or *propositions*. Usually the theorems are arranged in order, beginning with the ones that are proved immediately after the axioms and then placing the others so that they depend only on theorems already proved.

The idea of arranging things in a *logical system* or a *deductive system* was first carried out by a Greek mathematician called Euclid over 2000 years ago. Other sciences still try to imitate his system and that is why we bother about proofs.

Some proofs to try

Here are some more examples of things to prove, for you to try. Just as you can learn to play the piano or cricket or learn tennis only if you practise, so you can learn mathematics only if you work out proofs for yourself. These exercises will help you to do that. In some cases the proof is almost completely written out for you. In those cases, write out the argument, making sure you understand all the steps.

1 Look at the properties of circles in Some properties of the circle, *Book Y*, and write out proofs in the style of the one in this Interlude. As you are writing them out, make a note of each agreed starting point and then make a list of all the agreed starting points.

2 Imagine that you have a piece of paper with an arc of a circle drawn on it. Suppose the centre of the circle is off the paper. How would you draw the tangent at a point on the arc?

Prove that your construction works.

3 You have already met the theorem: *the angle between the tangent and the chord equals the angle in the alternate segment*. What is the hypothesis of that theorem? What is the conclusion? What is the converse theorem? Prove it! [*Hint*. When trying to prove a converse, it is often a good idea to use an indirect proof.]

4 In the diagram, the circle is inscribed in the quadrilateral $ABCD$.

Prove that $AB + CD = BC + DA$.

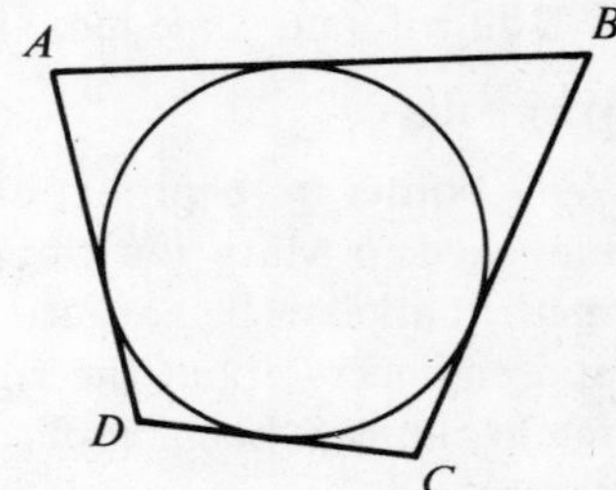

Fig. 5

5 What is the converse of the theorem in Question 4? Is it true? (In order to decide whether a theorem is true or not you must either prove it or find an example which shows that it cannot be true, like when we looked at equi-angular triangles to see if they had to have equal sides.)

Exercises 6, 7 and 8 show that a theorem may have more than one converse, some of which are true and some false.

6 In Figure 6, XY is drawn through the mid-point X of the side AB of the triangle ABC, parallel to the side BC. Prove that Y is the mid-point of the side AC and that XY is half of BC.

The information and things to prove can be summarized as follows.

There is a triangle ABC, with points X, Y on AB, AC respectively. The following statements are made about X, Y etc.

(1) X is the mid-point of AB;
(2) XY is parallel to BC;
(3) $XY = \frac{1}{2}BC$;
(4) Y is the mid-point of AC.

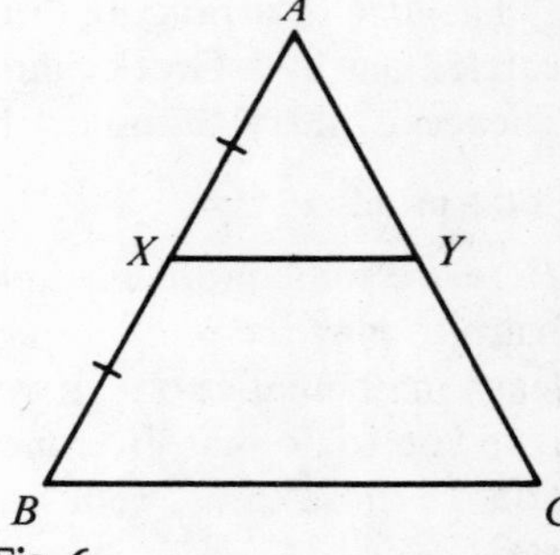

Fig. 6.

The question then asks you to prove that statements (1) and (2) (the hypotheses) imply statements (4) and (3) (the conclusions).

[*Hint*. Take as an agreed starting point the fact that an enlargement preserves parallel lines. Consider an enlargement with scale-factor 2 and centre A.

What happens to X? What happens to Y? (Why?) So what can you say about AY and YC?]

7 One converse of the theorem in Exercise 6 is this: *the line joining the mid-points of two sides of a triangle is parallel to the third side and equal to one-half of it.* Is that true? (Either prove it or give an example which shows it to be false. Note that it says: statements (1) and (4) imply statements (2) and (3).)

8 Here is another converse of Exercise 6. *If X is the mid-point of AB and if Y is a point on AC such that XY is one-half of BC, then Y is the mid-point of AC.* Is that true? (That is, is it true that statements (1) and (3) imply statement (4)?)

9 (This question is quite hard, but remember that mathematicians like mathematics because it is hard!)
ABC is a triangle and X, Y, Z are points between A, B; A, C; B, C respectively. The circles through A, X, Y and B, X, Z meet at a point P. Prove that the circle through C, Y, Z also passes through P.

10 In the last question, does it matter that X, Y, Z are between the points as stated? Would the theorem still be true if X, Y, Z were any points on AB, AC, BC extended outside the triangle?

11 Here is a variation of the proof that 2 cannot be expressed as a^2/b^2. Suppose that $2b^2 = a^2$ where a and b are whole numbers that have no common factor. Since a^2 is twice b^2, a^2 must be even. Can you prove that a must be even? (If a is *not* even, then its square cannot be even. Put $a = 2m + 1$ and notice that $a^2 = 4(m^2 + m) + 1$, which must be odd. So a must be of the form $a = 2m$). Now put $a = 2m$. Can you see why $b^2 = 2m^2$? Show that b must also be even and deduce that a and b have 2 as a common factor. Why is that a contradiction?

12 Can you prove that 3 cannot be expressed as a square of an ordinary fraction? What about 5, 7, 11, 13, . . .?

13 Here is yet another variation. Take as agreed starting point the fact that every number can be written as a product of prime numbers in exactly one way. Does a^2 have an even number of prime factors or an odd number? What about $2b^2$ (or $3b^2$, $5b^2$, $7b^2$ . . .)? What do you deduce?

14 This exercise introduces you to an important method of proof called *proof by induction.* In *Book A* you met formulas like $1 + 3 = 2^2$, $1 + 3 + 5 = 3^2$, $1 + 3 + 5 + 7 = 4^2$, which you proved by counting dots. All those formulas are summarized in

$$1 + 3 + 5 + \cdots + (2n - 1) = n^2,$$

which you can also prove by counting dots. If you think about it, the dot-counting argument enables you to go from 'the dot pattern for $n - 1$' to 'the dot pattern for n'. In other words you say: 'the argument goes like this, and so on'. This method of proof is called *induction* (strictly speaking, *mathematical induction*) and it has two essential ingredients: (*a*) you must prove the formula for a starting value of n (usually $n = 1$); (b) you must prove that *if the formula is true for* $n = k$, *then it must be true for* $n = k + 1$. So if we know it is true for $n = 1$, then (b) tells us that it must be true for $n = 2$. But (b) then tells us that it must be true for $n = 3$, and so on.

Let us see how this works for the sum of the first n odd numbers. It is certainly true for $n = 1$, because $1 = 1^2$. Now suppose that it is true for $n = k$. So

$$1 + 3 + 5 + \cdots + (2k - 1) = k^2.$$

Now add on the next odd number, $2k + 1$. We get

$$1 + 3 + 5 + \cdots + (2k - 1) + (2k + 1) = k^2 + 2k + 1$$

But $k^2 + 2k + 1 = (k + 1)^2$; so

$$1 + 3 + 5 + \cdots + (2k + 1) = (k + 1)^2.$$

15 *The tower of Hanoi*. This is another exercise in induction. It concerns a problem which is supposed to have originated in a Buddhist Temple in Hanoi. You may have seen a version of it which is sold as a puzzle.

There are three vertical pegs A, B, C with n numbered discs on one of them, say A. The discs are numbered so that the largest is 1 and the numbers increase as the discs decrease. The rule of the game is that one may not place a larger disc on a smaller disc and the object of the game is to transfer all the discs to another peg, say B.

By trial and error it is easy to see that if you have one disc only ($n = 1$), then you need one move. If you have two discs ($n = 2$), you need 3 moves and if you have three discs ($n = 3$) you need 7. By more trials you might be led to guess that for n discs you need $2^n - 1$ moves. Let us see if we can prove that, by induction.

The case $n = 1$ has already been settled. For then $2^n - 1 = 2 - 1 = 1$ and one disc needs one move.

Suppose then that we know that k discs require $2^k - 1$ moves. Think of $k + 1$ discs on peg A. Keep the bottom, largest disc fixed. How many moves are required to transfer the remaining k discs to peg C? Imagine that done. You need 1 move to transfer the $(k + 1)$th disc to peg B. Now how many moves do you require to transfer the original k discs on top of the one on peg B? Is $(2^k - 1) + 1 + (2^k - 1) = 2^{k+1} - 1$? Does that do the trick?

16 Look at all the agreed starting points in your geometrical proofs and discuss whether or not they require proof. For example, why do the angles of a triangle add up to 180°? Why are the base angles of an isosceles triangle equal? Is it necessary to prove that two sides of a triangle are greater than the third?

Review chapters
7 Matrices

1 Storing information

Matrices can be used to store information in a concise manner. For example:

$$\mathbf{D} = \begin{array}{l} \text{Breakfasts} \\ \text{Lunches} \\ \text{Teas} \\ \text{Suppers} \end{array} \overset{\begin{array}{ccc} \text{Boys} & \text{Teaching staff} & \text{Domestic staff} \end{array}}{\begin{pmatrix} 35 & 3 & 4 \\ 65 & 7 & 8 \\ 48 & 7 & 0 \\ 35 & 5 & 4 \end{pmatrix}}$$

gives the daily numbers of meals required on weekdays in a small boarding school and

$$\mathbf{L} = \begin{array}{c} a \\ b \\ c \\ d \end{array} \overset{\begin{array}{cccc} a & b & c & d \end{array}}{\begin{pmatrix} 0 & 1 & 0 & 0 \\ 0 & 0 & 0 & 0 \\ 0 & 1 & 0 & 0 \\ 1 & 1 & 1 & 0 \end{pmatrix}}$$

describes the relation 'is longer than' on the sides a, b, c, d of a quadrilateral.

Which is the longest side of the quadrilateral? Which is the shortest side? Is there anything special about this quadrilateral?

D has 4 rows and 3 columns. We say that the order of **D** is 4×3 (read as 'four by three'). Remember that we always write the row number first.

What is the order of **L**?

2 Combining matrices

Whatever information matrices contain, they all obey the same laws of algebra.

Let $\qquad \mathbf{A} = \begin{pmatrix} 5 & ^-2 \\ ^-1 & 2 \\ 4 & 3 \end{pmatrix}$ and $\mathbf{B} = \begin{pmatrix} 2 & 0 \\ ^-5 & 1 \\ 4 & ^-1 \end{pmatrix}$.

(*a*) We add two matrices by adding the pairs of elements which have corresponding positions in the matrices:

$$\mathbf{A}+\mathbf{B}=\begin{pmatrix} 5+\ 2 & {}^-2+\ 0 \\ {}^-1+{}^-5 & 2+\ 1 \\ 4+\ 4 & 3+{}^-1 \end{pmatrix}=\begin{pmatrix} 7 & {}^-2 \\ {}^-6 & 3 \\ 8 & 2 \end{pmatrix}.$$

Since $\mathbf{A}-\mathbf{B}$ is the matrix which added to $\mathbf{B}$ gives $\mathbf{A}$, we can subtract $\mathbf{B}$ from $\mathbf{A}$ by subtracting the pairs of elements which have corresponding positions in the matrices.

Calculate $\mathbf{A}-\mathbf{B}$.

We can add or subtract two matrices only if they have the same order. Why? What can you say about the order of the answer?

Is matrix addition commutative? Is it associative?

(*b*) We can multiply a matrix by a scalar:

$$3\mathbf{A}=\begin{pmatrix} 3\times\ 5 & 3\times{}^-2 \\ 3\times{}^-1 & 3\times\ 2 \\ 3\times\ 4 & 3\times\ 3 \end{pmatrix}=\begin{pmatrix} 15 & {}^-6 \\ {}^-3 & 6 \\ 12 & 9 \end{pmatrix}.$$

Calculate $\mathbf{B}+\mathbf{B}+\mathbf{B}+\mathbf{B}$, or $4\mathbf{B}$.

(*c*) When we multiply two matrices, we combine each row of the first matrix with each column of the second matrix. If

$$\mathbf{C}=\begin{pmatrix} 3 & 0 & 1 \\ 2 & {}^-1 & 4 \end{pmatrix} \quad \text{and} \quad \mathbf{D}=\begin{pmatrix} 1 & {}^-1 \\ 3 & 1 \\ 2 & 5 \end{pmatrix},$$

the *second* row of $\mathbf{C}$ combines with the *first* column of $\mathbf{D}$ to give

$$(2 \quad {}^-1 \quad 4)\begin{pmatrix} 1 \\ 3 \\ 2 \end{pmatrix}=2\times 1+{}^-1\times 3+4\times 2=7$$

and this appears in the *second* row and *first* column of $\mathbf{CD}$:

$$\mathbf{CD}=\begin{pmatrix} * & * & * \\ 2 & {}^-1 & 4 \end{pmatrix}\begin{pmatrix} 1 & * \\ 3 & * \\ 2 & * \end{pmatrix}=\begin{pmatrix} * & * \\ 7 & * \end{pmatrix}.$$

Calculate $\mathbf{CD}$ and $\mathbf{DC}$. Is matrix multiplication commutative?

Is it associative?

We can work out $\mathbf{EF}$ only if the number of elements in each row of $\mathbf{E}$ is equal to the number of elements in each column of $\mathbf{F}$.

Remember the domino pattern:

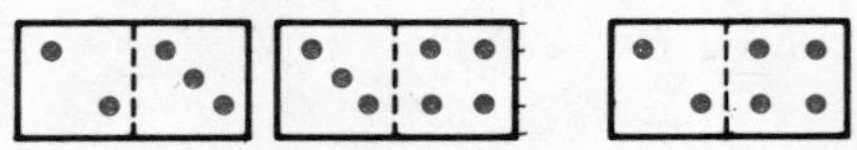

If $\quad \mathbf{E}=\begin{pmatrix} 1 & 0 & 1 \\ 1 & 1 & 0 \end{pmatrix} \quad \text{and} \quad \mathbf{F}=\begin{pmatrix} 1 & 0 & 1 & 1 \\ 1 & 1 & 0 & 1 \\ 1 & 1 & 1 & 0 \end{pmatrix},$

does $\mathbf{EF}$ have a meaning? Does $\mathbf{FE}$ have a meaning?

Exercise A

1 Calculate, where possible,

(*a*) $(3 \quad 2)\begin{pmatrix} ^-5 \\ 4 \end{pmatrix}$; (*b*) $(2 \quad ^-3 \quad 7)\begin{pmatrix} 4 \\ 2 \\ ^-5 \end{pmatrix}$;

(*c*) $(8 \quad 0 \quad 5)\begin{pmatrix} 2 \\ 7 \end{pmatrix}$; (*d*) $(6 \quad ^-5 \quad 7 \quad 0)\begin{pmatrix} ^-1 \\ 4 \\ 3 \\ 9 \end{pmatrix}$.

2 Calculate:

(*a*) $\begin{pmatrix} 3 & 6 \\ ^-2 & 5 \end{pmatrix}\begin{pmatrix} ^-4 \\ 1 \end{pmatrix}$; (*b*) $(4 \quad ^-7)\begin{pmatrix} 2 & ^-3 \\ 1 & 4 \end{pmatrix}$.

3 If $\mathbf{A} = \begin{pmatrix} 2 & 1 \\ 0 & ^-3 \\ 7 & 2 \end{pmatrix}$ and $\mathbf{B} = \begin{pmatrix} 1 & ^-3 \\ ^-2 & 6 \\ 8 & 5 \end{pmatrix}$, calculate:

(*a*) $\mathbf{A}+\mathbf{B}$; (*b*) $3\mathbf{A}$; (*c*) $4\mathbf{B}$; (*d*) $4\mathbf{B}-3\mathbf{A}$.

4 If $\mathbf{X} = \begin{pmatrix} 1 & 0 & 2 \\ ^-3 & 2 & 0 \\ ^-1 & 1 & 3 \end{pmatrix}$, $\mathbf{Y} = \begin{pmatrix} ^-3 & ^-2 & 2 \\ ^-9 & 7 & 3 \\ 2 & 1 & 0 \end{pmatrix}$ and $\mathbf{Z} = \begin{pmatrix} 2 & 0 \\ 0 & ^-1 \\ 3 & 2 \end{pmatrix}$,

calculate, where possible,

(*a*) $\mathbf{X}-\mathbf{Y}$; (*b*) $\mathbf{X}+\mathbf{Z}$; (*c*) $\mathbf{XY}$;
(*d*) $\mathbf{YZ}$; (*e*) $\mathbf{Y}^2$; (*f*) $\mathbf{YXZ}$.

5 Find p, q, r and s if

$$\begin{pmatrix} p & ^-1 \\ 6 & 2 \end{pmatrix} + \begin{pmatrix} 2 & q \\ r & ^-7 \end{pmatrix} = \begin{pmatrix} 5 & ^-5 \\ 3 & s \end{pmatrix}.$$

6 Find a and b if

$$\begin{pmatrix} ^-6 & 0 & 1 \\ 1 & 2 & b \end{pmatrix}\begin{pmatrix} a \\ 3 \\ 1 \end{pmatrix} = \begin{pmatrix} 1 \\ 1 \end{pmatrix}.$$

7 $\mathbf{A}$ is a 2×3 matrix.

(*a*) If $\mathbf{A}+\mathbf{B}=\mathbf{C}$, what can you say about the order of (i) $\mathbf{B}$; (ii) $\mathbf{C}$?
(*b*) If $\mathbf{AD}=\mathbf{E}$, what can you say about the order of (i) $\mathbf{D}$; (ii) $\mathbf{E}$?

3 Sets of matrices

(*a*) Consider $S_{2,3} = \{2 \times 3 \text{ matrices}\}$ under matrix addition. $S_{2,3}$ is closed under addition and contains the identity

$$\mathbf{0} = \begin{pmatrix} 0 & 0 & 0 \\ 0 & 0 & 0 \end{pmatrix}.$$

0 is often called the 'zero' matrix for $S_{2,3}$.
Matrix addition is associative.

Every 2×3 matrix has an additive inverse. For example, the inverse of

$$\begin{pmatrix} 1 & ^-3 & 0 \\ 4 & 2 & ^-7 \end{pmatrix} \text{ is } \begin{pmatrix} ^-1 & 3 & 0 \\ ^-4 & ^-2 & 7 \end{pmatrix}.$$

Thus $S_{2,3}$ forms a group under addition, that is, it possesses all four of the properties of closure, identity, inverse and associativity.

(*b*) Now consider $S_{2,2} = \{2 \times 2 \text{ matrices}\}$ under matrix multiplication.
Is $S_{2,2}$ closed under multiplication?
The identity is

$$\mathbf{I} = \begin{pmatrix} 1 & 0 \\ 0 & 1 \end{pmatrix}.$$

I is the 'unit' matrix for $S_{2,2}$.
Matrix multiplication is associative.
By calculating

$$\begin{pmatrix} 2 & ^-5 \\ ^-1 & 4 \end{pmatrix}\begin{pmatrix} \frac{4}{3} & \frac{5}{3} \\ \frac{1}{3} & \frac{2}{3} \end{pmatrix} \text{ and } \begin{pmatrix} \frac{4}{3} & \frac{5}{3} \\ \frac{1}{3} & \frac{2}{3} \end{pmatrix}\begin{pmatrix} 2 & ^-5 \\ ^-1 & 4 \end{pmatrix}$$

check that the multiplicative inverse of

$$\begin{pmatrix} 2 & ^-5 \\ ^-1 & 4 \end{pmatrix} \text{ is } \begin{pmatrix} \frac{4}{3} & \frac{5}{3} \\ \frac{1}{3} & \frac{2}{3} \end{pmatrix}.$$

Does every 2×2 matrix have a multiplicative inverse?
Does $S_{2,2}$ form a group under multiplication?

(*c*) A method for finding the multiplicative inverse of a 2×2 matrix **M** is indicated by the flow diagram in Figure 1 and is applied to the matrix

$$\begin{pmatrix} 2 & ^-5 \\ ^-1 & 4 \end{pmatrix}.$$

What happens when the method is applied to the matrix

$$\begin{pmatrix} 4 & 2 \\ 6 & 3 \end{pmatrix}?$$

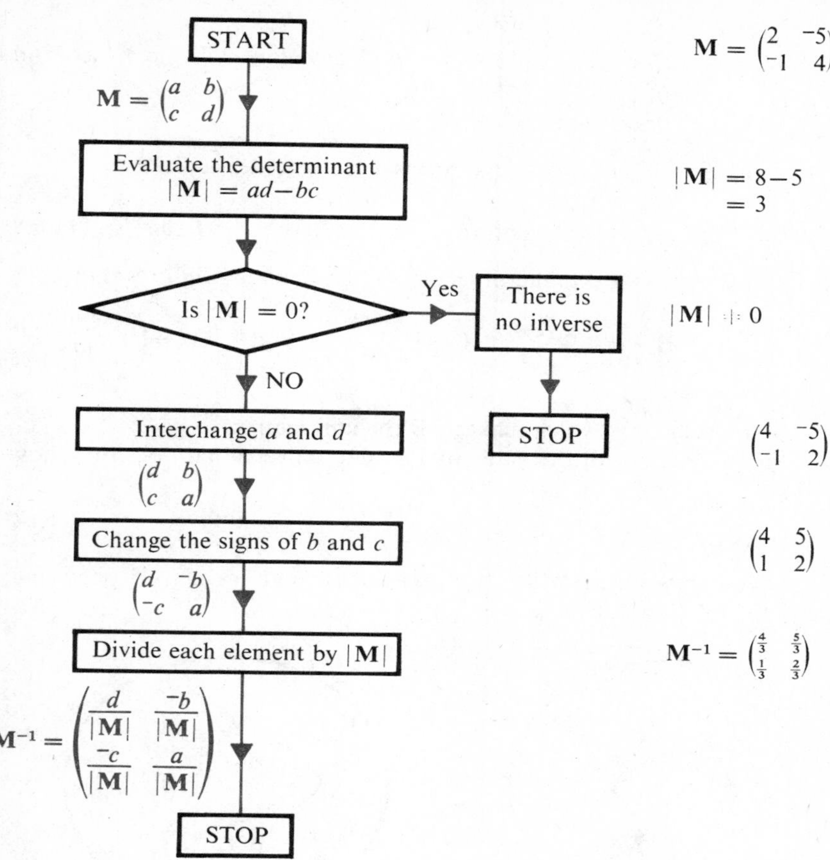

Fig. 1

Exercise B

1 Find the multiplicative inverse, if it exists, of each of the following matrices:

(*a*) $\begin{pmatrix} 5 & 4 \\ 6 & 5 \end{pmatrix}$; (*b*) $\begin{pmatrix} 7 & 2 \\ 13 & 4 \end{pmatrix}$; (*c*) $\begin{pmatrix} 5 & 8 \\ ^-2 & ^-3 \end{pmatrix}$;

(*d*) $\begin{pmatrix} 3 & 7 \\ 7 & 16 \end{pmatrix}$; (*e*) $\begin{pmatrix} 1 & 0 \\ 1 & ^-1 \end{pmatrix}$; (*f*) $\begin{pmatrix} 0{\cdot}6 & ^-0{\cdot}8 \\ 0{\cdot}8 & 0{\cdot}6 \end{pmatrix}$;

(*g*) $\begin{pmatrix} 6 & 24 \\ 2 & 8 \end{pmatrix}$; (*h*) $\begin{pmatrix} 2 & 5 \\ 1 & ^-7 \end{pmatrix}$; (*i*) $\begin{pmatrix} \frac{1}{2} & ^-\frac{1}{4} \\ \frac{1}{3} & \frac{1}{3} \end{pmatrix}$.

2 Write down the additive inverses of the following matrices:

(*a*) $\begin{pmatrix} 1 & ^-5 & 2 \\ 0 & 3 & ^-7 \\ 4 & ^-1 & 9 \end{pmatrix}$; (*b*) $\begin{pmatrix} 6 \\ ^-8 \\ 11 \end{pmatrix}$; (*c*) $\begin{pmatrix} 4 & ^-9 \\ 3 & 2 \\ ^-8 & 0 \end{pmatrix}$.

3 State the unit and zero matrices for the set of 4 × 4 matrices.

4 If $\mathbf{A}=\begin{pmatrix}2 & 1 & 1\\ 0 & 3 & ^-2\\ 1 & 2 & 0\end{pmatrix}$ and $\mathbf{B}=\begin{pmatrix}4 & 2 & ^-5\\ ^-2 & ^-1 & 4\\ ^-3 & ^-3 & 6\end{pmatrix}$, calculate **AB** and hence find the multiplicative inverse of **B**.

5 Explain why $S_{2,1}=\{2\times 1 \text{ matrices}\}$ forms a group under addition.

6 If S is the set of matrices of the form $\begin{pmatrix}a & ^-b\\ b & a\end{pmatrix}$ where $a, b\in\{\text{reals}\}$ and a and b are not both zero, show that every member of S has a multiplicative inverse.

7 $\mathbf{A}=\begin{pmatrix}1 & 2\\ 0 & 3\end{pmatrix}$, $\mathbf{B}=\begin{pmatrix}1 & ^-\frac{2}{3}\\ 0 & \frac{1}{3}\end{pmatrix}$, $\mathbf{C}=\begin{pmatrix}2 & 0\\ 4 & 1\end{pmatrix}$ and $\mathbf{D}=\begin{pmatrix}\frac{1}{2} & 0\\ k & 1\end{pmatrix}$.

(*a*) Evaluate **AB**.
(*b*) Find the value of k which makes **CD** the unit matrix.
(*c*) Simplify **CABD** with the value of k found above. What does this show about the inverse of **CA**?

4 Relation matrices

(*a*) Figure 2(*a*) shows the relation 'is on the left of' on a set of six dancers and Figure 2(*b*) the relation 'is opposite to' on the same set.

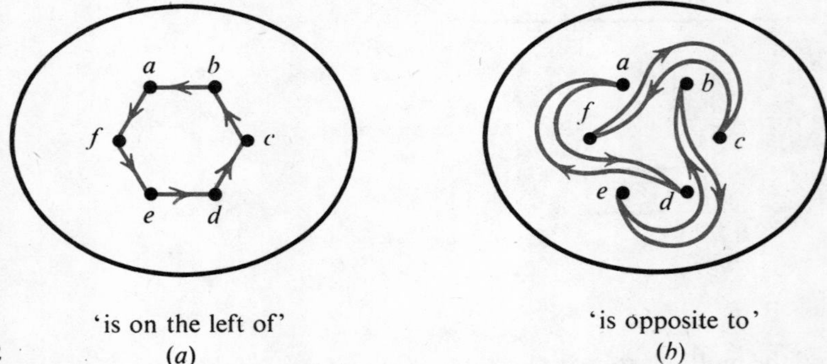

Fig. 2 'is on the left of' (*a*) 'is opposite to' (*b*)

The matrix which represents the relation 'is on the left of' is

$$\mathbf{L}=\begin{pmatrix}0 & 0 & 0 & 0 & 0 & 1\\ 1 & 0 & 0 & 0 & 0 & 0\\ 0 & 1 & 0 & 0 & 0 & 0\\ 0 & 0 & 1 & 0 & 0 & 0\\ 0 & 0 & 0 & 1 & 0 & 0\\ 0 & 0 & 0 & 0 & 1 & 0\end{pmatrix}.$$

Another matrix can be formed from **L** by interchanging the rows and columns, that is, the first row of **L** becomes the first column of the new matrix and so on. This matrix is called the *transpose* of **L**; it is denoted by **L′** and represents the inverse relation 'has on the left' or 'is on the right of'. How would a diagram showing the relation which **L′** represents differ from the one in Figure 2(*a*)?

Write down the matrix **O** which represents the relation 'is opposite to'. Now write down the transpose of **O**. What relation does **O**′ represent? What happens when the directions of the arrows in Figure 2(*b*) are reversed?

Calculate **LO**. Draw a diagram to show the relation which **LO** represents and check that this relation is 'is on the left of the dancer opposite to'.

What relations do you think **OL**, $\mathbf{L}^2$ and $\mathbf{O}^2$ represent?

(*b*) The matrices **X** and **Y**, where

$$\mathbf{X} = \begin{array}{c} \\ A \\ B \\ C \\ D \end{array} \begin{array}{c} \begin{array}{ccc} \text{1sts} & \text{2nds} & \text{3rds} \end{array} \\ \begin{pmatrix} 2 & 3 & 1 \\ 2 & 2 & 2 \\ 1 & 2 & 3 \\ 3 & 1 & 2 \end{pmatrix} \end{array} \quad \text{and} \quad \mathbf{Y} = \begin{array}{c} \\ \text{1sts} \\ \text{2nds} \\ \text{3rds} \end{array} \begin{array}{c} \begin{array}{cc} (a) & (b) \end{array} \\ \begin{pmatrix} 4 & 4 \\ 2 & 3 \\ 1 & 1 \end{pmatrix} \end{array}$$

give respectively the results of four schools *A*, *B*, *C* and *D* competing in an athletics match and the points awarded for 1sts, 2nds and 3rds under two possible scoring schemes (*a*) and (*b*).

$$\mathbf{XY} = \begin{array}{c} \\ A \\ B \\ C \\ D \end{array} \begin{array}{c} \begin{array}{ccc} \text{1sts} & \text{2nds} & \text{3rds} \end{array} \\ \begin{pmatrix} 2 & 3 & 1 \\ 2 & 2 & 2 \\ 1 & 2 & 3 \\ 3 & 1 & 2 \end{pmatrix} \end{array} \begin{array}{c} \\ \text{1sts} \\ \text{2nds} \\ \text{3rds} \end{array} \begin{array}{c} \begin{array}{cc} (a) & (b) \end{array} \\ \begin{pmatrix} 4 & 4 \\ 2 & 3 \\ 1 & 1 \end{pmatrix} \end{array} = \begin{array}{c} \\ A \\ B \\ C \\ D \end{array} \begin{array}{c} \begin{array}{cc} (a) & (b) \end{array} \\ \begin{pmatrix} 15 & 18 \\ 14 & 16 \\ 11 & 13 \\ 16 & 17 \end{pmatrix} \end{array},$$

that is, the relations

'schools → positions' and 'positions → points'

represented by **X** and **Y** are combined to obtain the relation

'school → points'

represented by **XY**.

(*c*) A one-stage route matrix gives the number of direct routes between the nodes of a network.

If a network has no one-way routes then

(i) its one-stage route matrix is symmetrical about the leading diagonal (top left to bottom right);

(ii) the numbers on the leading diagonal are even.

Multiplying a one-stage route matrix by itself gives the two-stage route matrix.

Copy and complete the one-stage and two-stage route matrices for the networks in Figures 3 and 4.

The one-stage route matrix is

$$\mathbf{L} = \text{from}\ \begin{matrix} \\ \\ A \\ B \\ C \end{matrix}\!\!\begin{matrix} \text{to} \\ \begin{matrix} A & B & C \end{matrix} \\ \begin{pmatrix} 1 & 1 & 0 \\ 0 & & \\ 1 & & \end{pmatrix} \end{matrix},$$

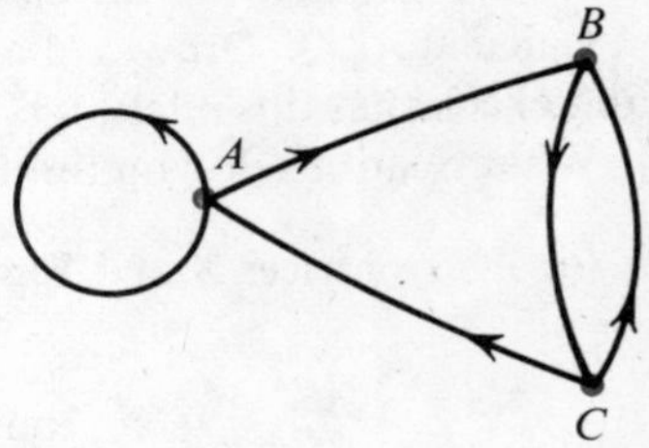

and the two-stage matrix is

$$\mathbf{L}^2 = \begin{pmatrix} 1 & 1 & 0 \\ 0 & & \\ 1 & & \end{pmatrix}\begin{pmatrix} 1 & 1 & 0 \\ 0 & & \\ 1 & & \end{pmatrix} = \begin{pmatrix} 1 & & \\ & & \\ & & \end{pmatrix}.$$

Fig. 3

The one-stage route matrix is

$$\mathbf{M} = \text{from}\ \begin{matrix} \\ \\ A \\ B \\ C \end{matrix}\!\!\begin{matrix} \text{to} \\ \begin{matrix} A & B & C \end{matrix} \\ \begin{pmatrix} 2 & 1 & 1 \\ 1 & & \\ 1 & & \end{pmatrix} \end{matrix},$$

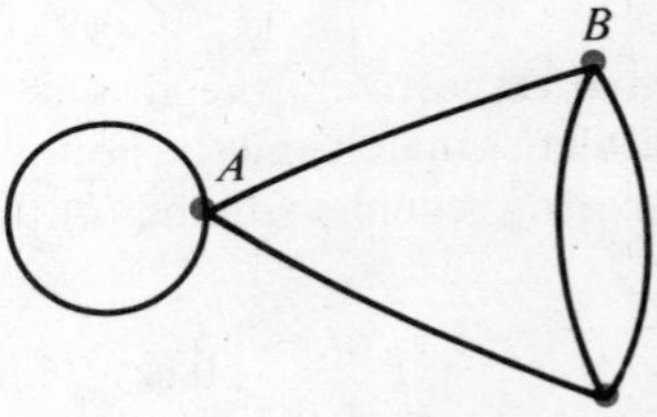

and the two-stage route matrix is

$$\mathbf{M}^2 = \begin{pmatrix} 2 & 1 & 1 \\ 1 & & \\ 1 & & \end{pmatrix}\begin{pmatrix} 2 & 1 & 1 \\ 1 & & \\ 1 & & \end{pmatrix} = \begin{pmatrix} 6 & & \\ & & \\ & & \end{pmatrix}.$$

Fig. 4

Calculate $\mathbf{L}^3$ and $\mathbf{M}^3$. What information do these matrices give?

(*d*) For a network, incidence matrices relate nodes, arcs and regions. For example

$$\mathbf{R} = \begin{matrix} \\ A \\ B \\ C \\ D \end{matrix}\!\!\begin{matrix} \begin{matrix} 1 & 2 & 3 & 4 & 5 \end{matrix} \\ \begin{pmatrix} 1 & 0 & 0 & 0 & 0 \\ 1 & 1 & 1 & 0 & 0 \\ 0 & 1 & 1 & 1 & 0 \\ 0 & 0 & 0 & 1 & 2 \end{pmatrix} \end{matrix}$$

shows the incidence of nodes on arcs for the network in Figure 5.

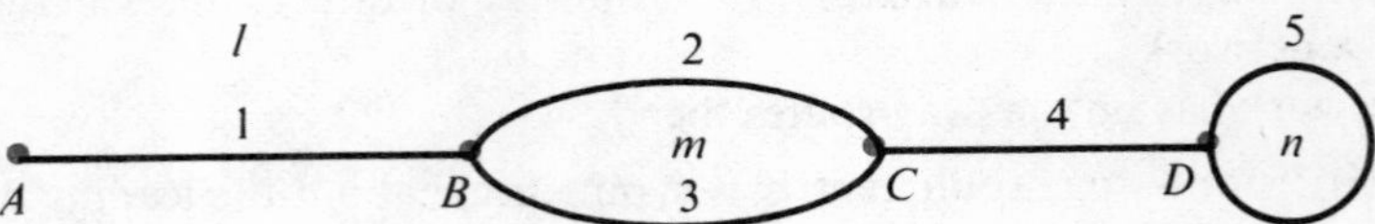

Fig. 5

What does $\mathbf{R}'$ show?

Verify that $\mathbf{RR}'$ differs from the one-stage route matrix only on the leading diagonal.

Copy and complete **S** and **T** to show the incidence of arcs on regions and the incidence of nodes on regions respectively.

$$\mathbf{S} = \begin{matrix} \\ 1 \\ 2 \\ 3 \\ 4 \\ 5 \end{matrix}\begin{matrix} \begin{matrix} l & m & n \end{matrix} \\ \begin{pmatrix} 2 & 0 & 0 \\ 1 & 1 & 0 \\ & & \\ & & \\ & & \end{pmatrix} \end{matrix}; \qquad \mathbf{T} = \begin{matrix} \\ A \\ B \\ C \\ D \end{matrix}\begin{matrix} \begin{matrix} l & m & n \end{matrix} \\ \begin{pmatrix} 1 & 0 & 0 \\ 2 & 1 & 0 \\ & & \\ & & \end{pmatrix} \end{matrix}.$$

Verify that **RS** = 2**T**.

(*e*) Matrices can be used to display probabilities. Suppose that if it is fine today the probability that it will be fine tomorrow is $\frac{2}{3}$ and that if it is wet today the probability that it will be fine tomorrow is $\frac{1}{2}$.

This information is illustrated in Figure 6 and the matrix of probabilities is

$$\mathbf{P} = \text{Tomorrow}\begin{cases}\text{Fine} \\ \text{Wet}\end{cases}\overbrace{\begin{matrix}\text{Fine} & \text{Wet}\end{matrix}}^{\text{Today}} \begin{pmatrix} \frac{2}{3} & \frac{1}{2} \\ \frac{1}{3} & \frac{1}{2} \end{pmatrix}.$$

P shows the relation between fine and wet days in terms of the probabilities that one will follow the other.

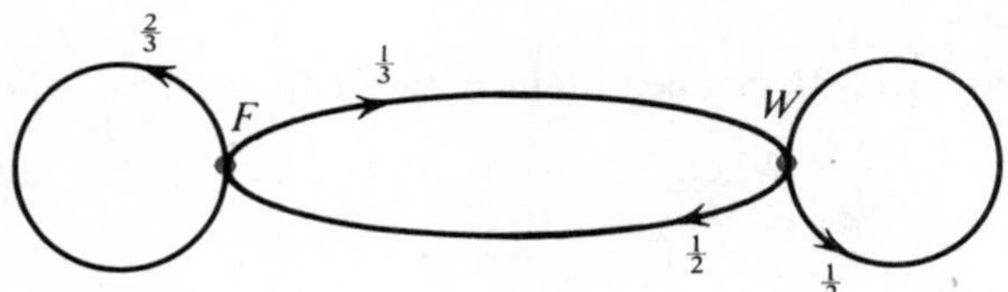

Fig. 6

The 'two-stage routes' in Figure 6 which start at *F* and finish at *F* are:

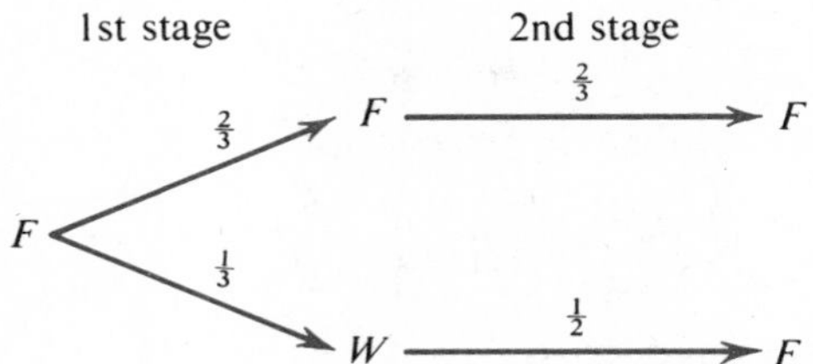

Therefore if today is fine the probability that the day after tomorrow will be fine is

$$\tfrac{2}{3} \times \tfrac{2}{3} + \tfrac{1}{3} \times \tfrac{1}{2} = \tfrac{11}{18} = 0{\cdot}611 \text{ to 3 D.P.}$$

By considering 'two-stage routes', find the remaining elements of

$$\mathbf{Q} = \text{Day after tomorrow}\begin{cases}\text{Fine}\\ \text{Wet}\end{cases}\overbrace{\begin{pmatrix} \frac{11}{18} & \\ & \end{pmatrix}}^{\substack{\text{Today}\\ \text{Fine} \quad \text{Wet}}} \approx \begin{pmatrix} 0{\cdot}611 & \\ & \end{pmatrix}.$$

Calculate $\mathbf{P}^2$ and compare $\mathbf{P}^2$ with $\mathbf{Q}$.

Calculate $\mathbf{P}^3$. What information does $\mathbf{P}^3$ give?

Calculate the probability matrix for 4 days' time.

What information would $\mathbf{P}^7$ give?

You should find that the sequence of matrices $\mathbf{P}$, $\mathbf{P}^2$, $\mathbf{P}^3$, $\mathbf{P}^4$, . . . approaches the matrix

$$\begin{pmatrix} 0{\cdot}6 & 0{\cdot}6 \\ 0{\cdot}4 & 0{\cdot}4 \end{pmatrix}.$$

The sequence is said to *tend to a limit*.

Verify that
$$\mathbf{P}\begin{pmatrix} 0{\cdot}6 & 0{\cdot}6 \\ 0{\cdot}4 & 0{\cdot}4 \end{pmatrix} = \begin{pmatrix} 0{\cdot}6 & {\cdot}06 \\ 0{\cdot}4 & 0{\cdot}4 \end{pmatrix}$$

and try to explain how this shows that in the long run 0·6 of all days are fine and 0·4 are wet.

The limiting matrix, $\mathbf{L}$, can be obtained by the following method.

Suppose $\mathbf{L} = \begin{pmatrix} a & b \\ c & d \end{pmatrix}$. When $\mathbf{L}$ is premultiplied by $\mathbf{P}$ it will be unchanged, so

$$\begin{pmatrix} \frac{2}{3} & \frac{1}{2} \\ \frac{1}{3} & \frac{1}{2} \end{pmatrix}\begin{pmatrix} a & b \\ c & d \end{pmatrix} = \begin{pmatrix} a & b \\ c & d \end{pmatrix}.$$

That is

$$\begin{pmatrix} \frac{2}{3}a + \frac{1}{2}c & \frac{2}{3}b + \frac{1}{2}d \\ \frac{1}{3}a + \frac{1}{2}c & \frac{1}{3}b + \frac{1}{2}d \end{pmatrix} = \begin{pmatrix} a & b \\ c & d \end{pmatrix}.$$

Hence

$$\tfrac{2}{3}a + \tfrac{1}{2}c = a,$$
$$\tfrac{1}{3}a + \tfrac{1}{2}c = c,$$
$$\tfrac{2}{3}b + \tfrac{1}{2}d = b,$$
$$\tfrac{1}{3}b + \tfrac{1}{2}d = d.$$

The first two equations reduce to

$$\tfrac{1}{2}c = \tfrac{1}{3}a$$
$$c = \tfrac{2}{3}a.$$

But $a + c = 1$. Why?

Therefore

$$a + \tfrac{2}{3}a = 1$$
$$\tfrac{5}{3}a = 1$$
$$a = \tfrac{3}{5}$$

and so

$$c = \tfrac{2}{3} \times \tfrac{3}{5} = \tfrac{2}{5}.$$

Similarly the last two equations give $b = \frac{3}{5}$ and $d = \frac{2}{5}$.
Therefore

$$\mathbf{L} = \begin{pmatrix} \frac{3}{5} & \frac{3}{5} \\ \frac{2}{5} & \frac{2}{5} \end{pmatrix} = \begin{pmatrix} 0{\cdot}6 & 0{\cdot}6 \\ 0{\cdot}4 & 0{\cdot}4 \end{pmatrix}.$$

Exercise C

1 Copy and complete Figure 7 to show the relation 'is a prime factor of' on the set {2, 3, 4, 6, 8}.

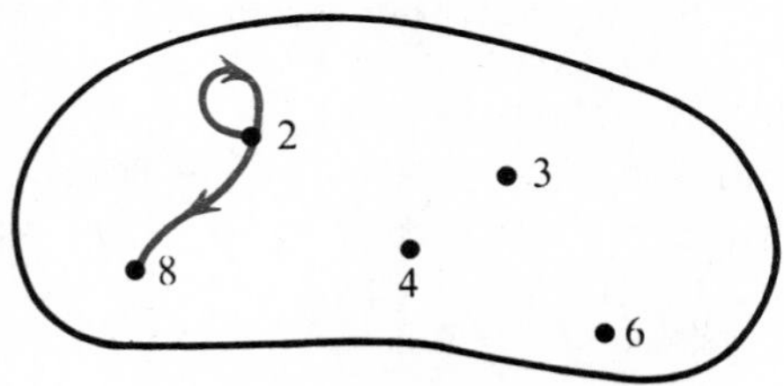

'is a prime factor of'

Fig. 7

Write down the matrix which represents this relation.

2 Figure 8 shows the relation 'is a parent of' on the set $\{a, b, c, d\}$.

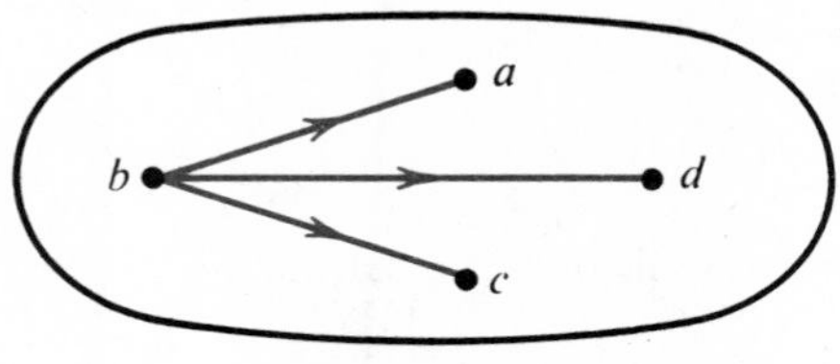

'is a parent of'

Fig. 8

(*a*) Write down the matrix **P** which describes this relation.
(*b*) Write down the transpose of **P**. What relation does it represent?

3 The petrol sales in litres at two garages A and B are given by the matrix

$$\begin{array}{c} \\ A \\ B \end{array} \begin{array}{cc} \text{Four star petrol} & \text{Two star petrol} \\ \end{array}$$
$$\begin{matrix} A \\ B \end{matrix}\begin{pmatrix} 300 & 100 \\ 200 & 200 \end{pmatrix}.$$

(*a*) Four star petrol is sold at 10p per litre and two star petrol at 9p per litre. Write this information as a 2 by 1 matrix.

(*b*) Multiply the two matrices to form a 2 by 1 matrix $\begin{pmatrix} x \\ y \end{pmatrix}$.

(*c*) Write down the value of $x + y$. What information is conveyed by this number?

(*d*) Four star petrol gives a profit of 1p per litre and two star petrol gives a profit of 2p per litre. Multiply two matrices to find the profit made by each of the two garages. Give your answers in pounds.

4 The Trans-Can Airline has eight Viscounts, six Tridents and two Caravelles. East Atlantic Airways have nine Viscounts, one Trident and seven Caravelles, and the Il-Oil Company has two Viscounts, eleven Tridents and no Caravelles.

(*a*) Express this information as a 3×3 matrix **A**.

(*b*) The Viscount carries 50 passengers, the Trident carries 140 and the Caravelle 80. Write down a suitable product of two matrices which, when evaluated, will determine the number of passengers that each airline is able to carry when all its aircraft are full.

(*c*) Evaluate this product.

(*d*) Premultiply your matrix **A** by the row vector $(1 \quad 1 \quad 1)$. What information does the resulting matrix give?

5 Write down the one-stage route matrix for each of the networks in Figure 9.

Fig. 9 (*a*) (*b*)

6 Draw the network described by the matrix

$$\text{From}\ \begin{matrix} \\ A \\ B \\ C \\ D \end{matrix}\!\!\begin{matrix} \text{to} \\ \begin{matrix} A & B & C & D \end{matrix} \\ \begin{pmatrix} 0 & 1 & 2 & 1 \\ 1 & 0 & 1 & 1 \\ 2 & 1 & 0 & 1 \\ 1 & 1 & 1 & 0 \end{pmatrix} \end{matrix}.$$

7 Write down the one-stage route matrix for the directed network in Figure 10.

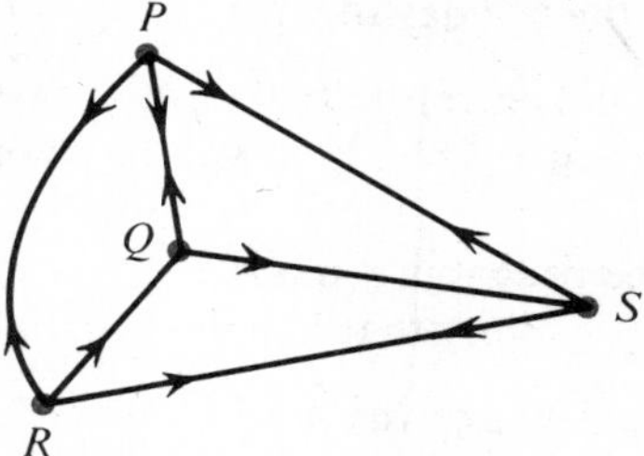

Fig. 10

Calculate the two-stage and three-stage route matrices for this network. How many three-stage routes are there from P to R? List these routes.

8 A network has nodes A, B, C and arcs 1, 2, 3. The relation between its nodes and arcs is given by the incidence matrix:

$$\begin{array}{c} \\ A \\ B \\ C \end{array} \begin{array}{c} \begin{array}{ccc} 1 & 2 & 3 \end{array} \\ \begin{pmatrix} 1 & 0 & 0 \\ 1 & 1 & 1 \\ 0 & 1 & 1 \end{pmatrix} \end{array}.$$

Sketch the network.

9 (*a*) Write down the matrices **R**, **S** and **T** which show the incidence of nodes on arcs, arcs on regions and nodes on regions respectively for the network in Figure 11.

(*b*) What information do **R**′, **S**′ and **T**′ give?

(*c*) Calculate **S**′**R**′ and compare **S**′**R**′ with **T**′.

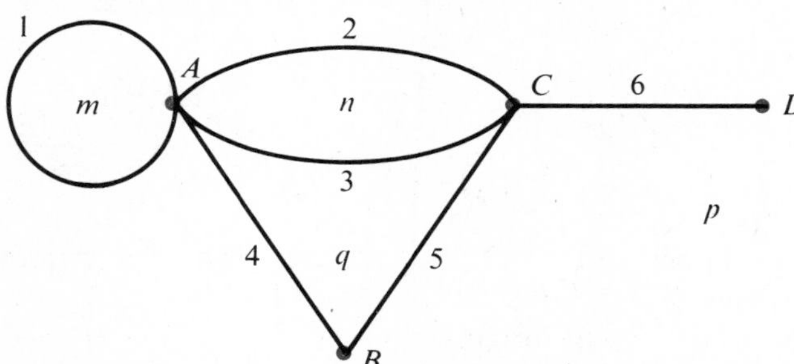

Fig. 11

10 If a boy oversleeps today, the probability that he will oversleep tomorrow is $\frac{1}{4}$; if he gets up in time to day, the probability that he will oversleep tomorrow is $\frac{1}{2}$.

(*a*) Copy and complete the following probability matrix:

$$\begin{array}{cc} & \overbrace{\begin{array}{cc}\text{Oversleeps} & \text{Wakes}\end{array}}^{\text{Today}} \\ \text{Tomorrow}\left\{\begin{array}{l}\text{Oversleeps} \\ \text{Wakes}\end{array}\right. & \begin{pmatrix} \frac{1}{4} & \phantom{\frac{1}{2}} \\ & \end{pmatrix}. \end{array}$$

(*b*) Find the probability matrices for the second and third days.

(*c*) If he oversleeps on the first day of term, what is the probability that he will oversleep on the last day of term?

11 The digits 0 and 1 should be transmitted through several stages of a computer. There is a probability of $\frac{1}{3}$ that a digit will be changed from 0 to 1 or from 1 to 0 at any stage.

(*a*) Write down the probability matrix.

(*b*) What is the probability that a digit is changed in a long calculation?

12 A salesman travels between towns A and B. If he is at town A today, he will certainly go to town B tomorrow; if he is at B today, the probability that he will stay at B tomorrow is $\frac{4}{5}$.

Write down the probability matrix and hence find the proportion of time he spends at A.

13 If a customer buys Brand X this week, the probability that he will buy Brand X next week is $\frac{1}{10}$. If he buys Brand Y this week, the probability that he will buy Brand X next week is $\frac{2}{3}$.

Write down the matrix of probabilities, and estimate the probabilities after one year.

5 Transformation matrices

(*a*) Transformations of the plane are mappings on ordered pairs of real numbers.

For example, consider the transformation **T** given by

$$\mathbf{T}: \begin{pmatrix} x \\ y \end{pmatrix} \to \begin{pmatrix} 2x - y \\ x + 3y \end{pmatrix}$$

or

$$\mathbf{T}: \begin{pmatrix} x \\ y \end{pmatrix} \to \begin{pmatrix} 2 & ^{-}1 \\ 1 & 3 \end{pmatrix}\begin{pmatrix} x \\ y \end{pmatrix}.$$

In practice, we can picture the geometrical effect of a transformation by considering what happens to the unit square $OIAJ$.

All 2×2 matrices map the origin onto itself, so

$$\begin{pmatrix} 0 \\ 0 \end{pmatrix} \xrightarrow{\mathbf{T}} \begin{pmatrix} 0 \\ 0 \end{pmatrix}.$$

The position vectors of I_1 and J_1, the images of $I(1, 0)$ and $J(0, 1)$ are given by the two columns of the matrix

$$\mathbf{M} = \begin{pmatrix} 2 & ^{-}1 \\ 1 & 3 \end{pmatrix}.$$

So,

$$\begin{pmatrix} 1 \\ 0 \end{pmatrix} \xrightarrow{\mathbf{T}} \begin{pmatrix} 2 \\ 1 \end{pmatrix} \quad \text{and} \quad \begin{pmatrix} 0 \\ 1 \end{pmatrix} \xrightarrow{\mathbf{T}} \begin{pmatrix} ^{-}1 \\ 3 \end{pmatrix}.$$

Therefore:

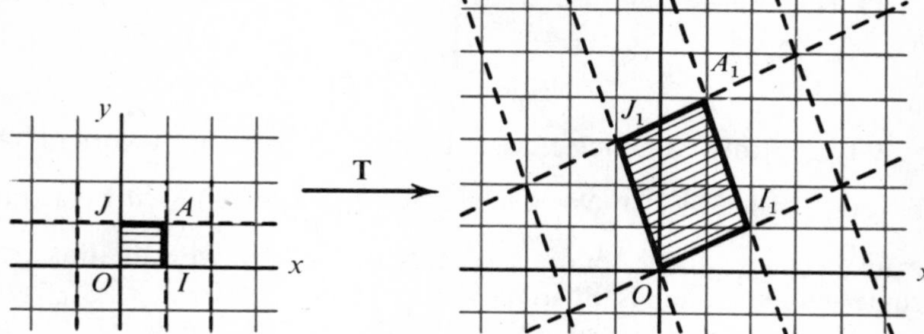

Fig. 12

Since $|\mathbf{M}| = 6 - {}^-1 = 7$, area $OI_1A_1J_1 = 7 \times$ area $OIAJ$.

Calculate $\mathbf{M}^{-1}$ and hence express the inverse transformation $\mathbf{T}^{-1}$ which maps $OI_1A_1J_1$ onto $OIAJ$ in the form

$$\mathbf{T}^{-1}: \begin{pmatrix} x \\ y \end{pmatrix} \rightarrow \begin{pmatrix} ax + by \\ cx + dy \end{pmatrix}.$$

(*b*) By considering what happens to the base vectors $\begin{pmatrix} 1 \\ 0 \end{pmatrix}$ and $\begin{pmatrix} 0 \\ 1 \end{pmatrix}$, verify that anticlockwise rotation of 90° about the origin is represented by

$$\mathbf{R} = \begin{pmatrix} 0 & {}^-1 \\ 1 & 0 \end{pmatrix}$$

and the shear with points on the line $y = 0$ invariant and shear factor 2 is represented by

$$\mathbf{S} = \begin{pmatrix} 1 & 2 \\ 0 & 1 \end{pmatrix}.$$

Figure 13 shows the effect on the unit square of *the shear represented by* **S** *followed by the rotation represented by* **R**.

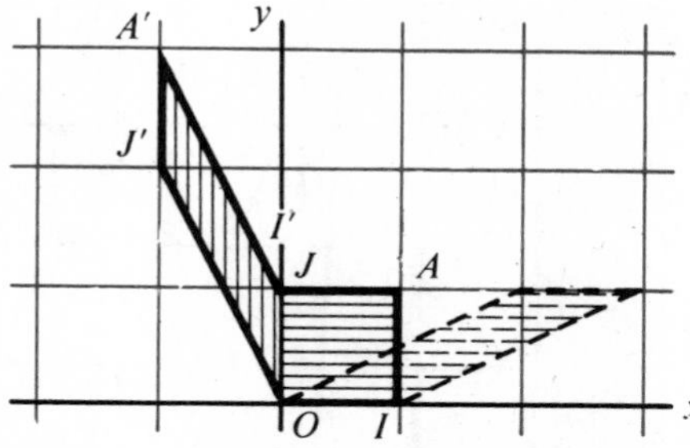

Fig. 13

The matrix for this combined transformation can be found by calculating **RS** (not **SR**!).

You may find it helpful to think of **R** as the matrix which 'gives the image under the rotation of', **S** as the matrix which 'gives the image under the shear of' and **RS** as the matrix which 'gives the image under the rotation of the image under the shear of'.

What combined transformation does **SR** represent?

(*c*) The transformation **U** given by

$$\mathbf{U}: \begin{pmatrix} x \\ y \end{pmatrix} \to \begin{pmatrix} 0 & ^-1 \\ 1 & 0 \end{pmatrix} \begin{pmatrix} x \\ y \end{pmatrix} + \begin{pmatrix} 4 \\ 2 \end{pmatrix}$$

may be regarded as anticlockwise rotation of 90° about the origin, O, followed by the translation $\begin{pmatrix} 4 \\ 2 \end{pmatrix}$. We know that a rotation followed by a translation is equivalent to a single rotation, so **U** is an anticlockwise rotation of 90° about some centre C which is different from O.

Let C be the point (p, q). Then, since C is invariant under **U**,

$$\begin{pmatrix} 0 & ^-1 \\ 1 & 0 \end{pmatrix} \begin{pmatrix} p \\ q \end{pmatrix} + \begin{pmatrix} 4 \\ 2 \end{pmatrix} = \begin{pmatrix} p \\ q \end{pmatrix}$$

$$\begin{pmatrix} ^-q+4 \\ p+2 \end{pmatrix} = \begin{pmatrix} p \\ q \end{pmatrix}$$

$$\begin{cases} ^-q+4 = p \\ p+2 = q \end{cases}$$

$$\begin{cases} p+q = 4 \\ ^-p+q = 2. \end{cases}$$

Adding,

$$2q = 6$$
$$q = 3$$

and hence

$$p = 1.$$

U is anticlockwise rotation of 90° about $C(1, 3)$. See Figure 14.

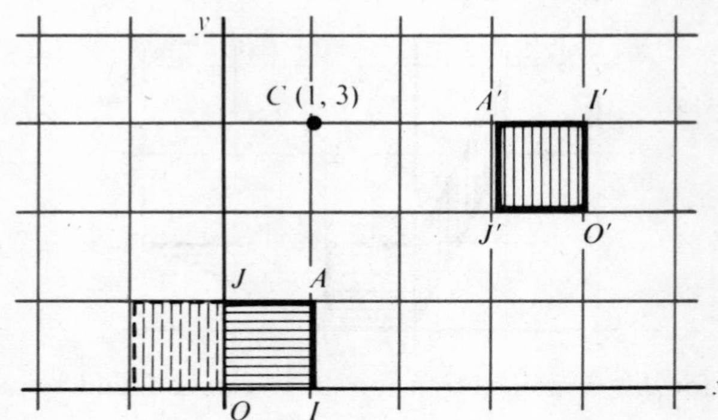

Fig. 14

Calculate the invariant point of the transformation

$$\begin{pmatrix} x \\ y \end{pmatrix} \to \begin{pmatrix} 3 & 0 \\ 0 & 3 \end{pmatrix} \begin{pmatrix} x \\ y \end{pmatrix} + \begin{pmatrix} ^-6 \\ 2 \end{pmatrix}$$

and describe the transformation.

(*d*) Consider the transformation

$$\begin{pmatrix}x\\y\end{pmatrix} \to \mathbf{M}\begin{pmatrix}x\\y\end{pmatrix} + \begin{pmatrix}e\\f\end{pmatrix} \quad \text{where } \mathbf{M} = \begin{pmatrix}a & b\\c & d\end{pmatrix}.$$

Note that this transformation maps the origin onto the point (e, f).

The numerical value of $|\mathbf{M}|$ gives the area scale factor of the transformation.

A negative determinant indicates that a reflection is involved, that is, if the vertices of the unit square are lettered O, I, A, J in an anticlockwise sense, then the images O', I', A', J' are in a clockwise sense.

If $|\mathbf{M}| \neq 0$, the transformation has a unique inverse given by

$$\begin{pmatrix}x\\y\end{pmatrix} \to \mathbf{M}^{-1}\left[\begin{pmatrix}x\\y\end{pmatrix} + \begin{pmatrix}^-e\\^-f\end{pmatrix}\right].$$

Exercise D

1 Describe, by considering the image of the unit square, the transformations represented by the following matrices:

(*a*) $\begin{pmatrix}0 & ^-1\\^-1 & 0\end{pmatrix}$; (*b*) $\begin{pmatrix}1 & ^-3\\0 & 1\end{pmatrix}$; (*c*) $\begin{pmatrix}^-2 & 0\\0 & ^-2\end{pmatrix}$;

(*d*) $\begin{pmatrix}3 & 0\\0 & \frac{1}{2}\end{pmatrix}$; (*e*) $\begin{pmatrix}1 & ^-1\\1 & 1\end{pmatrix}$; (*f*) $\begin{pmatrix}2 & 0\\1 & 3\end{pmatrix}$.

2 Write down the matrices for the transformations under which the images of the unit square are as shown in Figure 15.

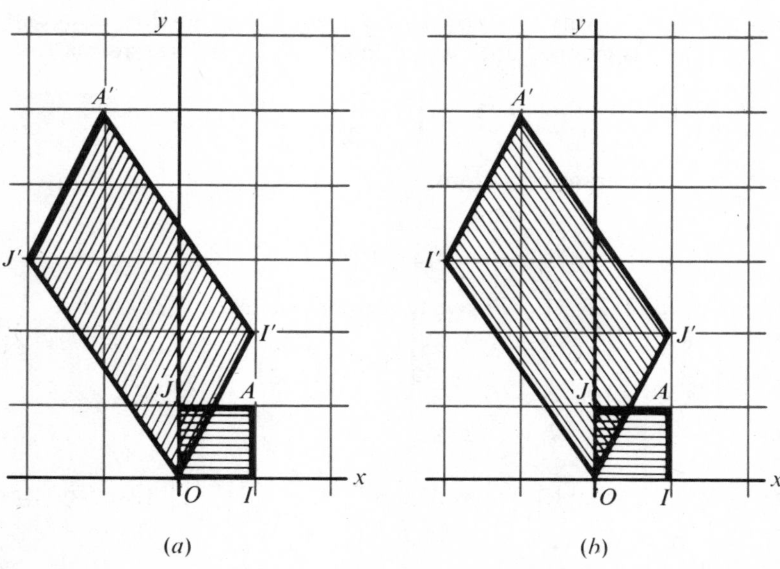

Fig. 15

3 Illustrate the effect on the unit square, and write down the matrix for the single transformation which is the result of:

(*a*) the shear $\begin{pmatrix}1 & ^-2\\0 & 1\end{pmatrix}$ followed by the one-way stretch $\begin{pmatrix}1 & 0\\0 & 3\end{pmatrix}$;

(*b*) the one-way stretch $\begin{pmatrix}1 & 0\\0 & 3\end{pmatrix}$ followed by the shear $\begin{pmatrix}1 & ^-2\\0 & 1\end{pmatrix}$;

(*c*) the enlargement $\begin{pmatrix}5 & 0\\0 & 5\end{pmatrix}$ followed by the rotation $\begin{pmatrix}0{\cdot}6 & ^-0{\cdot}8\\0{\cdot}8 & 0{\cdot}6\end{pmatrix}$;

(*d*) the transformation $\begin{pmatrix}1 & ^-2\\^-3 & 6\end{pmatrix}$ followed by the transformation $\begin{pmatrix}1 & \frac{1}{3}\\\frac{1}{2} & \frac{1}{6}\end{pmatrix}$.

In each case, find the area of the image of the unit square.

4 Find the invariant point for each of the following transformations and then describe the transformations:

(*a*) $\begin{pmatrix}x\\y\end{pmatrix} \to \begin{pmatrix}^-1 & 0\\0 & ^-1\end{pmatrix}\begin{pmatrix}x\\y\end{pmatrix} + \begin{pmatrix}^-2\\4\end{pmatrix}$;

(*b*) $\begin{pmatrix}x\\y\end{pmatrix} \to \begin{pmatrix}^-3 & 0\\0 & ^-3\end{pmatrix}\begin{pmatrix}x\\y\end{pmatrix} + \begin{pmatrix}^-20\\12\end{pmatrix}$.

6 Simultaneous equations

(*a*) Consider the equations

$$\begin{cases}3x - 2y = 25\\ x + 4y = {}^-1\end{cases} \quad\text{or}\quad \begin{pmatrix}3 & ^-2\\1 & 4\end{pmatrix}\begin{pmatrix}x\\y\end{pmatrix} = \begin{pmatrix}25\\^-1\end{pmatrix}.$$

The problem of solving these equations is equivalent to that of finding the point (x, y) which is mapped onto the point $(25, {}^-1)$ by the transformation

$$\begin{pmatrix}x\\y\end{pmatrix} \to \begin{pmatrix}3 & ^-2\\1 & 4\end{pmatrix}\begin{pmatrix}x\\y\end{pmatrix}.$$

Hence the inverse transformation

$$\begin{pmatrix}x\\y\end{pmatrix} \to \frac{1}{14}\begin{pmatrix}4 & 2\\^-1 & 3\end{pmatrix}\begin{pmatrix}x\\y\end{pmatrix}$$

applied to the point $(25, {}^-1)$ will give the required solution.

$$\begin{pmatrix}x\\y\end{pmatrix} = \frac{1}{14}\begin{pmatrix}4 & 2\\^-1 & 3\end{pmatrix}\begin{pmatrix}25\\^-1\end{pmatrix} = \frac{1}{14}\begin{pmatrix}98\\^-28\end{pmatrix} = \begin{pmatrix}7\\^-2\end{pmatrix},$$

and so $\qquad x = 7, y = {}^-2.$

Note that it is simpler to evaluate $\frac{1}{14}\left[\begin{pmatrix}4 & 2\\^-1 & 3\end{pmatrix}\begin{pmatrix}25\\^-1\end{pmatrix}\right]$ than to evaluate $\left[\frac{1}{14}\begin{pmatrix}4 & 2\\^-1 & 3\end{pmatrix}\right]\begin{pmatrix}25\\^-1\end{pmatrix}$.

(*b*) If $|\mathbf{M}| \neq 0$, then **M** has an inverse and the transformation which it represents is a one-to-one mapping. Therefore for all p and q

$$\mathbf{M}\begin{pmatrix} x \\ y \end{pmatrix} = \begin{pmatrix} p \\ q \end{pmatrix}$$

has a unique solution as in (*a*) above.

If $|\mathbf{M}| = 0$ then **M** has no inverse and the transformation which it represents is a many-to-one mapping. The points of the plane are mapped onto a single line or even onto a single point.

For all a, the point $(a, 2-3a)$ lies on the line $3x+y=2$. Verify that the transformation represented by

$$\begin{pmatrix} 6 & 2 \\ 3 & 1 \end{pmatrix}$$

maps $(a, 2-3a)$ onto $(4, 2)$.

In fact for all values of k this transformation maps all the points of the line $3x+y=k$ onto the point $(2k, k)$ and so maps all the points of the plane onto the points of the line $x=2y$. See Figure 16.

The equations $\begin{cases} 6x+2y=5 \\ 3x+y=1 \end{cases}$ or $\begin{pmatrix} 6 & 2 \\ 3 & 1 \end{pmatrix}\begin{pmatrix} x \\ y \end{pmatrix} = \begin{pmatrix} 5 \\ 1 \end{pmatrix}$

state that (x, y) is mapped onto $(5, 1)$. But no point can be mapped onto $(5, 1)$ since it does not lie on the line $x=2y$. Hence the solution set is $\varnothing$.

The equations $\begin{cases} 6x+2y=4 \\ 3x+2y=2 \end{cases}$ or $\begin{pmatrix} 6 & 2 \\ 3 & 1 \end{pmatrix}\begin{pmatrix} x \\ y \end{pmatrix} = \begin{pmatrix} 4 \\ 2 \end{pmatrix}$

state that (x, y) is mapped onto $(4, 2)$. *All* the points of the line $3x+y=2$ are mapped onto $(4, 2)$. So the solution set is

$$\{(x, y): 3x+y=2\}.$$

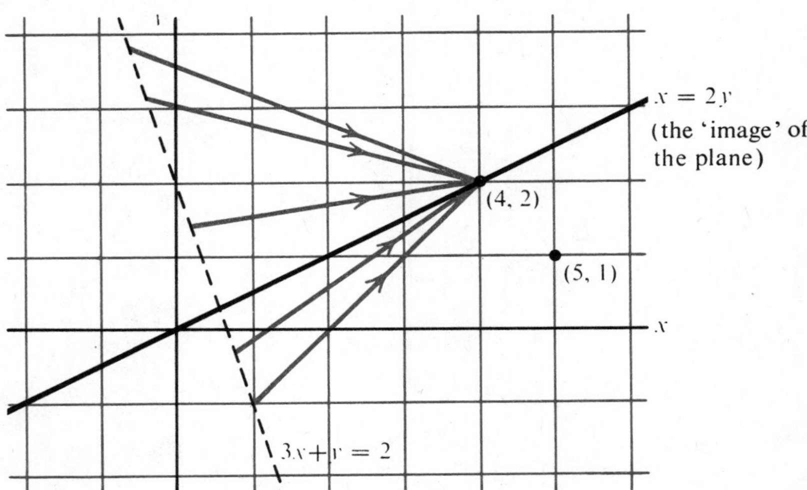

Fig. 16

State the solution of (i) $\begin{cases} 6x+2y={}^-2 \\ 3x+y={}^-1 \end{cases}$; (ii) $\begin{cases} 6x+2y={}^-3 \\ 3x+y=7. \end{cases}$

Exercise E

1 Solve the following pairs of simultaneous equations:

(*a*) $\begin{cases} 2x+y=3 \\ 3x+2y=4; \end{cases}$ (*b*) $\begin{cases} 5x-4y=7 \\ 4x-3y=6; \end{cases}$

(*c*) $\begin{cases} 4x+5y=9 \\ 2x+3y=7; \end{cases}$ (*d*) $\begin{cases} 7x-9y=8 \\ 4x-5y=3; \end{cases}$

(*e*) $\begin{cases} 6x-y=17 \\ 3x+2y={}^-4; \end{cases}$ (*f*) $\begin{cases} x-0{\cdot}3y=0{\cdot}2 \\ x-0{\cdot}7y=0{\cdot}5; \end{cases}$

(*g*) $\begin{cases} 5x+11y=13 \\ 3x+7y=8; \end{cases}$ (*h*) $\begin{cases} 3x+y+11=0 \\ 9x-7y+13=0; \end{cases}$

(*i*) $\begin{cases} 0{\cdot}2x+0{\cdot}3y=0{\cdot}4 \\ {}^-0{\cdot}9x+0{\cdot}4y=1; \end{cases}$ (*j*) $\begin{cases} 13x+6y-1=0 \\ 4y+9x+1=0. \end{cases}$

2 Find the solution set for the following pairs of simultaneous equations:

(*a*) $\begin{cases} 13x+6y=12 \\ 39x+18y=35; \end{cases}$ (*b*) $\begin{cases} 3x-2y=10 \\ 5x+4y=13; \end{cases}$

(*c*) $\begin{cases} 2x+y=6 \\ 4x+2y=12; \end{cases}$ (*d*) $\begin{cases} {}^-3x+5y=9 \\ 6x-10y={}^-18. \end{cases}$

3 The graph of $y = ax + bx^2$ passes through the points (2, 5) and (3, 12). Find the values of a and b and hence find y when $x = 4$.

4 What point is mapped onto (15, 23) by the transformation

$$\begin{pmatrix} x \\ y \end{pmatrix} \to \begin{pmatrix} 2 & 3 \\ 4 & 7 \end{pmatrix}\begin{pmatrix} x \\ y \end{pmatrix} + \begin{pmatrix} 5 \\ -3 \end{pmatrix}?$$

8 Geometry

1 Symmetries

(*a*) *Line symmetry*. A plane figure is symmetrical about a line m if for every point P of the figure there is a corresponding point P' such that m is the *mediator* of PP' (see Figure 1). This means that a reflection in m will map the figure onto itself. Line symmetry is also known as bilateral symmetry.

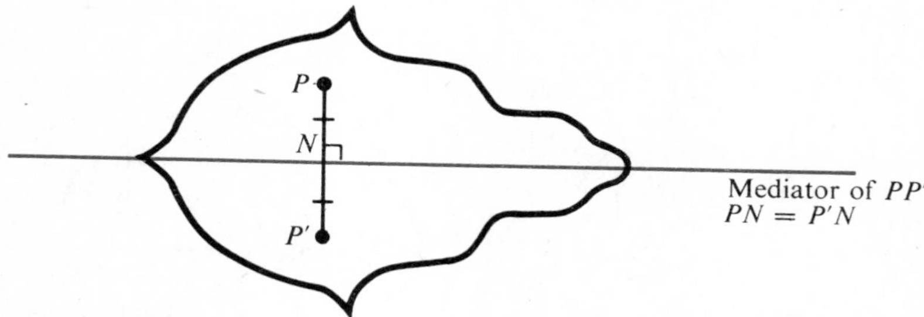

Fig. 1

Some plane figures have more than one line of symmetry – a rhombus, for example, has two; whilst others have no lines of symmetry (e.g. a parallelogram).

(*b*) *Rotational symmetry*. A plane figure has rotational symmetry of order n about a point O (the centre of rotational symmetry) if $1/n$th of a full turn maps the figure onto itself. The shape in Figure 2 has order of symmetry 4. Note that the shape will be mapped onto itself by a $\frac{1}{4}$ turn, a $\frac{1}{2}$ turn and a $\frac{3}{4}$ turn – all plane figures will be mapped onto themselves by a full turn about *any* point in the plane.

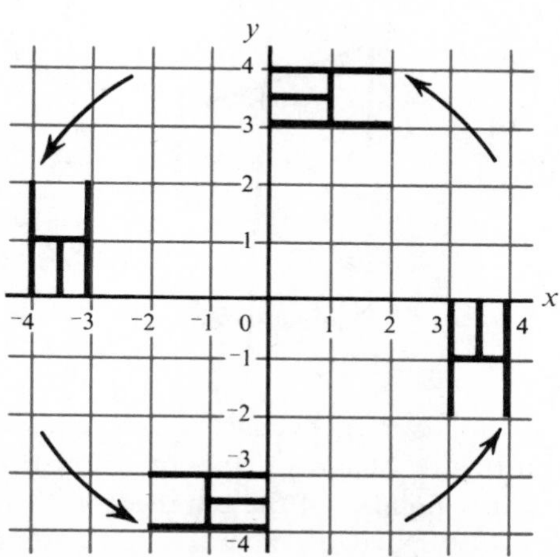

Fig. 2

Half-turn symmetry (rotational symmetry of order 2) is also known as point symmetry. If a figure is mapped onto itself by an enlargement scale factor $^{-}1$, then it has point symmetry because a half turn and an enlargement scale factor $^{-}1$ are equivalent transformations.

(*c*) Figure 3 shows some well-known polygons. The lines of symmetry of each figure are shown in red and the arrows attempt to represent the rotational symmetries. The regular pentagon, for example, has five lines of symmetry and rotational symmetry of order 5.

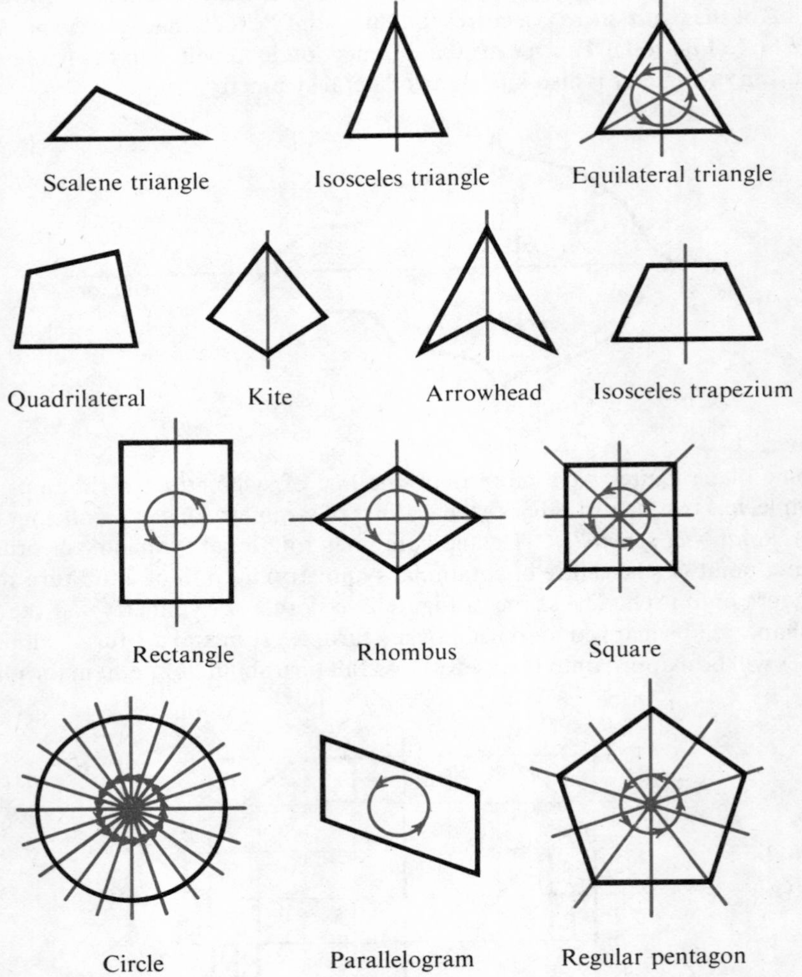

Fig. 3

(*d*) *Plane symmetry*. Solids may, or may not, have planes of symmetry. The regular tetrahedron in Figure 4 has six planes of symmetry – one is represented by the shading. For every point P of the tetrahedron there is a corresponding point P' such that P' is the image of P after a reflection in a plane of symmetry.

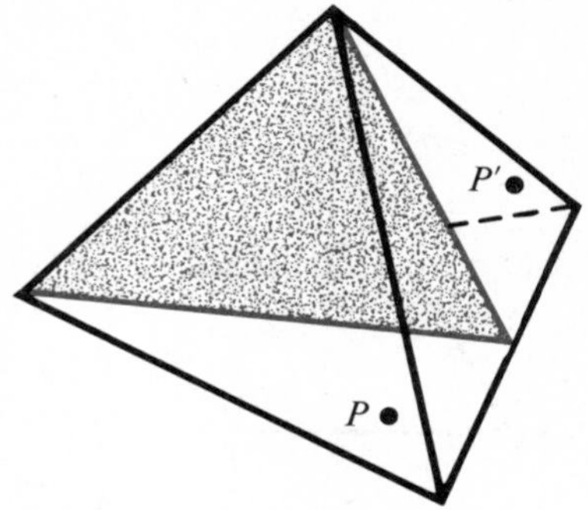

Fig. 4

(*e*) *Rotational symmetry about an axis.* The red line in Figure 5 is an *axis of symmetry* of the cuboid. The cuboid has three axes of symmetry and has rotational symmetry of order two about each axis.

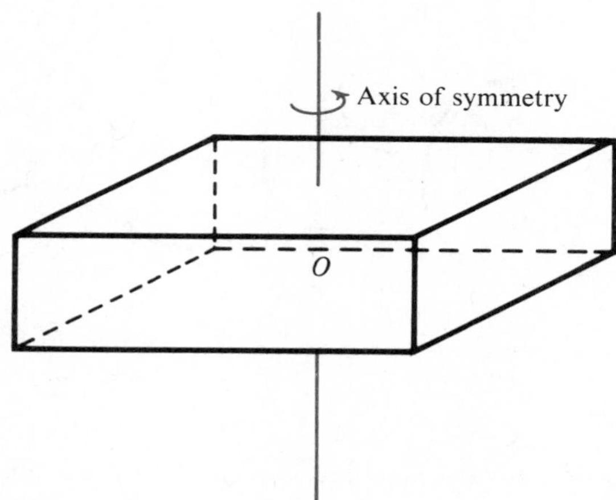

Fig. 5

(*f*) *Point symmetry.* A solid has point symmetry, centre O, if an enlargement S.F. $^{-}1$, centre O, maps the solid onto itself. The cuboid in Figure 5 has point symmetry, centre O.

(*g*) Figure 6 shows three solids with one plane of symmetry represented by shading, and one axis of symmetry represented by a red line. Both the cube and the regular octahedron have point symmetry.

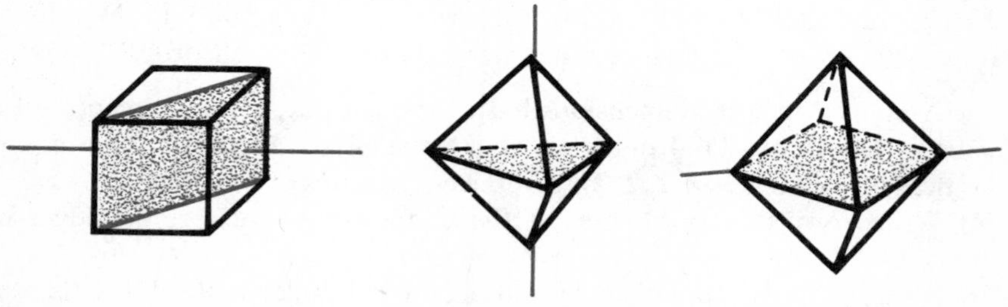

Fig. 6 Cube Bipyramid Regular octahedron

Exercise A

1 Describe, as accurately as you can, the symmetries of each of the shapes in Figure 7. Draw diagrams if necessary.

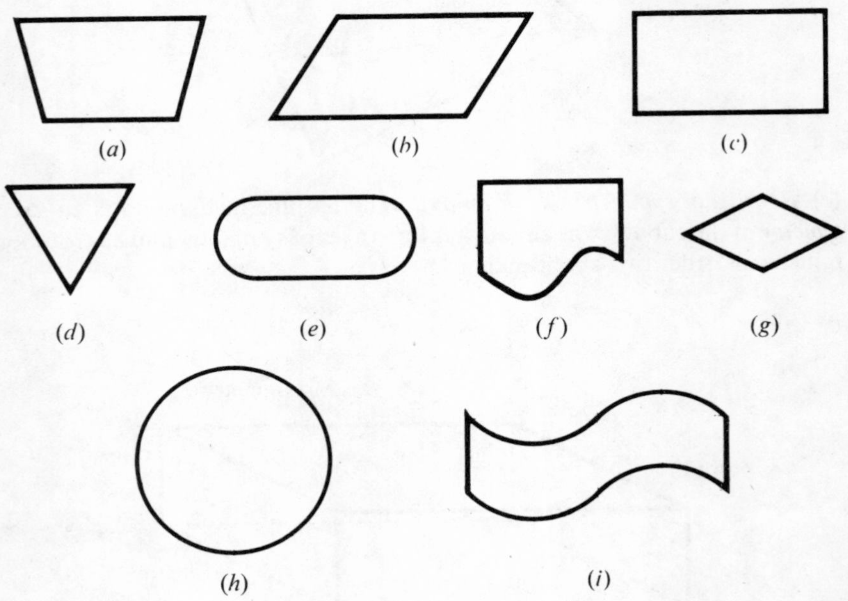

Fig. 7

2 (*a*) Draw a quadrilateral which has point symmetry but not line symmetry. (*b*) Draw two different types of quadrilaterals which have exactly one line of symmetry. Is it possible for a figure to have exactly one line of symmetry and point symmetry?

3 If $\mathscr{E} = \{\text{Quadrilaterals}\}$, $P = \{\text{Quadrilaterals with point symmetry}\}$, $Q = \{\text{Quadrilaterals with line symmetry}\}$, draw a quadrilateral which is a member of $P' \cap Q$.

4 How many (i) planes, (ii) axes of symmetry has (*a*) a cuboid; (*b*) a cube; (*c*) a cylinder; (*d*) a regular tetrahedron; (*e*) a regular octahedron?

5 (*a*) A point P in three dimensions has coordinates (x, y, z). The centre of a cube of length of side 4 units lies at the origin (see Figure 8). One of the vertices of the cube is $A(2, 2, 2)$. List the coordinates of the other vertices.
(*b*) Write down the coordinates of the midpoints of each of the edges of the cube.
(*c*) If (x_0, y_0, z_0) is a point on the surface of the cube, write down the coordinates, in terms of x_0, y_0 and z_0 of another point on the cube's surface.

(*d*) One plane of symmetry of the cube has equation $z = 0$. Copy Figure 8 and represent the plane by shading. Write the equations of the other eight planes of symmetry of the cube.

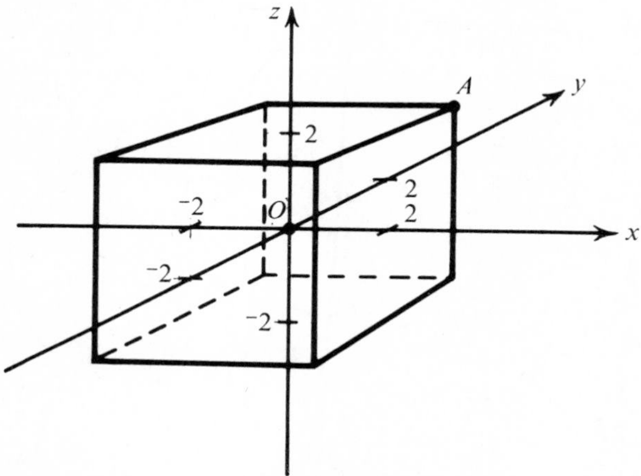

Fig. 8

6 A polygon has one vertex $A(2, 3)$. The figure has rotational symmetry of order 4 about (0, 0), and a line of symmetry $y = x$.

(i) Sketch such a polygon on squared paper and write down the coordinates of its vertices.

(ii) What other symmetries does the polygon necessarily possess?

7 (*a*) A plane figure is cut from a sheet of paper. The figure is then folded along one of its lines of symmetry. The new figure is folded again along a line of symmetry. The operation is repeated once more. If the final figure has the shape of Figure 9, sketch the original figure. Is there more than one possibility? Sketch as many possible shapes as you can.

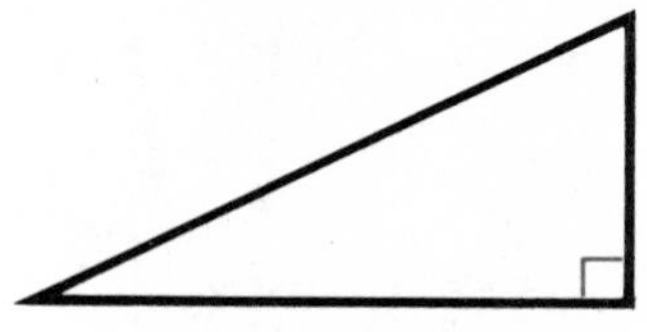

Fig. 9

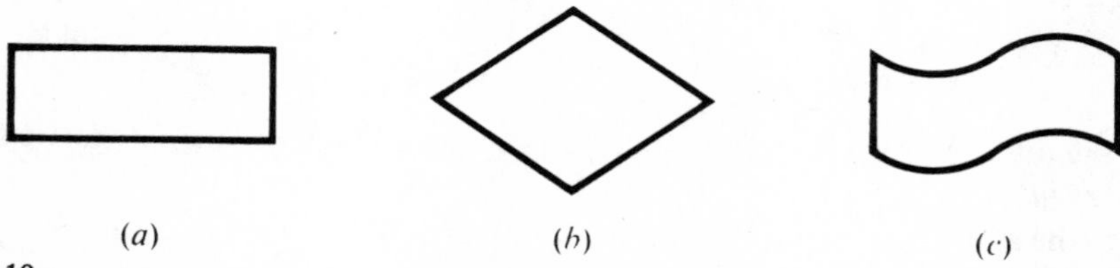

Fig. 10

(*b*) Answer the same question for the shapes in Figure 10.

8 By considering the symmetries of the following polygons, calculate (i) the exterior angle of each polygon; (ii) the interior angle.

(*a*) a regular hexagon; (*b*) a regular pentagon; (*c*) a regular octagon.

2 Polygons and polyhedra

(*a*) A convex polygon has each of its interior angles less than 180°. Figure 11 shows some examples of convex polygons, and Figure 12 examples of non-convex polygons.

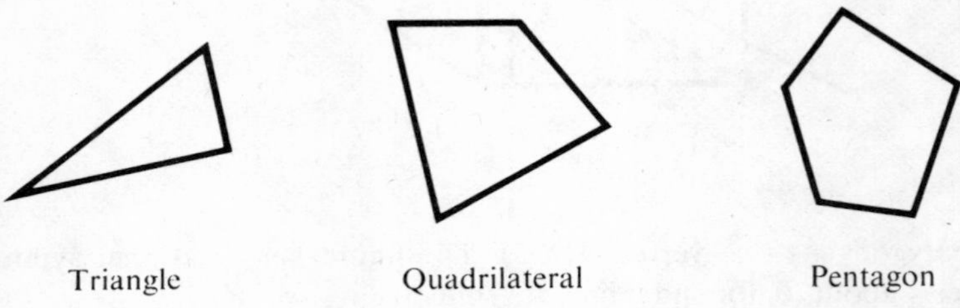

Fig. 11

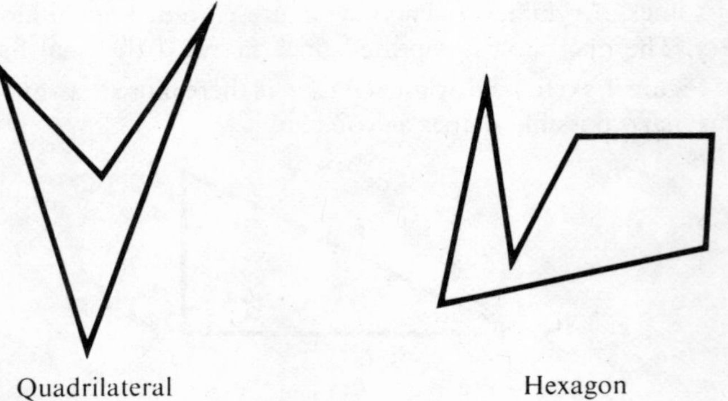

Fig. 12

If the interior angles and sides of a polygon are equal then the polygon is said to be *regular*.

(*b*) The sum of the exterior angles of any polygon is 360°, or one full turn (see Figure 13).

$$a^\circ + b^\circ + c^\circ + d^\circ + e^\circ = 180^\circ.$$

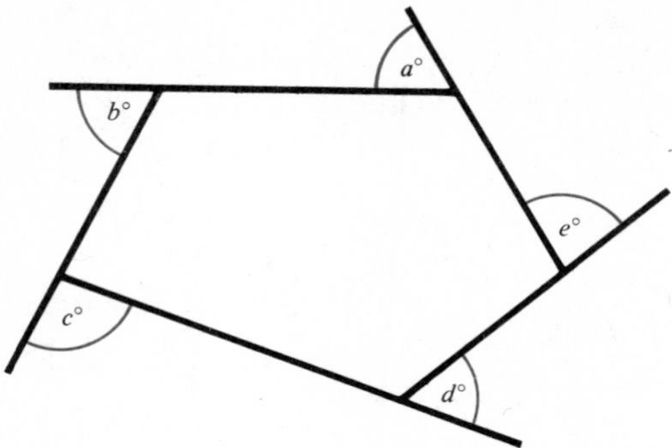

Fig. 13

A regular n-gon has exterior angles of $(360/n)°$. For example, the exterior angle of a regular hexagon is $(360/6)° = 60°$.

(*c*) The angle sum of a triangle is 180° (Figure 14).

$$a° + b° + c° = 180°.$$

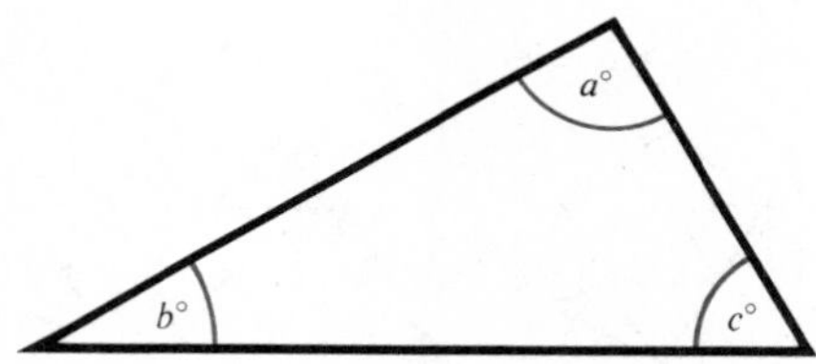

Fig. 14

(*d*) Since an n-sided polygon can be divided into $(n-2)$ triangles with a common vertex, then the sum of the interior angles of an n-gon is $\{(n-2) \times 180\}°$, or $(2n-4)$ right angles. Figure 15 shows how a pentagon can be divided into triangles in this way. The sum of its interior angles is $\{(5-2) \times 180\}° = 540°$.

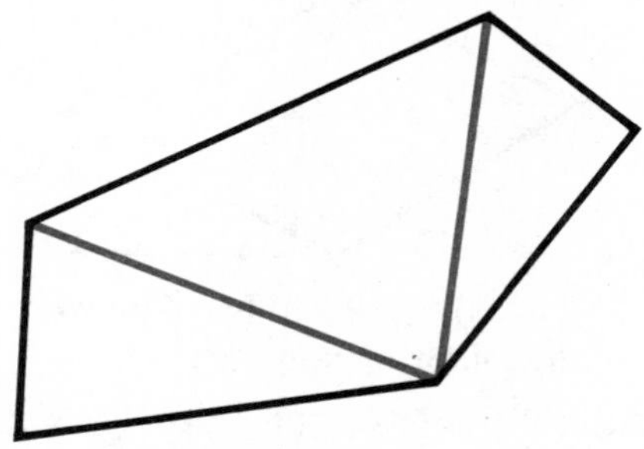

Fig. 15

(*e*) Polyhedra can be constructed from nets. Figure 16 shows the five regular polyhedra whose nets are made up of regular polygons.

All polyhedra obey the relation (Euler's relation):

Number of Faces + Number of Vertices − Number of Edges = 2

that is

$$F+V-E=2.$$

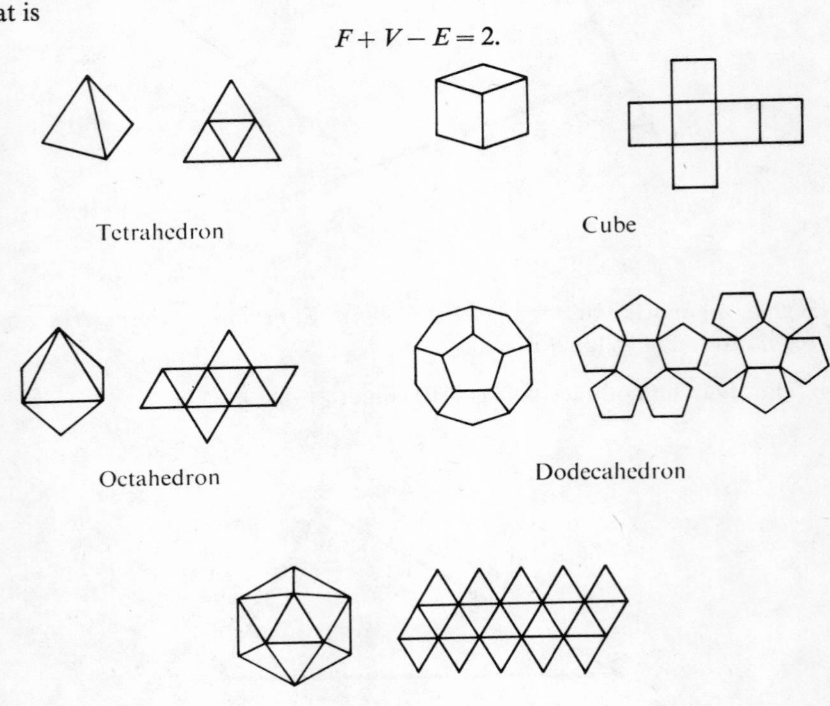

Fig. 16

Exercise B

1 In the solid in Figure 17 all faces are either rectangles or triangles. Draw a net for the solid, indicating any lengths which need to be equal. How many (i) planes; (ii) axes of symmetry has the solid?

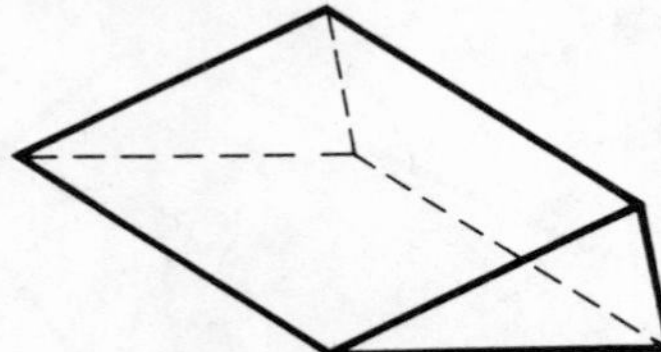

Fig. 17

2 Use the formula $N = \{(n-2) \times 180\}^\circ$ to help you write down the sum (N) of the interior angles of the following polyhedra:

(i) a pentagon; (ii) a hexagon; (iii) a 10-gon; (iv) a 20-gon.

If the polygons are regular, write down in degrees the size of (*a*) each interior angle; (*b*) each exterior angle.

3 Explain carefully why there is no regular polygon with interior angles of 164°.

4 (*a*) A regular polygon has an interior angle of 140°. How many sides has it?
(*b*) Can a regular polygon have an interior angle less than (i) 90°; (ii) 60°? Explain your answers.

5 A regular hexagon with centre (0, 0) has one vertex $A(2, 0)$. Write down the coordinates of each of the other vertices.

6 An isosceles triangle has one angle $x°$ ($x° > 90°$). Write down, in terms of x, the size of the other two angles.

7 Calculate, in Figure 18,

(i) $a° + b° + c°$;
(ii) $p° + q° + r°$.

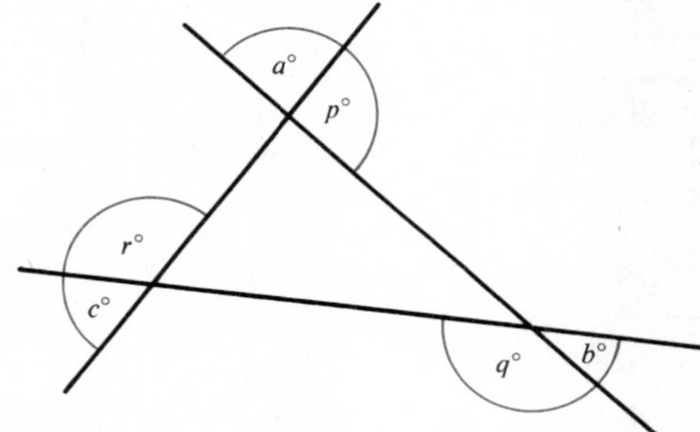

Fig. 18

8 In Figure 19, calculate

(i) $a° + b° + c° + d° + e°$;
(ii) $p° + q° + r° + s° + t°$.

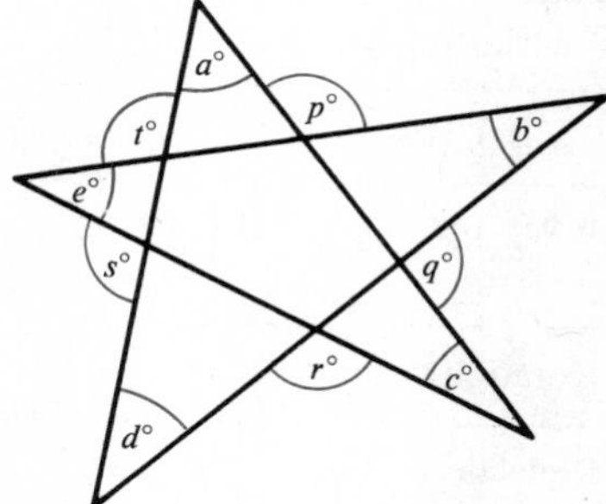

Fig. 19

3 Transformations

(*a*) Transformations of the plane are classified according to properties of invariance (that is, properties such as angle, shape, size, . . ., which do not change under the transformation). For example, the isometries – reflection, rotation, translation and glide-reflection – leave shape and size (and consequently length, area and angle) invariant.

The similarities constitute the class of transformations leaving shape invariant. Thus enlargements, and the isometries, are similarities.

Table 1 shows which properties of geometrical figures are invariant under the transformations we have studied.

	Length	Angle	Ratio	Area	Sense	Parallelism	Order of points	Node order
Rotation (i)	✓	✓	✓	✓	✓	✓	✓	✓
Translation (ii)	✓	✓	✓	✓	✓	✓	✓	✓
Reflection (iii)	✓	✓	✓	✓		✓	✓	✓
Glide reflection (iv)	✓	✓	✓	✓		✓	✓	✓
Enlargement (v)		✓	✓		✓	✓	✓	✓
Shearing (vi)				✓	✓	✓	✓	✓
Stretching (vii)						✓	✓	✓
Topology (viii)							✓	✓

TABLE 1

(*b*) Table 2 shows to which class each transformation belongs, and the minimum invariance requirements of each class. Notice that each and every transformation we have studied is a topology, and that this is the most 'general' transformation requiring only that 'order of points' and 'order of nodes' be invariant.

Class	Topology	Affine	Similarities	Isometrics
Invariants	Order of nodes Order of points	Parallelism	Shape	Shape and size
Rotation	✓	✓	✓	✓
Reflection	✓	✓	✓	✓
Translation	✓	✓	✓	✓
Glide reflection	✓	✓	✓	✓
Enlargement	✓	✓	✓	
One-way stretch	✓	✓		
Two-way stretch	✓	✓		
Shearing	✓	✓		
Topology	✓			

TABLE 2

When we study transformations we often consider simple geometrical figures and their images under the transformation to help us understand its effect. It is important to remember, however, that a transformation is applied *to the whole* (x, y) *plane* and consequently that each and every point of the plane has an image under the transformation.

4 Transformations – isometries

(*a*) *Rotation.* A rotation is defined by a given amount of turn about a centre of rotation (the only invariant point of the transformation). Figure 20 shows the effect of three different rotations on figures in the (x, y) plane. The centre of rotation is marked in red. Note that a positive turn is in an anticlockwise direction, and that if the transformation is denoted by **N**, then the image of a figure P under the transformation is denoted by **N**(P).

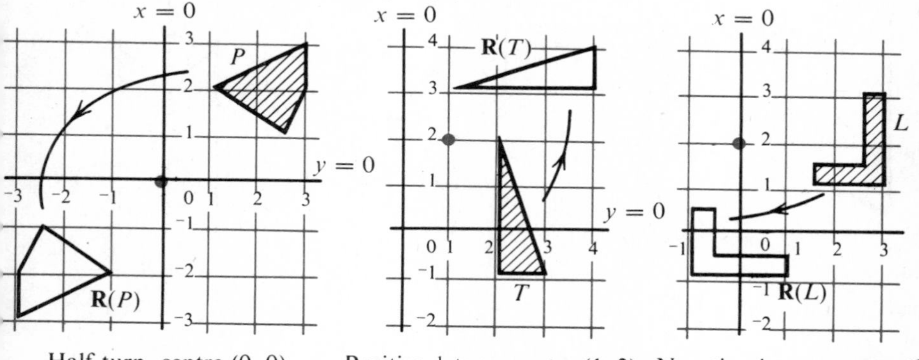

Half turn, centre (0, 0)

Positive $\frac{1}{4}$ turn, centre (1, 2) (or, negative $\frac{3}{4}$ turn, centre (1, 2))

Negative $\frac{1}{4}$ turn, centre (0, 2) (or positive $\frac{3}{4}$ turn, centre (0, 2))

Fig. 20 (*a*) (*b*) (*c*)

Given an object figure and its image under a particular rotation we can determine the centre of rotation by constructing the mediators of two points and their images (the mediators, for example, of AA' and BB' in Figure 21). The point in which these mediators meet (O) is then the centre of rotation. The reason for this is that A and A' are equidistant from O, and so are B and B'. A therefore

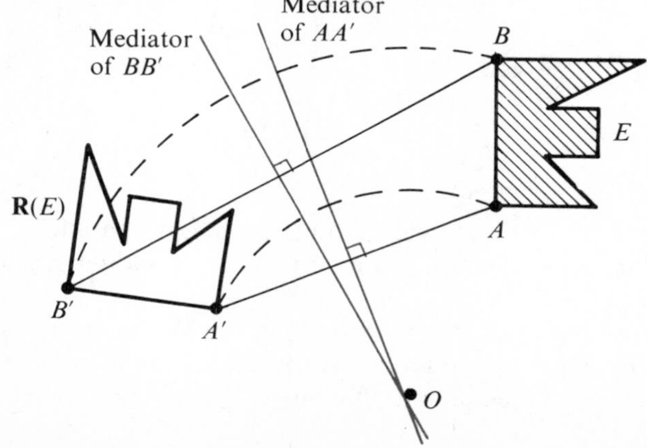

Fig. 21

moves on the circumference of a circle, centre O, under the rotation, and is mapped to A'. Simultaneously, B is also mapped to B'. The angle of rotation for this particular transformation is given by $\angle AOA'$ (or $\angle BOB'$).

(*b*) *Reflection.* A reflection leaves all points on a particular line (the mirror line) invariant and maps all other points P onto images P' so that the invariant line is the mediator of PP'. Figure 22 shows the effect of three different reflections on figures in the (x, y) plane. Note that lines perpendicular to the mirror line (for example, the lines $x = c$ in Figure 22(*a*)) are mapped onto themselves, although the points on these lines are not (other than the point of $x = c$, which also lies on the mirror line).

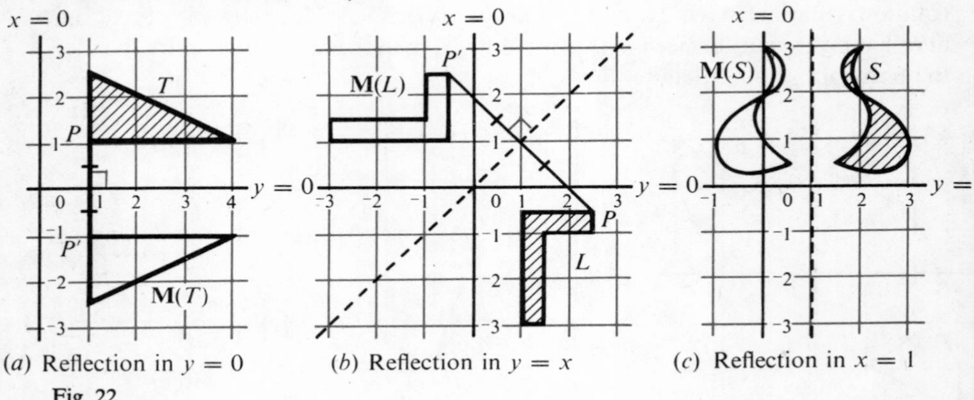

(*a*) Reflection in $y = 0$ (*b*) Reflection in $y = x$ (*c*) Reflection in $x = 1$

Fig. 22

(*c*) *Translation.* A translation is completely defined by the displacement of any one point of the (x, y) plane. Figure 23 represents a translation of the quadrilateral $ABCD$. The translation is completely defined by any displacement equivalent to $\mathbf{PP}'$ (for example, $\mathbf{AA}'$, $\mathbf{BB}'$, . . .).

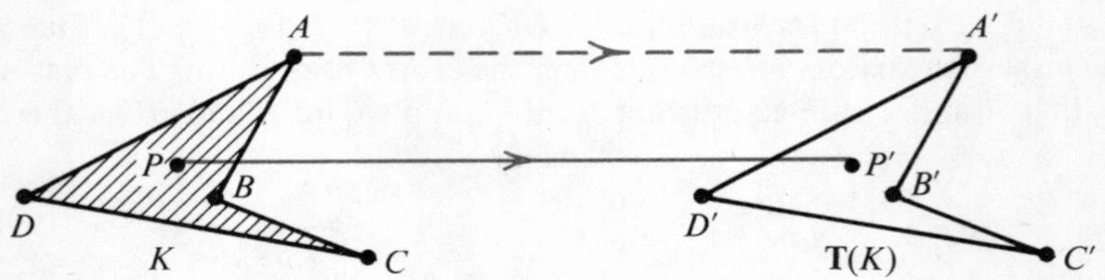

Fig. 23

There are no invariant points under a translation. Lines parallel to the direction of translation are, however, mapped onto themselves, although points on these lines are not.

(*d*) *A glide-reflection* is a combination of a translation and a reflection in a line *parallel to the direction of translation* (see Figure 24). No points are invariant, but the glide line is mapped onto itself under the transformation.

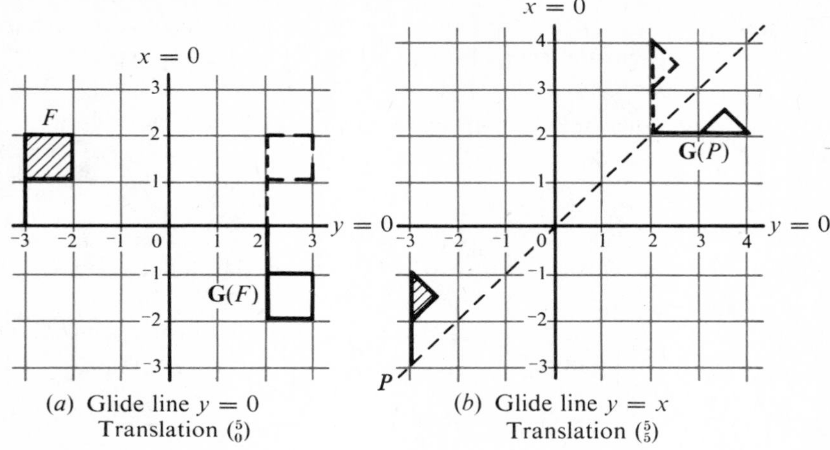

(a) Glide line $y = 0$
Translation $\binom{5}{0}$

(b) Glide line $y = x$
Translation $\binom{5}{5}$

Fig. 24

(e) *Congruence.* Figures which can be mapped onto each other by direct isometries (rotations, and translations) are said to be *directly congruent*. Those which can be mapped onto each other by opposite isometries (reflections, glide reflections) are said to be *oppositely congruent*. An opposite isometry changes the sense of a figure – in layman's terms this means that the object figure must be 'flipped over' before it can be fitted exactly over its image. For example, in Figure 25 L_1 and L_3 are directly congruent, but L_2 is oppositely congruent to both. L_1 can be mapped onto L_3 by a rotation, and onto L_2 by a glide-reflection. There is always a *single simple transformation* which will map congruent figures onto each other.

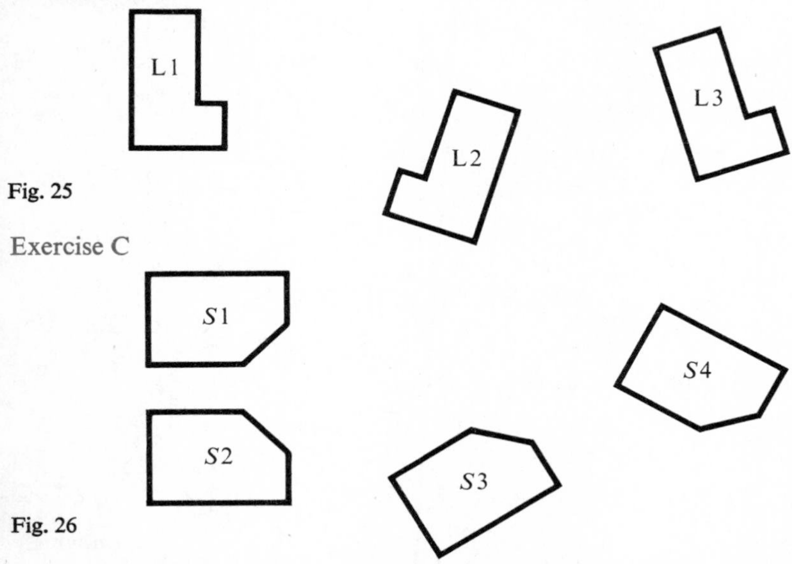

Fig. 25

Exercise C

Fig. 26

1 Trace Figure 26 and explain carefully how each can be mapped onto the others by single transformations.

2 A triangle has vertices $A(2, 0)$, $B(4, 1)$, $C(3, 5)$. Sketch the triangle on a grid and draw its image after each of the following transformations:

(i) a reflection in $y = 0$;
(ii) a half turn about (0, 0);
(iii) a positive quarter turn about (0, 0);
(iv) a translation $\begin{pmatrix}3\\1\end{pmatrix}$;
(v) a glide-reflection – translation $\begin{pmatrix}2\\2\end{pmatrix}$, followed by a reflection in $y = x$.

3 Describe accurately the transformation which will map triangle ABC onto triangle $AB'C'$ in Figure 27.

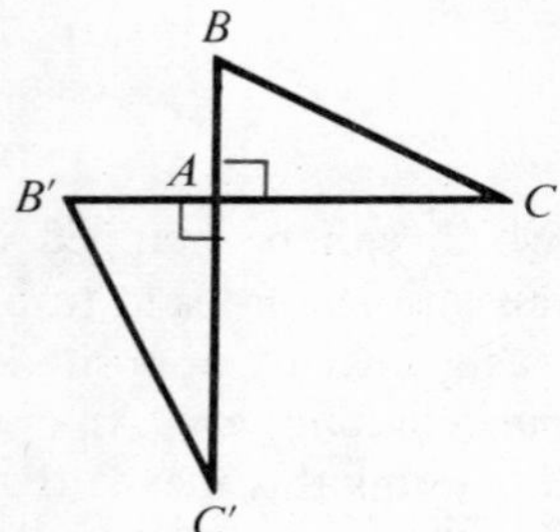

Fig. 27

4 The point P in Figure 28 is reflected in m_1, then its image in m_2, and finally this image in m_3. (m_1, m_2, m_3 are parallel.) If P' is the final image, state the distance PP'.

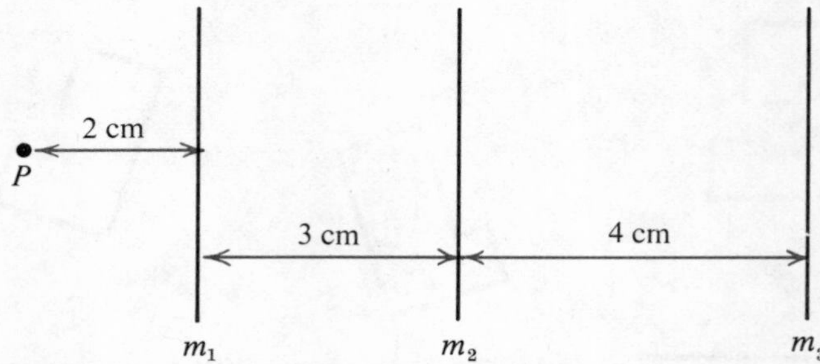

Fig. 28

State the distance PP' if the order of reflections is

(i) m_1, m_3, m_2; (ii) m_3, m_2, m_1; (iii) m_2, m_3, m_1.

Questions 5, 6, and 7 refer to Figure 29.

5 Describe the single transformation, if it exists, which maps:

(*a*) △4 onto △8; (*b*) △5 onto △6; (*c*) △4 onto △1;
(*d*) △5 onto △7; (*e*) △4 onto △7.

6 △3 is given a rotation of 180° about ($^-1$, 1). Specify a transformation which will map its image onto:

(*a*) △8; (*b*) △2.

7 In Figure 29, $\triangle 1$ is reflected in the line $y = {}^-x$. Specify a single transformation which will map its image onto:

(*a*) $\triangle 8$; (*b*) $\triangle 4$.

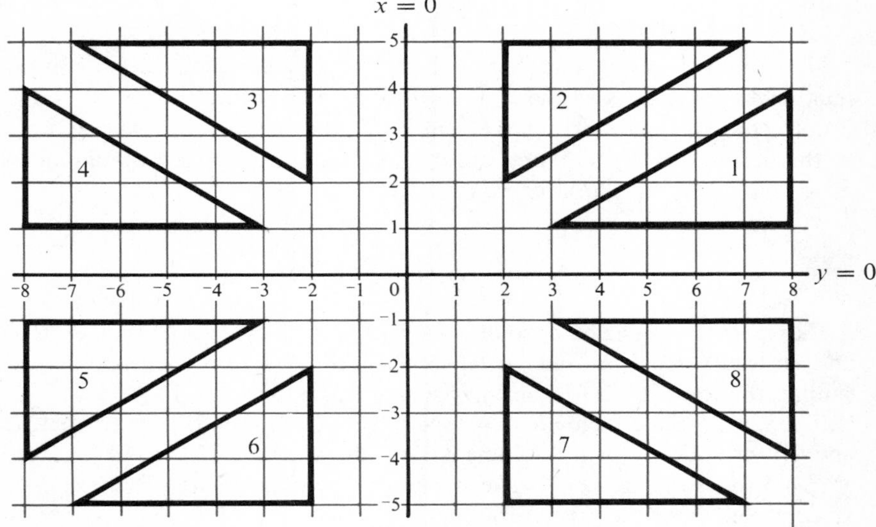

Fig. 29

5 Transformations – enlargements and stretches

(*a*) Enlargements belong to the class of transformations known as the similarities. This class also includes the isometries, and is made up of those transformations which leave shape (and consequently angle) invariant.

Under an enlargement the (x, y) plane, and in particular, geometrical figures are 'enlarged' by a constant scale factor K, with one point of the plane, called the centre of enlargement, invariant. Figure 30(*a*) represents an enlargement S.F.2, centre $({}^-1, 0)$, and Figure 30(*b*) an enlargement S.F. $^-2$, centre O.

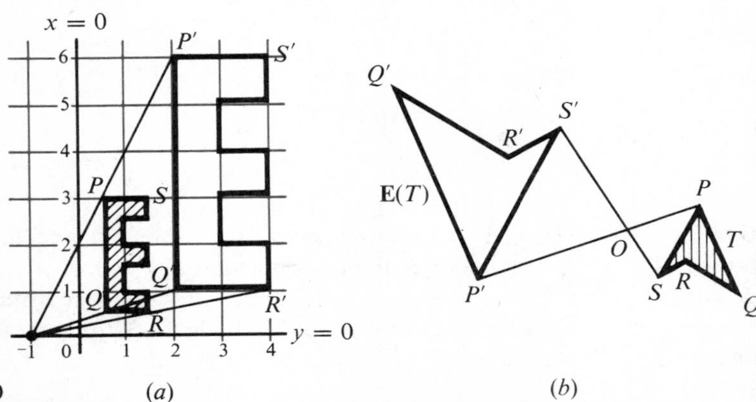

Fig. 30 (*a*) (*b*)

In Figure 30(a) $\frac{OP'}{OP} = \frac{OQ'}{OQ} = \frac{OR'}{OR} = \frac{OS'}{OS} = 2$

and in Figure 30(b) $\frac{OP'}{OP} = \frac{OQ'}{OQ} = \frac{OR'}{OR} = \frac{OS'}{OS} = 2.$

Note that all lines passing through the centre of enlargement are mapped onto themselves, but points on these lines (other than O) are not.

As a consequence of an enlargement S.F. K, the lengths of corresponding sides of the object and image figures are in the ratio $1:K$, and corresponding angles are equal. In Figure 30(b), for example,

$$\frac{P'Q'}{PQ} = \frac{Q'R'}{QR} = \frac{R'S'}{RS} = \frac{S'P'}{SP} = 2.$$

Shapes which can be mapped onto each other by a combination of an isometry and an enlargement are said to be similar. All congruent shapes are therefore similar, the scale factor for the enlargement being 1 (or $^-1$).

If the linear scale factor for an enlargement is K, then (i) the area S.F. is K^2 and (ii) the scale factor for volume is K^3. Thus the area of **E**(T) in Figure 30(b) is $2^2 = 4$ times the area of T. The two cans in Figure 31 are similar and one is an enlargement of the other with linear scale factor 2. The volume of the larger is therefore $2^3 = 8$ times that of the smaller.

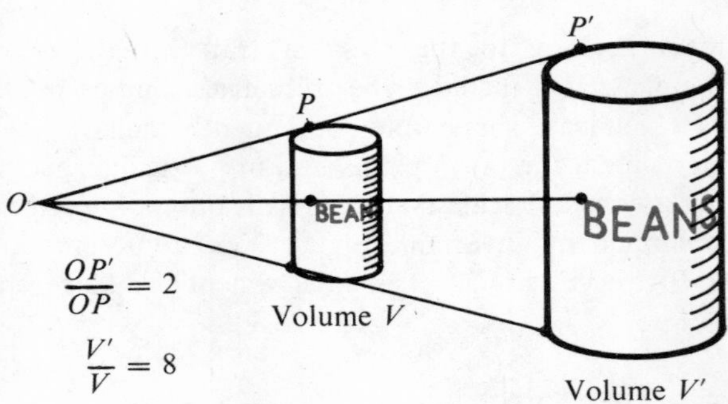

Fig. 31

(b) *One-way stretch.* Under a one-way stretch all points on one line are invariant whilst others move in a direction perpendicular to this line. If the scale factor for the stretch is K then each point is mapped onto an image distance K times the original distance from the invariant line. Figure 32(a) represents a stretch S.F 2 with $x = 0$ invariant, and Figure 32(b) a stretch S.F. $^-3$ with $y = 0$ invariant.

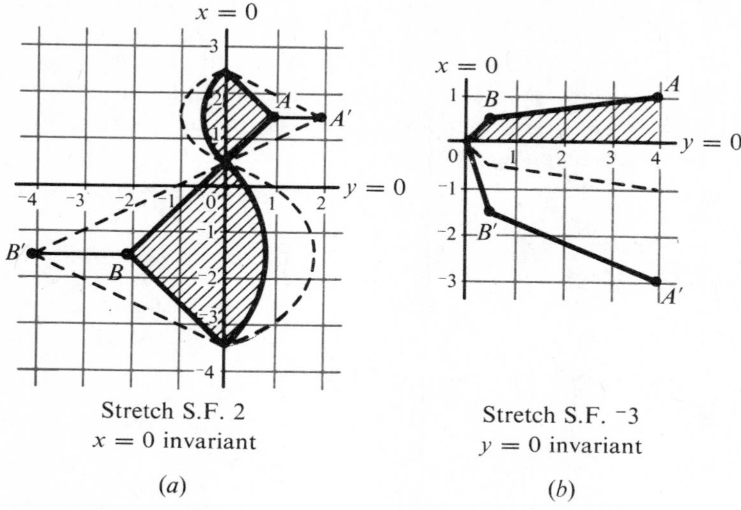

Stretch S.F. 2
$x = 0$ invariant

(*a*)

Stretch S.F. $^-3$
$y = 0$ invariant

(*b*)

Fig. 32

If the linear scale factor for a stretch is K, the area scale factor is also K.

(*c*) *A two-way stretch* is a combination of two one-way stretches with the invariant lines meeting in a right angle. The only invariant point under such a transformation is thus the point in which the lines meet. Figure 33 represents a two-way stretch S.F. 2 parallel to the x axis and S.F. $^-3$ parallel to the y axis (i.e. the combination of the transformations represented by Figures 32(*a*) and (*b*)).

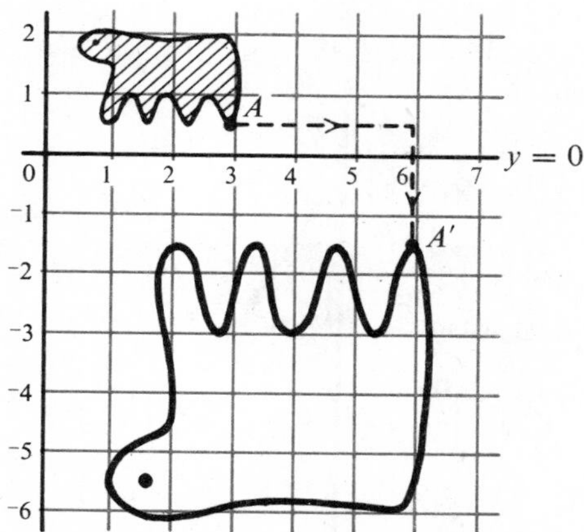

Fig. 33

If the linear scale factors for a two-way stretch are k and m, then the area scale factor is km. When the two scale factors are equal, the transformation is equivalent to an enlargement.

6 Transformations – shearing and topology

(*a*) *Shearing*. Under a shear one line of points remains invariant and all other points move parallel to this line, so that area is preserved. Figure 34(*a*) represents a shear with $y = 0$ invariant mapping $(0, 1) \rightarrow (1, 1)$, and Figure 34(*b*) a shear with $x = 1$ invariant mapping $(0, 1) \rightarrow (0, 0)$.

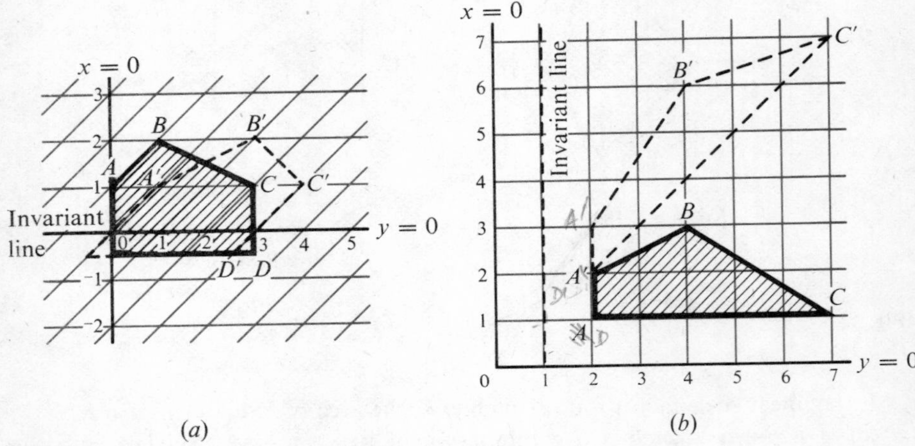

Fig. 34

The red grid in Figure 34(*a*) represents the image of the (x, y) plane under this shear.

Notice that points equidistant from the invariant line (for example, A and C in Figure 34(*a*)) remain the same distance apart after the transformation as they were before it – that is $AC = A'C'$. By studying Figure 35 you can also see that $(PP'/QQ') = (PO/QO)$. This means that the distance a point moves is proportional to its distance from the invariant line.

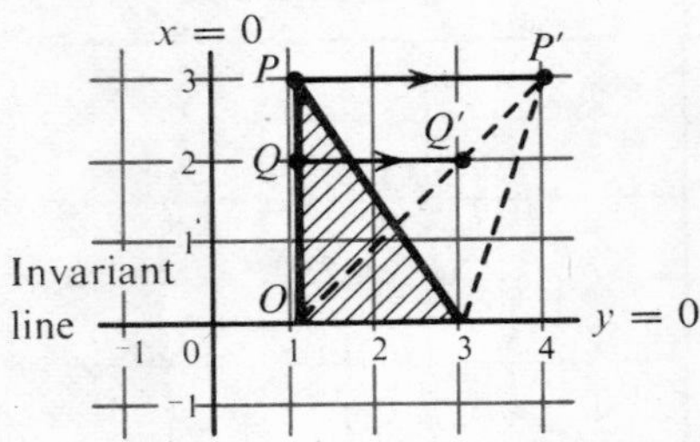

Fig. 35

(*e*) *Topology*. Topological transformations are the most 'general' transformations we study in this course. All other transformations have parallelism as an invariant property (that is, parallel lines are mapped to parallel lines) and are called *affine* transformations. For a transformation to be a topology, all we require is that order of points on a line, and order of nodes, be invariant. Affine transformations are therefore necessarily topologies.

The circle in Figure 36 can be distorted into the other shapes by stretching in the plane of the paper – we are allowed to 'flip' the plane over before distortion so that a change in sense occurs. Notice that in each of the figures, B lies between A and C (order of points on a line is invariant). A change in sense has occurred in Figure 36(c). These three shapes are said to be topologically equivalent.

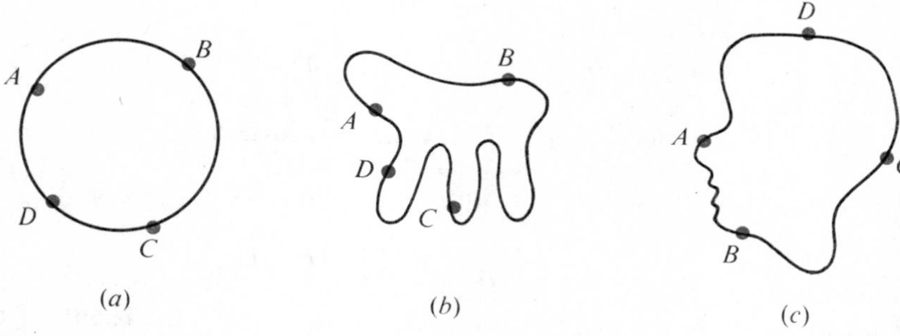

Fig. 36

The shapes in Figure 37 can be transformed into each other by stretching and are therefore topologically equivalent. Note that each figure has four three nodes and one four node (order of nodes is invariant). Without actually tearing the plane and sticking pieces together none of these shapes could be transformed into those in Figure 36. The two sets of shapes therefore belong to different equivalence classes.

Fig. 37

(f) *Inverse transformations.* The transformation **A** is the inverse of **B** if $\mathbf{AB} = \mathbf{BA} = \mathbf{I}$ where **I** is the identity transformation. We write the inverse of **B** as $\mathbf{B}^{-1}$. Hence if $\mathbf{AB} = \mathbf{BA} = \mathbf{I}$, $\mathbf{A} = \mathbf{B}^{-1}$ (and $\mathbf{B} = \mathbf{A}^{-1}$).

$\mathbf{A}^{-1}$ maps the image of the plane under the transformation **A** back into its original position. Hence if **R** is a quarter positive turn, centre (0, 0) then $\mathbf{R}^{-1}$ is a quarter negative turn centre (0, 0) (or a $\frac{3}{4}$ positive turn, centre (0, 0)).

Here are some more examples of transformations and their inverses:

Transformation	*Inverse transformation*
Reflection in a line m	Reflection in a line m
Enlargement S.F. k, centre (p, q)	Enlargement S.F $1/k$, centre (p, q)
Shear mapping $(p, q) \to (p, s)$ with $y = 0$ invariant	Shear mapping $(p, s) \to (p, q)$ with $y = 0$ invariant
Translation $\begin{pmatrix} a \\ b \end{pmatrix}$	Translation $\begin{pmatrix} -a \\ -b \end{pmatrix}$

Exercise D

1 State the image of the point (3, 1) under the following transformations of the (x, y) plane:

(i) enlargement S.F. 2 centre (0, 0);
(ii) enlargement S.F. $^{-}2$ centre (0, 0);
(iii) enlargement S.F. 2 centre (1, 1);
(iv) shear mapping $(0, 1) \to (1, 1)$ with $y = 0$ invariant;
(v) shear mapping $(1, 0) \to (1, 1)$ with $x = 0$ invariant;
(vi) shear mapping $(0, 1) \to (1, 2)$ with $x = y$ invariant;
(vii) one-way stretch S.F. 3 with $y = 0$ invariant;
(viii) two-way stretch S.F. 2 with $x = 0$ invariant, S.F. $^{-}3$ with $y = 0$ invariant.

2 Draw the image of the triangle $A(0, 0)$; $B(3, 0)$; $C(0, 2)$ under each of the transformations in Question 1. For each image state: (i) the length of the side of the image triangle corresponding to AB in the object triangle; (ii) the area of the image triangle.

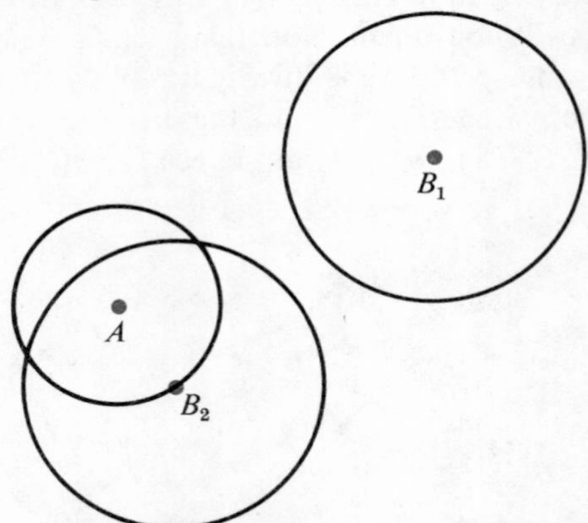

Fig. 38

3 Copy Figure 38 and construct the centres of the enlargements which map the circle centre A onto each of the circles centres B_1 and B_2.

4 In Figure 39, AE bisects $\angle BAC$, $AB = AD$ and FD is parallel to BC.

(i) State a single transformation that maps B onto D and E onto F. Name a line segment equal to BE.
(ii) State a single transformation that maps D onto C and F onto E. Copy and complete the statement $\frac{}{EC} = \frac{}{AC}$.
(iii) Given that $AB = 5$ cm, $AC = 7$ cm and $BC = 6$ cm, calculate EC.

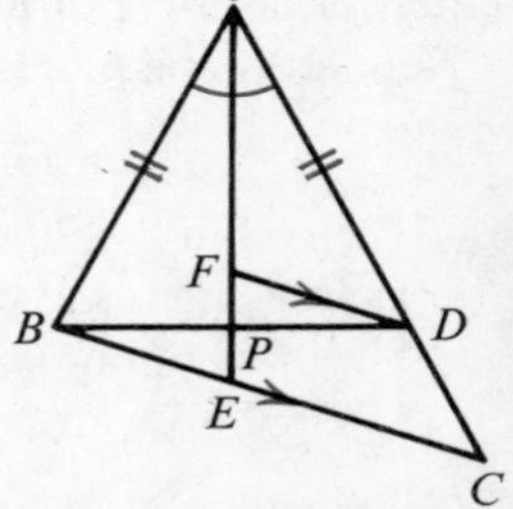

Fig. 39

5 It takes me 10 puffs to blow up a balloon to a radius 8 cm. How many puffs would you expect me to take to blow up the balloon to a radius 16 cm?

6 Which of the diagrams in Figure 40 are topologically equivalent?

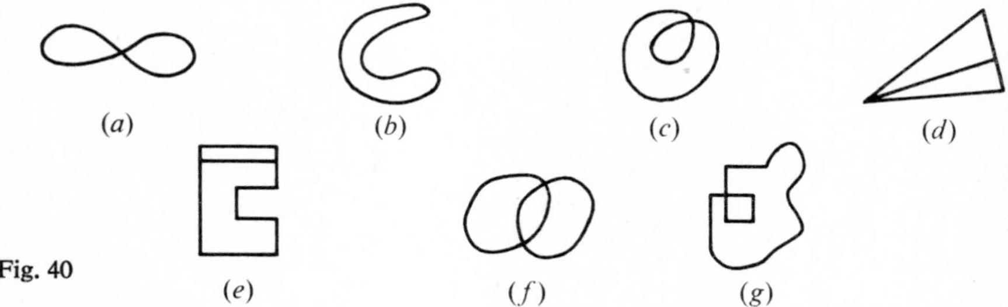

Fig. 40

7 In each of the pairs of figures in Figure 41, A has been transformed to B by stretching. Mark in a possible position for the points P and R on Figure B in each case.

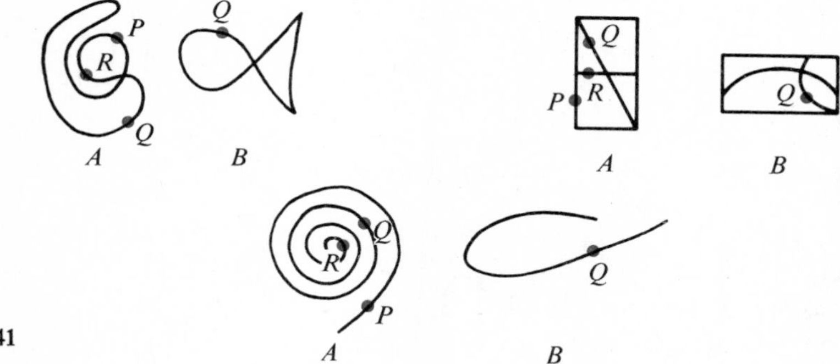

Fig. 41

7 Combination of transformations

(*a*) *Two rotations.* A turn through $\theta°$ about A followed by a turn through $\alpha°$ about B (B might be the same point as A) is equivalent to a turn through $(\theta + \alpha)°$ as suggested by Figure 42.

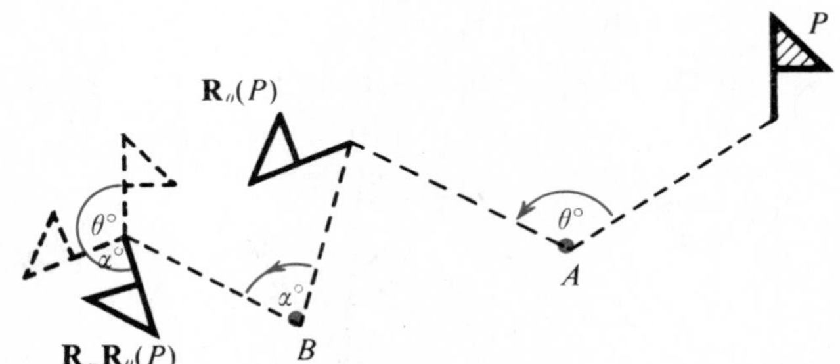

Fig. 42

Note that the final image is written $\mathbf{R}_\alpha\mathbf{R}_\theta(P)$. Since $\mathbf{R}_\alpha\mathbf{R}_\theta \neq \mathbf{R}_\theta\mathbf{R}_\alpha$ in this case, we must be careful to write the combined transformation correctly.

(*b*) *Reflection in intersecting mirrors.* Figure 43 shows that two reflections in intersecting mirror lines is equivalent to a rotation through twice the angle between the mirror lines about the point of intersection of the lines. Again $\mathbf{M}_1\mathbf{M}_2 \neq \mathbf{M}_2\mathbf{M}_1$.

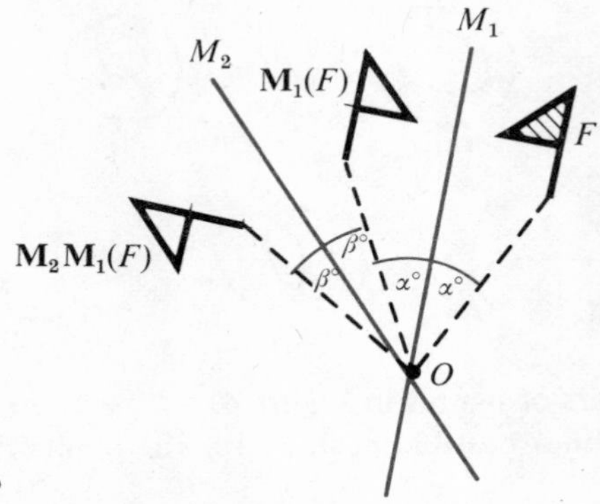

$F \rightarrow \mathbf{M}_2\mathbf{M}_1(F)$ by a rotation of $2(\alpha+\beta)°$ about O.

Fig. 43

(*c*) Two reflections in parallel mirror lines are equivalent to a translation through twice the distance between the lines (Figure 44).

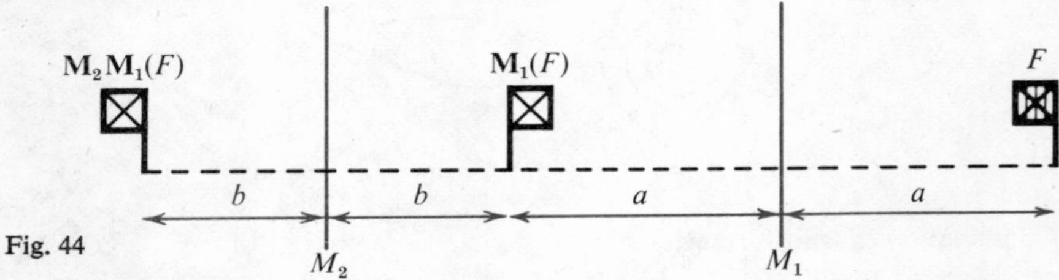

Fig. 44

Is $\mathbf{M}_1\mathbf{M}_2 = \mathbf{M}_2\mathbf{M}_1$ in this case?

(*c*) *Two translations.* A translation $\begin{pmatrix} a \\ b \end{pmatrix}$ followed by a translation $\begin{pmatrix} p \\ q \end{pmatrix}$ is equivalent to a single translation $\begin{pmatrix} a+p \\ b+q \end{pmatrix}$. In Figure 45 $\mathbf{AA}' + \mathbf{A}'\mathbf{A}'' = \mathbf{AA}''$, that is $\mathbf{T}_2\mathbf{T}_1 = \mathbf{T}_3$. In this case the commutative law holds: $\mathbf{T}_2\mathbf{T}_1 = \mathbf{T}_1\mathbf{T}_2$.

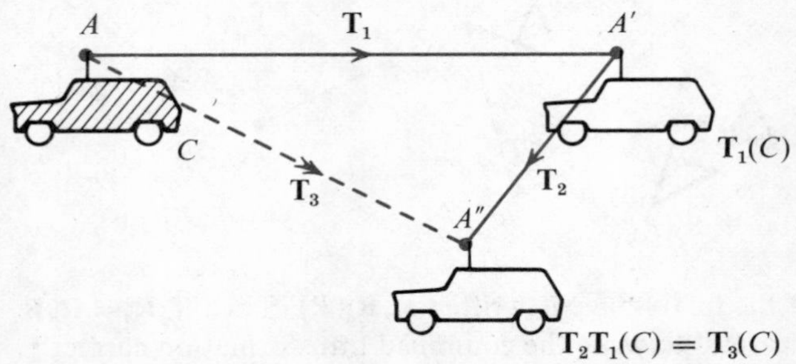

Fig. 45

(*d*) Two enlargements scale factors k and m are equivalent to a single enlargement S.F. km. In Figure 46 the enlargements have scale factors 2 and 3 with centres (0, 0) and (1, 4) respectively. These are equivalent to a single enlargement S.F. 6 centre $(\frac{2}{5}, \frac{8}{5})$.

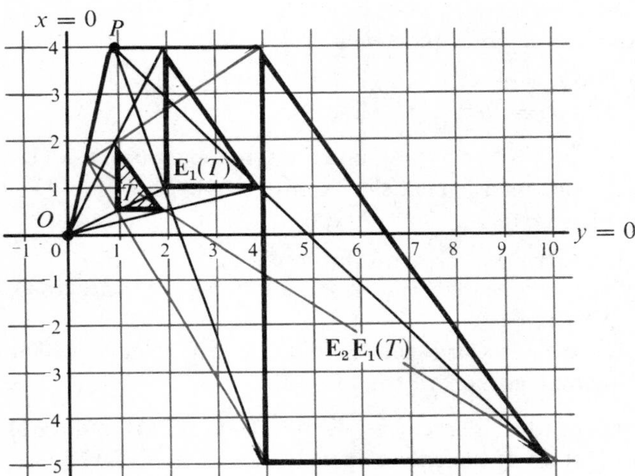

Fig. 46

The centre of enlargement lies on the line segment OP and divides it internally in the ratio 2:3.

(*e*) Other combinations of transformations, for example, a rotation and an enlargement cannot be reduced to a single simple transformation. Exercise E contains some questions on more complicated combinations of transformations.

(*f*) *Inverse transformations.* To map $\mathbf{M}_2\mathbf{M}_1(F)$ back to F in Figure 44 we apply first the inverse transformation of $\mathbf{M}_2$ and then the inverse transformation of $\mathbf{M}_1$ so that $(\mathbf{M}_2\mathbf{M}_1)^{-1} = \mathbf{M}_1^{-1}\mathbf{M}_2^{-1}$. Similarly, in Figure 42 we can see that $(\mathbf{R}_\alpha\mathbf{R}_\theta)^{-1} = \mathbf{R}_\theta^{-1}\mathbf{R}_\alpha^{-1}$. This means that to find the inverse of a combined transformation we first obtain the two inverse transformations and then combine them in the reverse order.

Exercise E

1 **X** denotes a reflection in the x-axis, **Y** a reflection in the y-axis, **R** a half-turn centre (0, 0). Copy and complete this combination table:

	X	**Y**	**R**
X	**I**		
Y			
R			**I**

Is the set {**X**, **Y**, **R**} closed under the operation 'followed by'?
Is the operation 'followed by' commutative for this set of transformations?

2 $\mathbf{X}$ denotes a reflection in $y=0$, $\mathbf{E}$ an enlargement S.F. 2 centre (0, 0), $\mathbf{T}$ a translation $\begin{pmatrix}2\\1\end{pmatrix}$, P is the point (2, 1). Find

(i) $\mathbf{XE}(P)$; (ii) $\mathbf{EX}(P)$; (iii) $\mathbf{TE}(P)$; (iv) $\mathbf{ET}(P)$; (v) $\mathbf{XET}(P)$.

3 If, with the notation of Question 2, $\mathbf{TE}(Q)=(5, 1)$ what are the coordinates of Q? Answer the same question if

(i) $\mathbf{T}^2(Q)=(4, 2)$; (ii) $\mathbf{E}^2(Q)=(7, 2)$;
(iii) $\mathbf{XE}(Q)=(0, 0)$; (iv) $\mathbf{XET}(Q)=(0, 0)$.

4 $\mathbf{S}$ represents a shear with $y=0$ invariant, mapping $(0, 1)\rightarrow(1, 1)$.
$\mathbf{Q}$ represents an enlargement S.F. 2 centre (1, 1).
A triangle has vertices $A(0, 0)$, $B(2, 0)$, $C(^-2, 3)$.

(*a*) Draw the triangle with vertices $\mathbf{S}(A)$, $\mathbf{S}(B)$, $\mathbf{S}(C)$.
(*b*) On the same diagram draw the triangle with vertices $\mathbf{QS}(A)$, $\mathbf{QS}(B)$, $\mathbf{QS}(C)$.
(*c*) Describe in words the combined transformation which will map the final image back to triangle ABC.

5 The centres of two enlargements $\mathbf{E}_1$ and $\mathbf{E}_2$ are (2, 1) and (6, 1) respectively. The scale factors of the enlargements are both 2. $\mathbf{E}_2\mathbf{E}_1(T)$ is a triangle with vertices (3, 1), (5, 1), (5, 5). Describe accurately the single transformation which will map $T\rightarrow\mathbf{E}_2\mathbf{E}_1(T)$.

6 $\mathbf{G}$ is a glide reflection, being a combination of a translation $\begin{pmatrix}3\\3\end{pmatrix}$ and a reflection in the line $x=y$. Describe $\mathbf{G}^{-1}$.
$\mathbf{H}$ is another glide reflection with translation $\begin{pmatrix}^-3\\3\end{pmatrix}$ and glide-line $x+y=0$. Describe $\mathbf{H}^{-1}$.
T is the triangle $A(0, 0)$, $B(2, 0)$, $C(2, 3)$. Write down the coordinates of the vertices of (i) $\mathbf{G}(T)$; (ii) $\mathbf{H}(T)$; (iii) $\mathbf{GH}(T)$.
Is there a single transformation which will map $T\rightarrow\mathbf{GH}(T)$?
Which single transformation will map $\mathbf{GH}(T)\rightarrow\mathbf{H}^{-1}\mathbf{G}^{-1}(T)$?

7 (i) ABC is any triangle. The enlargement with centre B and scale factor 2 maps A onto H, C onto K and K onto P. Name two pairs of parallel lines.
(ii) An 'enlargement' with centre C maps P onto K and H onto X. State the scale factor of this enlargement. Identify the position of X, justifying your answer by reference to the properties of enlargement, without any appeal to accurate drawing.
(iii) Describe the simplest transformation that maps A onto K while leaving X unaltered.

8 Solid geometry

(*a*) *Angle between skew lines.* It is relatively easy to calculate the angle between the intersecting line segments AB and BD of the cube in Figure 47 by using trigonometry:

$$\tan \angle ABD = \frac{AD}{BD} = \frac{\sqrt{(2a^2)}}{a} = \sqrt{2}; \qquad \text{(why is } AD = \sqrt{(2a^2)}\text{ ?)}$$

$$\angle ABD \approx 54{\cdot}7^\circ.$$

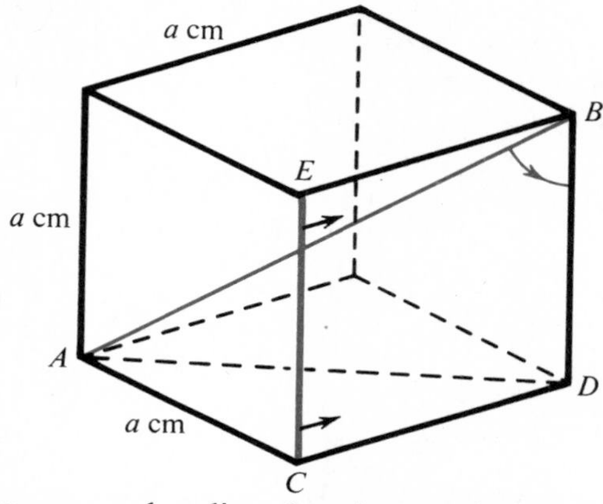

Fig. 47

AB and *EC*, however, are skew lines (non-intersecting). To calculate the angle between them we first translate one until it intersects the other. By translating *EC* to *BC* we do not change the angle between *AB* and *EC*, and the angle is therefore equal to angle *ABD* i.e. 54·7°.

(*b*) The angle between a line and a plane is defined as the angle between the line and its orthogonal projection onto the plane. The angle between the line through *AB* and the plane π in Figure 48 is therefore the angle *ABX*.

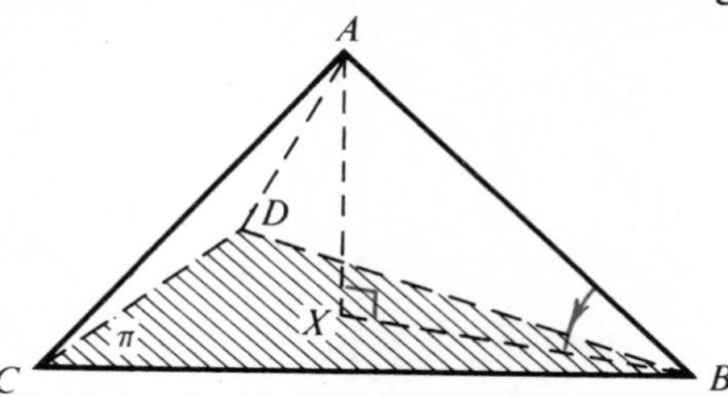

Fig. 48

(*c*) *Angle between two planes.* The angle between two planes is defined as the angle between any line in the first plane which meets the line of intersection of the two planes in a right angle, and its orthogonal projection onto the second plane. In Figure 49 the planes *VBC*, *BCD* meet in the line *BC*. *AP* is the projection of *VA* onto the plane *BCD*. The angle between the two planes is therefore represented by angle *VAP*.

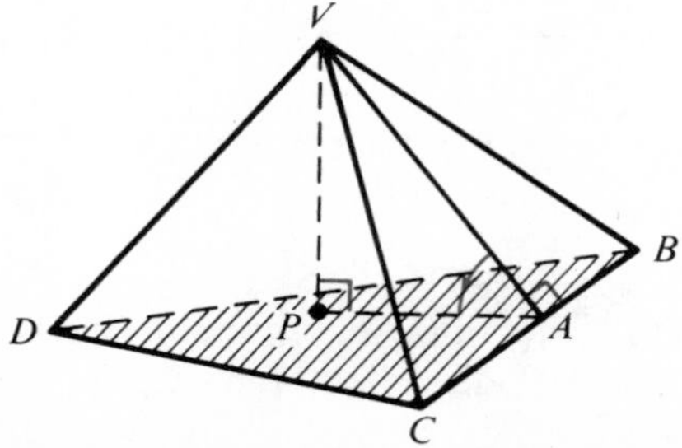

Fig. 49

(*d*) *Shortest distance between two points measured along the surface of a solid.* A useful way of calculating the shortest distance between two points measured along the surface of a solid is to first of all draw the net of the solid and then calculate the straight line distance along the net. For example, the shortest distance from A to B along the curved surface of the cone in Figure 50 is given by the distance AB on the net.

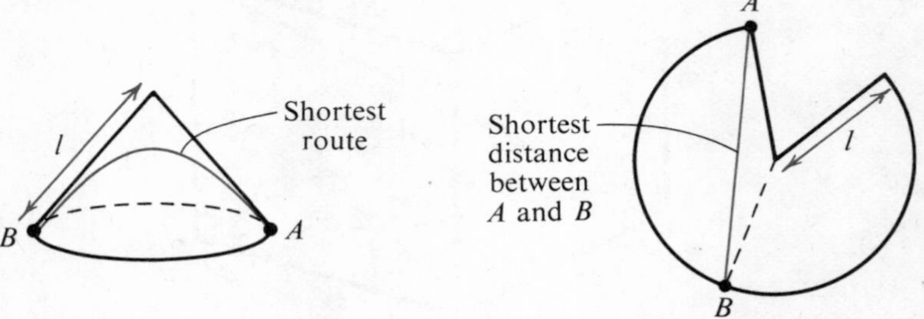

Fig. 50

9 Locus

(*a*) The locus of the pendulum bob P, of the clock in Figure 51 is an arc of a circle of radius l. The locus of the point A on the rim of the wheel in Figure 52 as the wheel rolls along a horizontal road is represented by the dotted line. The locus of the centre of the wheel is a straight line through X, X'.

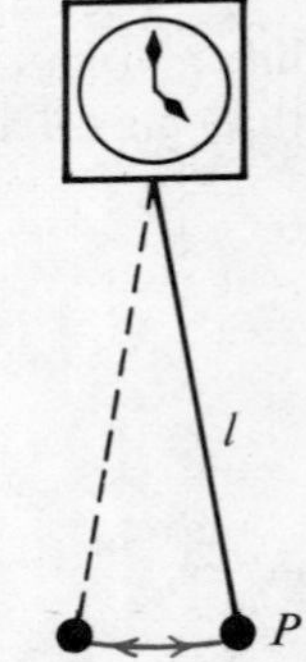

Fig. 51

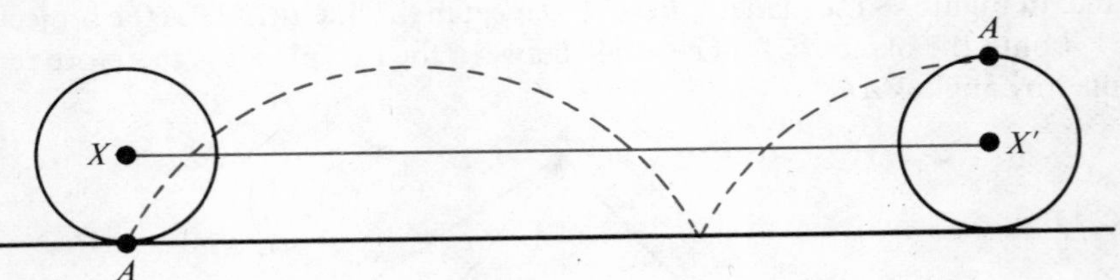

Fig. 52

The locus of the main blades of the helicopter in Figure 53 as it rises vertically is a cylinder, and of the tail blades a flat 'cigar' shape.

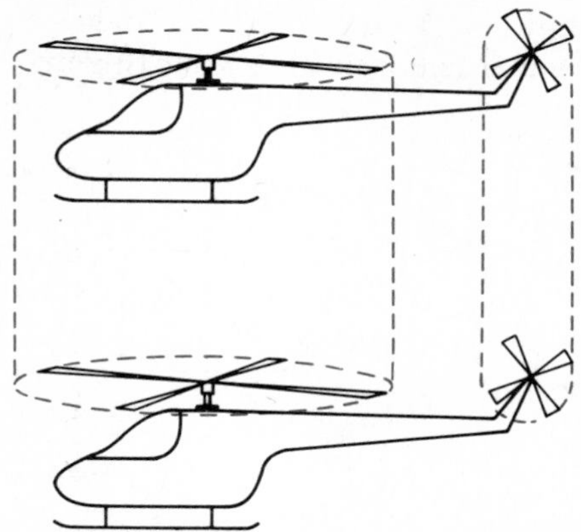

Fig. 53

Locus is the word used to describe a set of points which obey a given law. We have already seen that a locus can comprise a set of points in one, two or three dimensions.

(*b*) The locus of a point P which moves in a plane so that its distances from two fixed points A and B are equal, is the mediator of the two points i.e. $\{P: PA = PB\}$, (Figure 54).

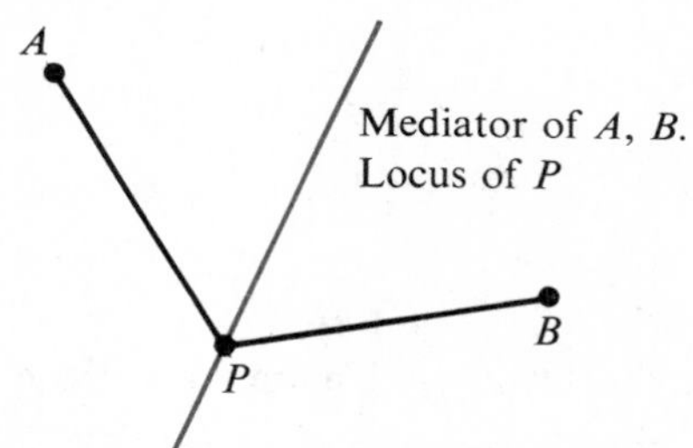

Fig. 54

The locus of points in a plane equidistant from two fixed lines is (i) the bisector of the angles between the lines or (ii) a line parallel to both lines and midway between them (see Figures 55(*a*) and (*b*)).

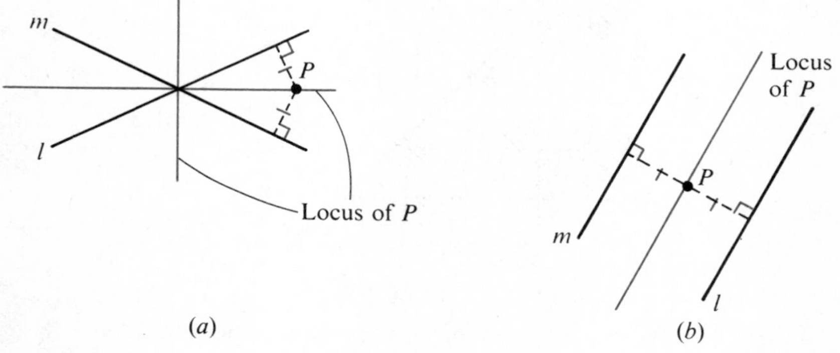

Fig. 55

(*c*) In 3-D Fred must fly in a plane parallel to π_1 and π_2 and midway between them if he is to remain equidistant from each plane (Figure 56).

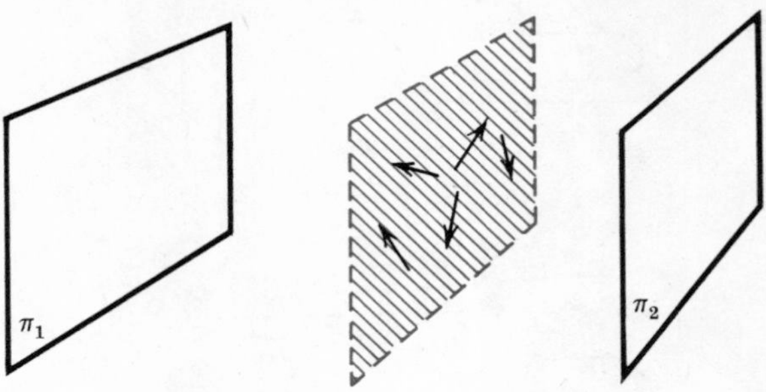

Fig. 56

If he flies so that he is always a distance p units from a fixed point O in space his locus is a sphere of radius p i.e. $\{P: PO = p\}$. (See Figure 57.)

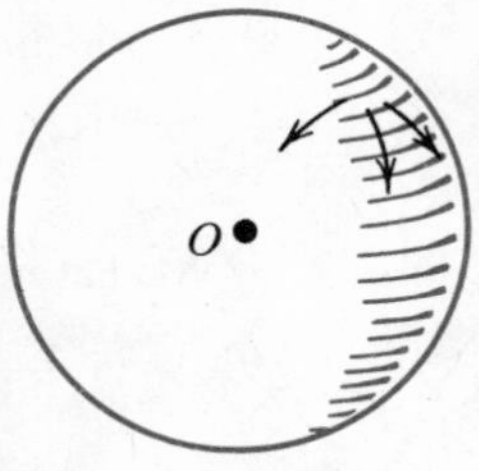

Fig. 57

(*d*) We can describe the locus of a point in a plane, or in 3-D by using coordinates:

$\{(x, y): x + y = 4\}$ is the set of points in a plane whose locus is represented by the line l in Figure 58.

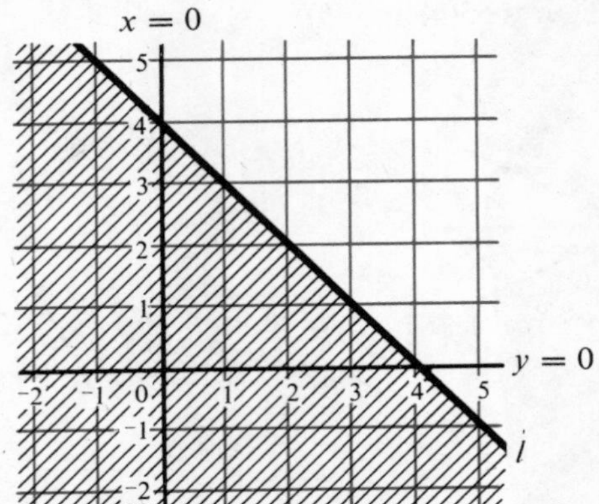

Fig. 58

$\{(x, y): x + y \leqslant 4\}$ describes the shaded region of the plane in Figure 58. The locus of the points (x, y) which obey the law $x + y \leqslant 4$ is therefore represented by the shading.

Exercise F

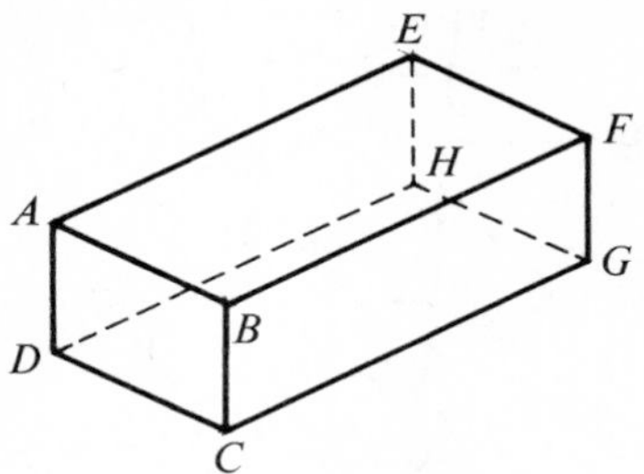

Fig. 59

1 Calculate the angles between

(i) BC and BF; (ii) BC and EH; (iii) BC and BG;
(iv) BC and BH; (v) BC and AG;

for the cuboid in Figure 59. ($AB = 4$ cm, $BC = 3$ cm, $BF = 10$ cm.)

2 $VABC$ is a right pyramid vertex V, with a square base of side 4 cm. The sloping edges are of length 6 cm. Calculate:

(i) the height of the pyramid;
(ii) $\angle VCB$;
(iii) the angle between the line segment VA, and the base of the pyramid;
(iv) the angle between the face VBC and the base.

3 Describe carefully the locus in (*a*) two (*b*) three dimensions of the following:

(i) the set of points equidistant from a fixed point;
(ii) the set of points equidistant from three fixed points;
(iii) the set of points equidistant from a fixed line;
(iv) the set of points equidistant from two fixed lines;
(v) the set of points more than 2 cm but less than 3 cm from a fixed line.

4 A, B are two points fixed in a plane. P moves so that angle $APB = 60°$. Describe the locus of P.

5 A, B are points fixed in a plane. Draw diagrams to show the locus of P for each of the following:

(i) $\{P: PA = PB\}$;
(ii) $\{P: PA \geqslant PB\}$;
(iii) $\{P: PA \leqslant PB\}$;
(iv) $\{P: PA = 2 \text{ units}\}$.

6 (*a*) Sketch the locus $X = \{(x, y): x^2 + y^2 = 25\}$.
(*b*) On the same diagram sketch the locus $Y = \{(x, y): x + y = 5\}$.
(*c*) Describe accurately $X \cap Y$.

7 Figure 60 shows a rigidly fixed cotton reel around which is wound a length of thread. Assuming that P moves in the plane of the paper sketch the locus of P as the thread is unwound – the thread is taut at all times. Calculate the length of unwound thread after P has made one full turn. ($C = 2\pi r$.)

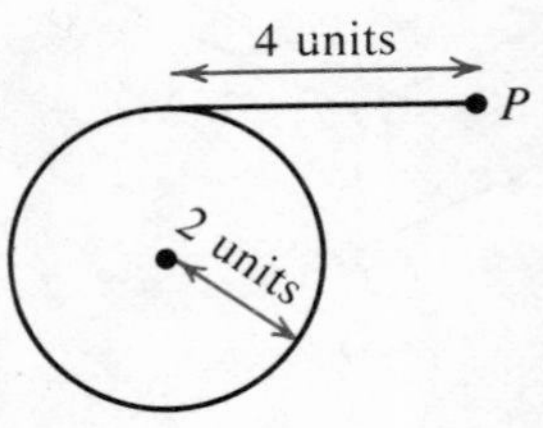

Fig. 60

8 Answer Question 7 for the cotton reel in Figure 61 with square cross-section.

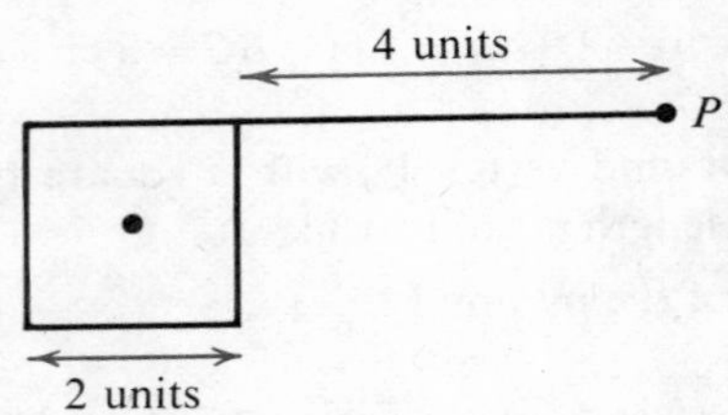

Fig. 61

9 Statistics

1 Pictorial representation

(a) Bar charts

A bar chart is drawn when one axis cannot have a numerical scale or when no physical meaning can be attached to the intermediate points on a numerical axis. See Figures 1 and 2.

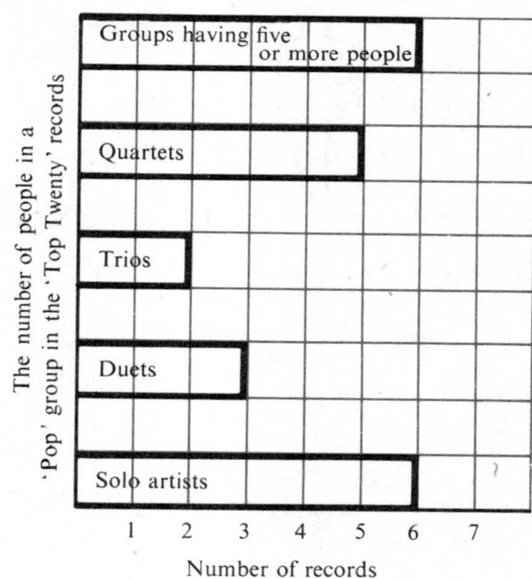

Fig. 1

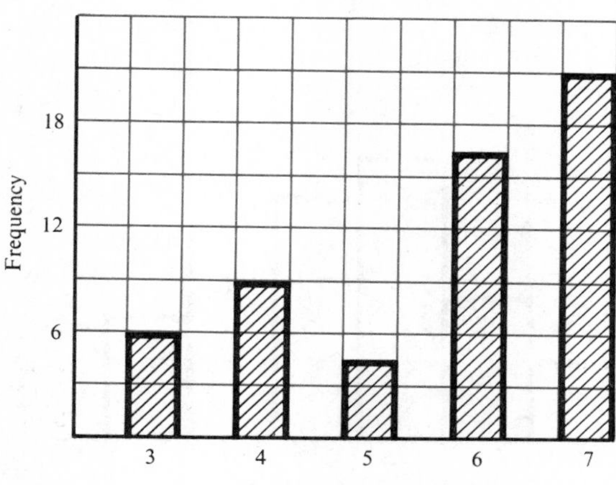

Fig. 2

A bar chart may be drawn with either vertical or horizontal columns. The columns are of constant width, though they can be just straight lines (see Figure 8).

(*b*) Line graphs

Figure 3 shows two line graphs. In graph (*a*) the points have been joined by a dotted line to emphasize the rise and fall in profits although no meaning can be given to the intermediate points.

In graph (*b*) the recorded values of temperatures have been joined by a smooth curve. In this case, an intermediate point is likely to be a reliable indication of temperature at that time.

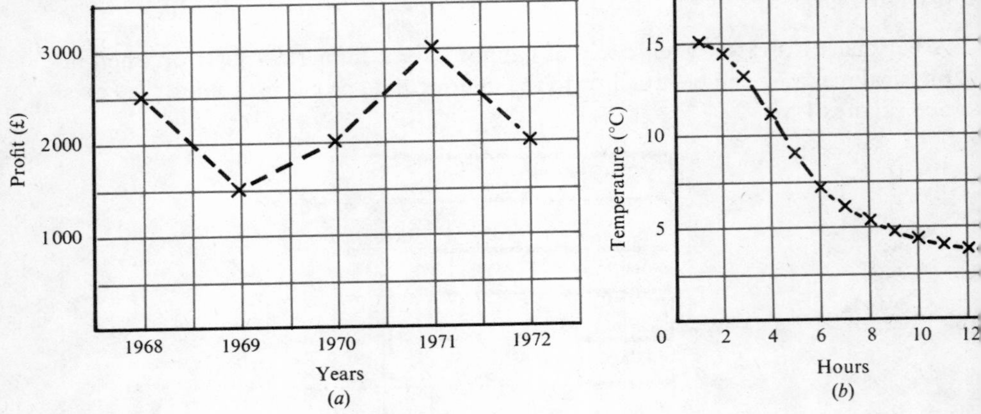

Fig. 3

(*c*) Frequency diagrams

These are diagrams where one axis (usually the vertical axis) represents the frequency of the values on the other axis (see Figure 4).

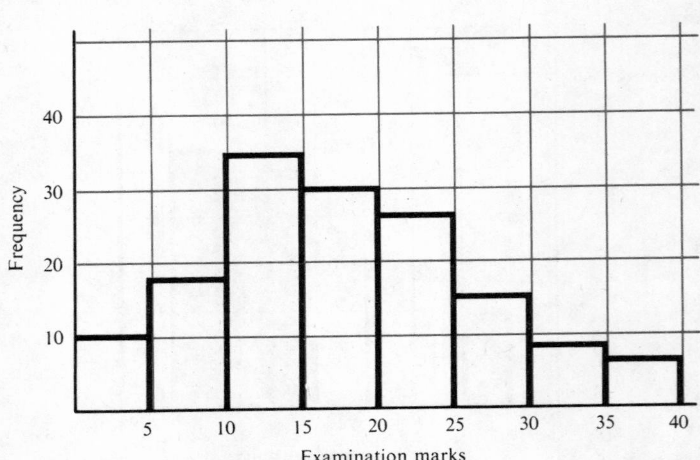

Fig. 4

Remember a cumulative frequency diagram is obtained from a frequency diagram and indicates the total frequency at or below a given value (see Figure 5).

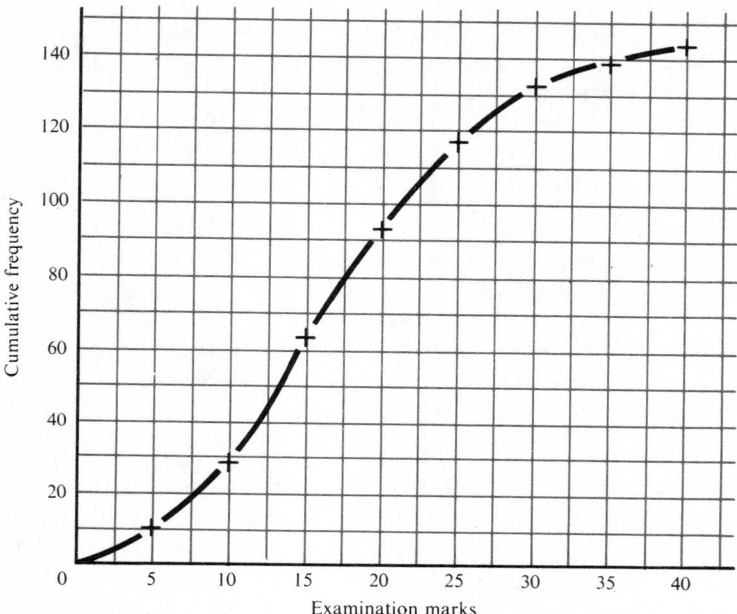

Fig. 5

When drawing these graphs it is important to remember the following:

(*a*) each graph should have a title;
(*b*) the values being plotted on each axis should be clearly labelled;
(*c*) choose easy scales: 1 cm to 1, 2 or 5 units are the best ones (or multiples of 10 of these three basic scales).

(*d*) Pictogram

These are a more attractive way of displaying data.

Strength of the armed Forces in 1968:

Royal Navy	90 thousands
Army	180 thousands
Royal Air Force	113 thousands

R.N.

R.A.F.

Army

One figure = 50 000 men

Fig. 6

(e) Pie charts

Another type of graph is the Pie Chart. This is used to show how a fixed quantity is divided amongst various uses.

Suppose the household budget of a man earning £20 per week is:

	£
Food	7.00
Rates	1.25
Mortgage	6.50
Clothing	1.00
Heating – lighting	2.00
Other expenses	1.25
Savings	1.00
Total:	20.00

The area representing FOOD on the pie chart will be $\frac{7}{20}$th of the total area. The angle at the centre will therefore be $\frac{7}{20}$ of 360° which is 126°. The angle for the RATES will be $\frac{5}{80}$ of $360° = \frac{1}{16} \times 360° = 22\frac{1}{2}°$.

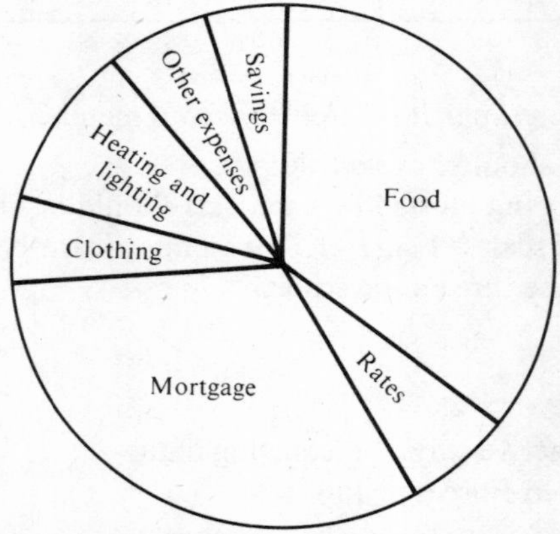

Fig. 7 *A pie chart showing a household budget*

Exercise A

In Questions 1–5 represent the data in a suitable pictorial form.

1 The estimated quantity of crops and grass harvested in 1968, in thousands of tonnes:

Wheat	3515	Rye (grain)	11
Barley	8274	Sugar beet	7006
Oats	1212	Potatoes	6738
Mixed corn	151		

2 The area of inland water in square kilometres:

England	712	Scotland	1577
Wales	125	N. Ireland	630

3 Population in thousands in 1966 of some English cities:

Bristol	429	Liverpool	712
Birmingham	1103	Manchester	625
Leeds	508	Sheffield	486

4 Production of beef in thousands of tonnes from 1958–68:

1958	1959	1960	1961	1962	1963	1964	1965	1966	1967	1968
783	693	789	870	884	913	851	810	842	893	880

5 Age distribution in Scotland in thousands in 1968:

5–9	237	35–39	150	60–64	130
10–14	210	40–44	155	65–69	96
15–19	204	45–49	151	70–74	62
20–24	185	50–54	147	75–79	38
25–29	152	55–59	151	80–84	20
30–34	144				

6 A Unit Trust invested its funds as follows:

In UK Equities	£3 600 000
In Fixed Interest Stocks	£800 000
Overseas	£1 000 000

Display this data clearly on a pie chart marking in the angle of each sector.

7 A pie chart is designed to show how an average family spends its income. If the total income is £24 per week and the angle of the sector representing expenditure on food is 105° find how much is spent on food per week.

8 The expenses of a small printing firm are as follows:

Printing costs	£5 000
Wages and salaries	£10 000
Maintenance of capital	£3 000
Miscellaneous	£2 000

(*a*) Illustrate these figures on a pie chart of radius 5 cm.

(*b*) In the following year printing costs go down by 20 % and wages and salaries go up by 10 %. What should be the area of the new pie chart, with reference to the old one, if it is to represent the proportional change in costs?

(*c*) What should be the size of the angle of the new sector devoted to wages and salaries?

2 Averages

(*a*) A research scientist who is working on the development of a new fertilizer for wheat decides to divide an area of land into 100 square plots of 10 m^2. The new fertilizer is applied to 50 plots.

Why are some plots left untreated?

How do you think he decided which plots were to be treated with the fertilizer?

To be able to say something about the effectiveness of the fertilizer, the yields of wheat in kilograms per plot have to be measured and some form of average found.

TABLE 1 *The yields in kilograms of the 50 plots which were treated with the fertilizer*

15	29	22	15	3	30	22	16	5	2
22	13	20	25	42	25	20	38	12	29
14	21	26	13	21	27	13	21	11	18
10	18	24	24	36	34	23	18	10	9
17	23	33	8	16	23	31	16	23	40

What information can the scientist get out of these numbers? As they stand, they convey very little, since there is no order in them. The next step is therefore to write them out in order.

TABLE 2 *The yields in kilograms of the 50 plots which were treated with the fertilizer arranged in order*

2	3	5	8	9	10	10	11	12	13
13	13	14	15	15	16	16	16	17	17
18	18	18	20	20	21	21	21	22	22
23	23	23	23	24	24	25	25	26	27
29	29	30	31	33	34	36	38	40	42

This table is much more useful as the yields extend from 2 kg to 42 kg and most quantities are between 15 and 25 kg.

(*b*) *Median.* We can use Table 2, arranged in order, to obtain one type of average. We simply take the middle number in the list. As there are 50 numbers, the middle number will be halfway between the 25th and 26th number, that is, between 20 and 21. This average is known as the *median.* In this case, the median yield is 20·5 kg.

Figure 8 shows the results in the form of a frequency diagram. As there are

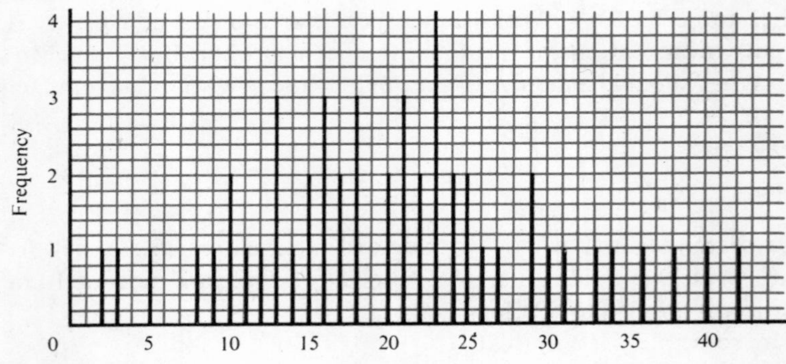

Fig. 8

so many columns to be drawn over a large range it is more convenient to use straight lines to represent frequency.

This only gives us some idea of his results, for there are far too many lines to pick out the main idea quickly. Also the lines are very irregular in height. To obtain a clearer picture we place the data into classes. What size of class is best? It is a matter of trial and error.

By taking classes of size 3, 5, 10, 20 the following frequency diagrams in Figures 9–12 are obtained.

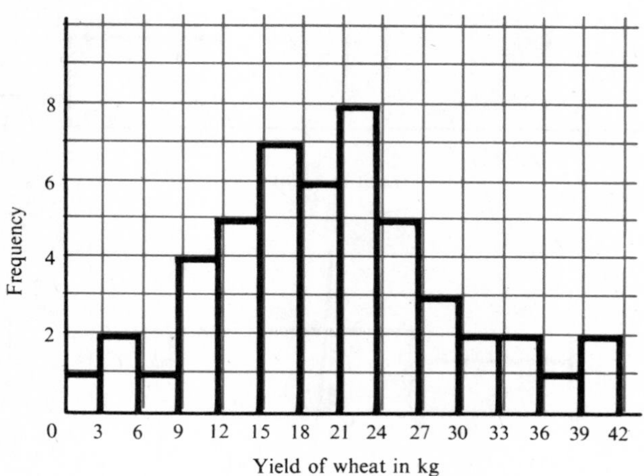

Fig. 9

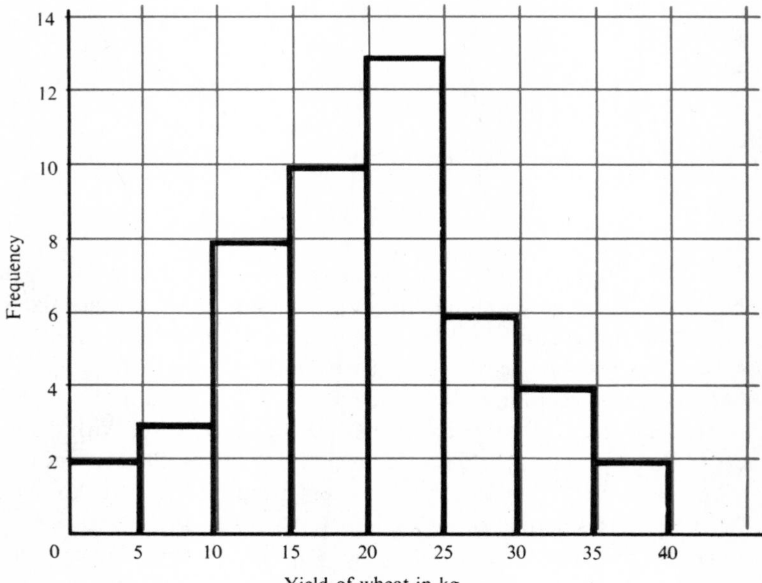

Fig. 10

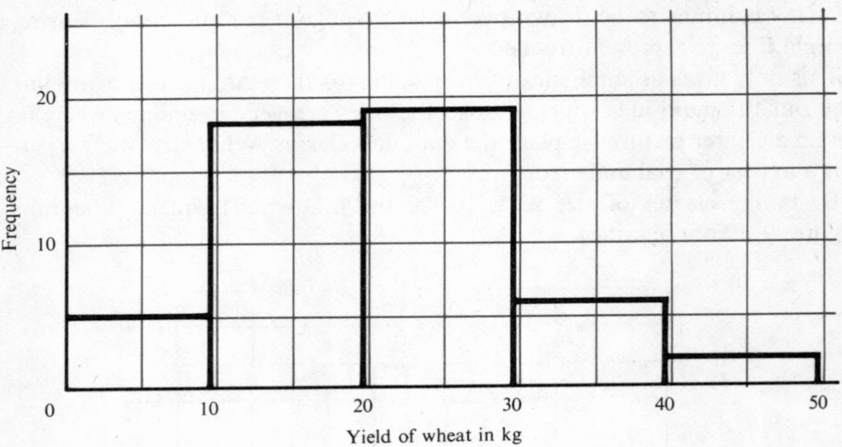

Fig. 11

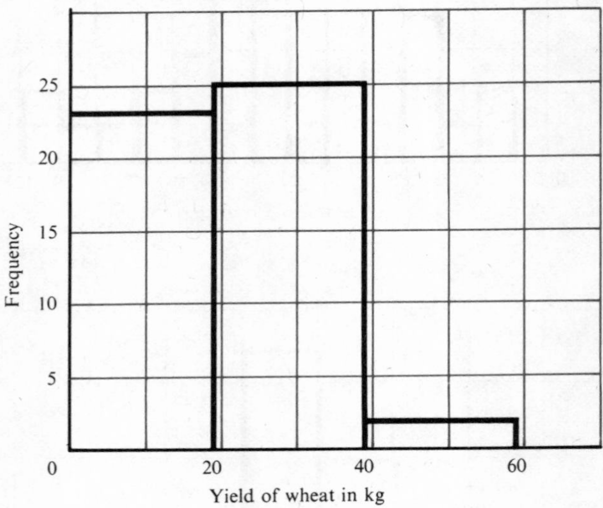

Fig. 12

As the class size increases, the frequency diagram at first assumes a more definite shape, then loses its shape and looks like a block. We choose a class size so that its shape is clearly shown, and this is more important than the individual measurements. A class interval of 5 seems to fit these data best.

(*c*) *Mode.* We have seen how the frequency rises to a peak and then falls off. This peak suggests another type of average and this is the value with the greatest frequency. In this case when the data is formed into classes, the average is known as the MODAL class. When the original numbers are used then it is known as the MODE. The modal class here is 20–24.

(*d*) *Mean.* A third form of average is the MEAN. It is probably the most commonly used and the most important of the averages. This is simply the sum of all the numbers divided by the number of separate numbers.

In this case, the sum of the individual numbers is 1021 so the mean is 20·4 correct to 1 decimal place. It is usual to calculate the mean to one more significant figure than there were in the original numbers.

When a large number of values are available this method of finding the mean is very laborious. An easier method is illustrated in columns 3 and 4 of Table 3. This method is not exact but is sufficiently accurate for most purposes.

TABLE 3

Class interval	Frequency	Class mid-mark		Deviation from working mean		Cumulative frequency
	f	x	fx	$d=x-22$	fd	
0–4	2	2	4	⁻20	⁻40	2
5–9	3	7	21	⁻15	⁻45	5
10–14	8	12	96	⁻10	⁻80	13
15–19	10	17	170	⁻5	⁻50	23
20–24	13	22	286	0	0	36
25–29	6	27	162	5	⁺30	42
30–34	4	32	128	10	⁺40	46
35–39	2	37	74	15	⁺30	48
40–44	2	42	84	20	⁺40	40
	50		1025		⁻75	

Class interval $= 5$.
Estimated mean $= \frac{1025}{50} = 20{\cdot}5$ or $22 - \frac{75}{50} = 20{\cdot}5$.

By grouping the data and using the middle of the class as the representative we introduce an error, but most of the errors we make in underestimating will be balanced by those we make in overestimating. The mean by this method is $\frac{1025}{50} = 20{\cdot}5$, which is an error of only 0·1.

The computation can be made easier if we use a working mean. In this case we take 22 as our working mean. The method is shown in columns 5 and 6 of Table 3. We work out the total yield above the working mean and the total below it, taking the values to be positive and negative respectively. If the sum of the two values is zero then the working mean is also the mean. The sum is negative in this case and the working mean is too high by an amount $\frac{75}{50}$.

That is: estimated mean $= 22 - \dfrac{(75)}{50} = 22 - 1{\cdot}5 = 20{\cdot}5$.

In general:

$$\text{estimated mean} = \text{working mean} + \frac{\text{total of } fd \text{ columns}}{\text{total frequency}}.$$

3 Comparison of averages

For this set of numbers:

Median = 20·5
Mode = 22 (mid-point of modal class)
Mean = 20·4

Which of these three averages should the scientist take? There is no fixed rule to answer this question. Part of the skill of using statistics lies in being able to choose which one gives the clearest summary of the original members. In our example, since it is advisable to have a fertilizer which will assist the wheat to produce a uniform yield, the median is probably the best. Although in this case there is very little to choose between the mean and median.

4 Spread

(*a*) Range

The scientist would like the fertilizer to yield a good, even crop. It is not enough to say that the median yield is 20·5 kg/plot for we need some measure of reliability of the fertilizer.

The most obvious way is to measure the spread of the yield. This is the difference between the highest and lowest yield and this measure is known as the RANGE. It is 42 − 2 = 40 kg.

It has one main disadvantage. A single extreme result can make the range misleading and we therefore need a simple measure which does not take too much account of values.

(*b*) Inter-quartile range

The median divides the population, arranged in order of magnitude, into two halves. We can also divide the population into four quarters. The dividing lines are known as quartiles.

These are the first (or lower) quartile, the median (second quartile), and the third (or upper) quartile.

The first quartile will be just after the twelfth number, which is between 13 kg and 14 kg.

The twelfth number is 13 kg and the thirteenth number is 14 kg. The first quartile is the $\frac{1}{4}(50+1)$th number = $12\frac{3}{4}$th number which is 13·75 kg.

Check that the third quartile is 25·25 kg.

The middle half of the population lies between 25·25 kg and 15·75 kg. The difference between these yields is called the INTER-QUARTILE range and is 11·5 kg.

This quantity tells us something about the majority of the yields and is not much affected by the exceptional yields.

(*c*) Cumulative Frequency

Usually we do not have the original numbers but we only have the grouped frequency distribution. A reliable way of estimating the median and quartiles is to draw a cumulative frequency diagram.

This is done in Table 3 and shown in Figure 13.

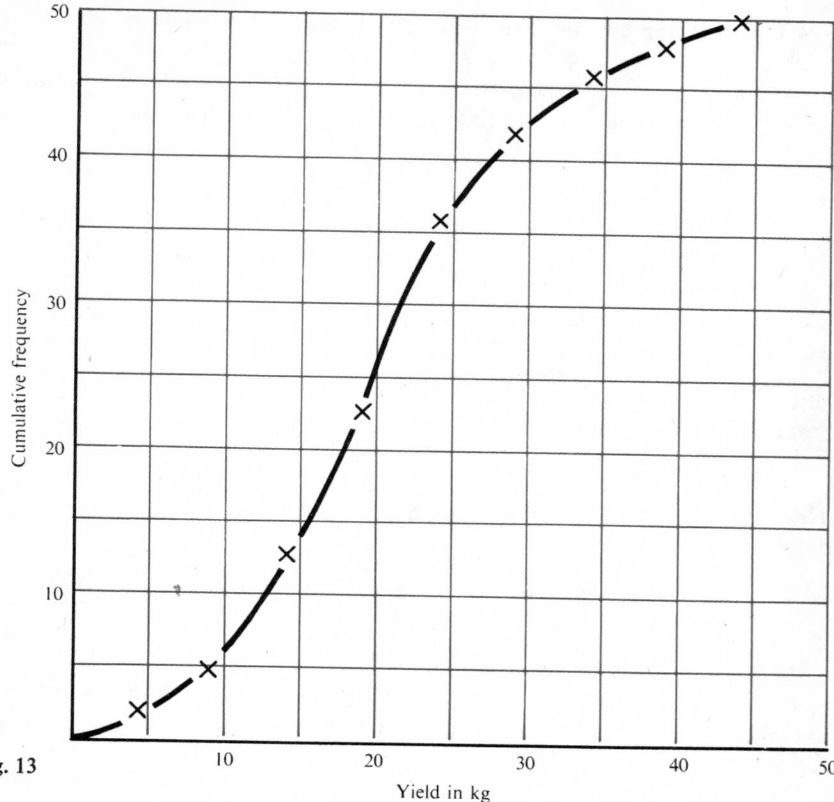

Fig. 13

It is important here to remember that we used the end of the class interval and not the class mid-mark.

To find the median, look on the cumulative frequency axis for $25\frac{1}{2}$, which is the middle number. The value corresponding to it on the other axis will be the median. The first and third quartiles are obtained in a similar way.

In our example:

First quartile is 14·5,
Second quartile is 21,
Third quartile is 25·8,
INTER-QUARTILE RANGE = 25·8 − 14·5 = 11·3.

The values from the grouped data may not agree exactly with the original numbers but the error will be small.

Here are the data of the yields of the plots which were not treated with the fertilizer:

23	16	16	17	22	3	10	10	8	14
16	5	24	16	32	23	15	18	9	21
4	24	5	24	15	2	15	25	17	29
33	39	16	17	2	15	17	17	26	13
26	11	18	19	12	20	27	12	28	22

Now find all three averages, the range, and inter-quartile range for both the original numbers and the numbers arranged into class intervals of 5.

Do you think the fertilizer has been successful? What other tests or evidence would you require to be sure that this fertilizer has some advantages?

Exercise B

1 Calculate the mean of the following:

(*a*) 2, 4, 6, 7, 8, 9, 16;
(*b*) 52, 54, 56, 57, 58, 59, 66;
(*c*) 122, 124, 126, 127, 128, 129, 136.

2 State the range, median and upper and lower quartiles of the following:

42, 39, 9, 4, 9, 18, 15, 16, 44, 16, 49, 36, 25, 44, 12.

3 The following are the number of hours of sunshine (to the nearest hour) per day for the month of May at a certain school:

3	1	4	8	11	12	9	7	5	3	4	9	8	4	7	9
11	12	1	8	7	4	5	6	6	6	5	7	6	9	8	

(*a*) Find the mean number of hours of sunshine per day.

(*b*) Classes work outside on sufficiently sunny days: in this case one-quarter of the days of the month. How many hours of sunlight are necessary for the class to work outside?

It rains on three-quarters of those days which have four hours of sunshine or less.

It rains on one-third of all the other days.

(*c*) Estimate the number of days on which it rained.

(*d*) Estimate the number of days on which the class got wet.

4 The table gives the number of planes landing per hour at Heathrow Airport:

Hour beginning	07.00	08.00	09.00	10.00	11.00	12.00	13.00	14.00
No. of planes	13	28	19	21	10	16	9	17
Hour beginning	15.00	16.00	17.00	18.00	19.00	20.00	21.00	
No. of planes	11	18	19	32	23	14	6	

(*a*) Illustrate these figures graphically.

(*b*) What is the modal number of planes arriving per hour?

(*c*) What is the mean number of planes arriving per hour?

(*d*) If four customs officers can clear the passengers through the customs in 15 minutes, approximately how many custom officers would be needed to ensure that no plane-load had to wait to be admitted to the customs hall? Make clear any assumptions you make in answering this question.

5 The number of days before fluorescent bulbs had to be replaced in a school were as follows:

Days	151–175	176–200	201–225	226–250
Bulbs	1	11	18	20
Days	251–275	276–300	301–325	326–350
Bulbs	25	16	7	2

Show these results graphically. Calculate the mean life of the bulbs.

6 At a recent golf competition 140 competitors took part and the results were as follows:

Score	65	66	67	68	69	70	71	72	73	74	75
Number of competitors having this score	1	2	5	3	8	17	20	31	22	21	10

(*a*) Represent this information graphically.
(*b*) Calculate the mean, the median and the interquartile range.

7 A manufacturer produces rods having a nominal length of 6 cm. A sample of 40 of these rods were taken and their lengths measured correct to two decimal places, the results are as follows:

6·02	6·01	5·94	5·94	5·97	5·96	5·98	6·01	5·98
6·00	6·03	5·99	5·97	5·98	6·00	6·03	5·95	6·00
5·92	5·93	5·99	5·99	6·00	5·95	6·00	5·97	5·96
6·03	6·01	5·98	5·99	6·04	6·00	6·02	5·97	5·96
5·99	6·00	5·97	5·98.					

(*a*) Calculate the mean length of the rods in the sample.
(*b*) Draw a cumulative frequency graph, and from it calculate the median and inter-quartile range.
(*c*) If the manufacturer allows a tolerance of 0·2 % from the nominal value of 6 cm, how many rods are acceptable from the sample?
(*d*) If the manufacturer requires 80 % of the sample to be acceptable, what are the tolerance limits of the length of the rod?

8 A record was kept of motor vehicles passing in one direction along a main road. The number of private cars in each group of ten vehicles was noted. When 60 successive groups had passed the frequencies were as follows:

No. of private cars in a group of ten	0	1	2	3	4	5	6	7	8	9	10
No. of groups	0	0	4	5	8	13	11	10	7	2	0

(*a*) Calculate the mean number of private cars in a group of ten vehicles.
(*b*) State the probability, at a given instant during the census, that the next vehicle to pass would be a private car.
(*c*) Taking the mean length of a private car to be 5 m and of a commercial vehicle to be 7 m and allowing 10 m between vehicles, calculate the length of road occupied by the modal group of vehicles.
(*d*) Represent the information in the above table in an appropriate graphical form.

10 Probability

1 What is probability?

(*a*) What is the probability of throwing an even number with a die? There are two ways of looking at this problem.

The experimental method

We can throw the die a large number of times and record the success fraction. If we perform the experiment of throwing the die 1000 times and succeed in obtaining an even number 513 times, we say that our success fraction is $\frac{513}{1000}$. This success fraction gives an estimate of the probability.

Repetitions of the experiment (either for 1000 or some other large number of trials) would give other success fractions, but the difference between any pair of these fractions would be small.

We assume that as the number of trials for an experiment increases indefinitely, the success fraction tends to a particular value and we call this value the probability of success. Thus the success fraction obtained from *any large number of trials* gives an estimate of the probability.

The theoretical method

When a die is thrown, the set, $\mathscr{E}$, of possible outcomes is

$$\mathscr{E} = \{1, 2, 3, 4, 5, 6\} \quad \text{and} \quad n(\mathscr{E}) = 6.$$

The set, S, of outcomes in which we are interested is

$$S = \{2, 4, 6\} \quad \text{and } n(S) = 3.$$

If the die is not loaded, then we may reasonably suppose that we are equally likely to obtain any one of the six possible outcomes.

The desired event 'the score on the die is even' is associated with three of these outcomes – the members of S. So we say that the probability of throwing an even number is

$$\frac{n(S)}{n(\mathscr{E})} = \frac{3}{6} = \frac{1}{2}.$$

If there is a universal set $\mathscr{E}$ of *equally likely* possible outcomes of a trial, then we say that the probability of an event **A** which is associated with a subset A of these outcomes is given by

$$p(\mathbf{A}) = \frac{n(A)}{n(\mathscr{E})}.$$

In the above example, we could have written $p(\mathrm{S}) = \frac{n(S)}{n(\mathscr{E})} = \frac{3}{6} = \frac{1}{2}$, where S denotes the event 'the score on the die is even' and $p(\mathrm{S})$ denotes the probability of the event S.

Before we use our definition we must always make sure that the members of our universal set *are* equally likely.

Suppose that we wish to find the probability of obtaining one head and one tail when we toss a coin twice. We could say quite correctly that the set of possible outcomes is {2 heads, 2 tails, one head and one tail} and that we are interested in the subset {one head and one tail}. But the required probability is not $\frac{1}{3}$ since the possible outcomes are not equally likely: one head and one tail is twice as likely as either two heads or two tails. In order to use our definition we must consider the set of possible outcomes {2 heads, 2 tails, a head and then a tail, a tail and then a head}. Since these are equally likely and since we are interested in the subset {a head and then a tail, a tail and then a head}, the required probability is $\frac{1}{2}$.

(*b*) Most of our work on probability is concerned with the theoretical method, but we must be careful not to underestimate the importance of the experimental approach. There are many problems for which the latter is the only appropriate method. For example, we cannot find from theoretical considerations the probability that the next child born in a given hospital will be left-handed. Neither can we forecast the number of times that an even number will turn up in a 1000 tosses of a particular biased die. There is no universal set of equally likely possible outcomes which we can consider.

Would you use theory or experiment in order to find:

(i) the probability of tossing two coins and obtaining two heads;
(ii) the probability of tossing two particular coins and obtaining two heads;
(iii) the probability that when a drawing pin is dropped it will land point up;
(iv) the probability of cutting a pack of cards and obtaining a king;
(v) the probability that the next child born in your county will be a boy?

Exercise A

1 Find the probability of each of the following events:

(i) throwing a die and obtaining a number greater than 4;
(ii) drawing a card from a pack of 52 cards and obtaining a spade;
(iii) tossing three coins and obtaining three heads;
(iv) throwing a die with one green, two blue and three red faces and obtaining a blue face;
(v) drawing a marble from a bag which contains three red marbles and two blue ones and obtaining a red marble.

Describe in each case the universal set of equally likely possible outcomes which you have used.

2 A die is to be tossed 1000 times. About how many times would you expect an odd number to turn up?

3 What conclusion, if any, would you draw if:

(i) you tossed a particular penny 10 times and obtained 1 tail;
(ii) you tossed a particular penny 100 times and obtained 23 tails;
(iii) you tossed a particular penny 1000 times and obtained 191 tails;
(iv) you tossed a particular penny 5000 times and obtained 1017 tails?

4 If you open a telephone directory at random and there are 1000 numbers on the double page facing you, how many would you expect to end in 3? Would you expect about the same number to begin with 3? Give your reasons.

2 'Both ... and ...' and 'either ... or ...'

(*a*) If a die is thrown the set of possible outcomes is

$$\mathscr{E} = \{1, 2, 3, 4, 5, 6\}.$$

Let the set of scores which are even numbers be S and the set of scores which are prime numbers be P. Then, remembering that 1 is not a prime number, we have

$$S = \{2, 4, 6\} \quad \text{and} \quad P = \{2, 3, 5\}.$$

The set of scores which are both prime and even is $\{2\}$. But this is the set $S \cap P$ (see Figure 1). So the probability of obtaining a score which is both prime and even is

$$\frac{n(S \cap P)}{n(\mathscr{E})} = \frac{1}{6}.$$

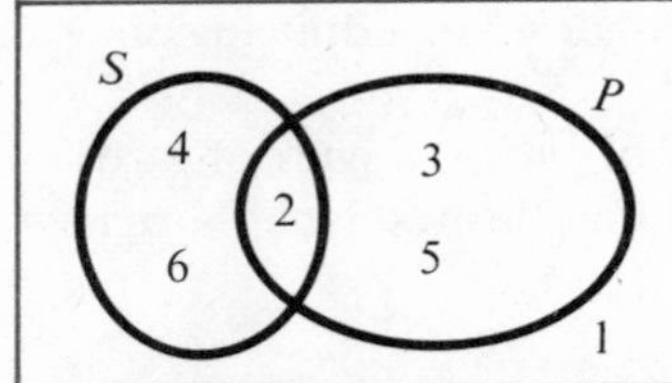

Fig. 1

Also, the set of scores which are either even or prime (or both) is $S \cup P = \{2, 3, 4, 5, 6\}$. So the probability of obtaining a score which is either even or prime is

$$\frac{n(S \cup P)}{n(\mathscr{E})} = \frac{5}{6}.$$

Thus we see that

$$p(\mathbf{S} \text{ and } \mathbf{P}) = \frac{n(S \cap P)}{n(\mathscr{E})}$$

and

$$p(\mathbf{S} \text{ or } \mathbf{P}) = \frac{n(S \cup P)}{n(\mathscr{E})}.$$

(*b*) If we draw a card from a pack of playing cards and record the suit, then a suitable universal set of equally likely outcomes is

$$\mathscr{E} = \{\text{spade, heart, diamond, club}\}.$$

Let **D** be the event 'the card is a diamond' and **H** be the event 'the card is a heart'. Then

$$p(\mathbf{D}) = \frac{n(D)}{n(\mathscr{E})} = \frac{1}{4} \quad \text{and} \quad p(\mathbf{H}) = \frac{n(H)}{n(\mathscr{E})} = \frac{1}{4}.$$

The probability of drawing either a heart or a diamond (that is, the probability of drawing a red card) is

$$p(\mathbf{D} \text{ or } \mathbf{H}) = \frac{n(D \cup H)}{n(\mathscr{E})} = \frac{2}{4} = \frac{1}{2}.$$

This is equal to $\frac{1}{4} + \frac{1}{4}$. So

$$p(\mathbf{D} \text{ or } \mathbf{H}) = p(\mathbf{D}) + p(\mathbf{H}).$$

Can we say that, for any events **A** and **B**,

$$p(\mathbf{A} \text{ or } \mathbf{B}) = p(\mathbf{A}) + p(\mathbf{B})\,?$$

(*c*) If we look again at the example in (*a*) we see that

$$p(\mathbf{S} \text{ or } \mathbf{P}) = \frac{n(S \cup P)}{n(\mathscr{E})} = \frac{5}{6}$$

and that

$$p(\mathbf{S}) + p(\mathbf{P}) = \tfrac{1}{2} + \tfrac{1}{2} = 1.$$

So

$$p(\mathbf{S} \text{ or } \mathbf{P}) \neq p(\mathbf{S}) + p(\mathbf{P}).$$

What is the difference between this example and the one in (*b*)? In the card-drawing example the sets H and D have no common element, that is $H \cap D = \varnothing$ and $n(H \cap D) = 0$. See Figure 2(*a*). If we draw a heart we can be sure that we have not drawn a diamond. Events such as **H** and **D** which cannot occur simultaneously are said to be *mutually exclusive*.

Fig. 2

(*a*) (*b*)

In the die-throwing example, the sets S and P intersect (see Figure 2(*b*)); they have the common element 2. If we throw an even number, we cannot be sure that we have not also thrown a prime number. The events **S** and **P** can occur simultaneously and are therefore not mutually exclusive.

We may *add* the probabilities of events *only if the events are mutually exclusive.*

Exercise B

1 A card is drawn from a pack of 52.

(i) Are drawing a seven and drawing a king mutually exclusive events?
(ii) Are drawing a seven and drawing a spade mutually exclusive events?
(iii) Find the probability of drawing either a seven or a king.
(iv) Find the probability of drawing either a seven or a spade.

2 Eight pieces of paper marked with the numbers 1, 2, 3, 4, 5, 6, 7 and 8 respectively are placed in a hat. A piece of paper is chosen at random. What is the probability that the number on the paper is (i) even, (ii) greater than 3, (iii) both even and greater than 3, (iv) either even or greater than 3?

3 In a group of 30 young people, 22 like pop music, 12 like classical music and 2 like neither. Draw a Venn diagram to illustrate this information. If a member of the group is chosen at random, what is the probability that he or she likes:

(i) pop music; (ii) classical music;
(iii) either pop or classical music; (iv) both pop and classical music?

Explain why your answer to (iii) is unequal to the sum of your answers to (i) and (ii).

3 Combined events

(*a*) What is the probability of obtaining first a '6' and then a number not greater than 2 with two throws of a die?

We could write out all the possible outcomes of throwing a die twice. For each of the six possible scores for the first throw there are six possible scores for the second: a total of 36 results, all equally likely. These 36 outcomes are shown in Figure 3.

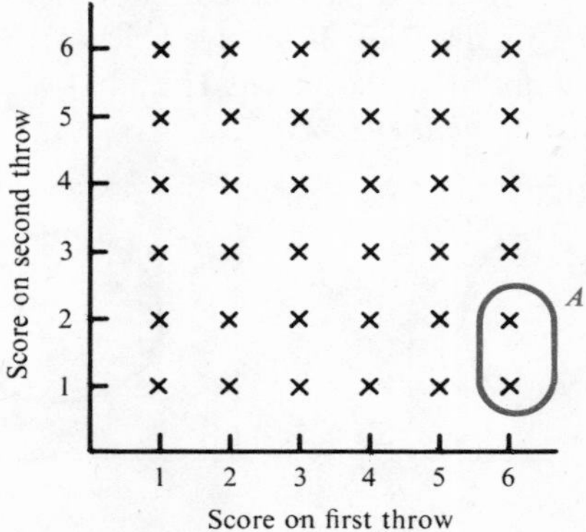

Fig. 3

The set A in which we are interested is

$$A = \{(6, 1), (6, 2)\} \quad \text{and} \quad n(A) = 2.$$

Therefore

$$p(\mathbf{A}) = \tfrac{2}{36} = \tfrac{1}{18}.$$

(*b*) Alternatively we could represent the 36 outcomes by the branches of a tree. The tree diagram for the first throw is shown in Figure 4; the red numbers on the branches are the probabilities of the possible scores.

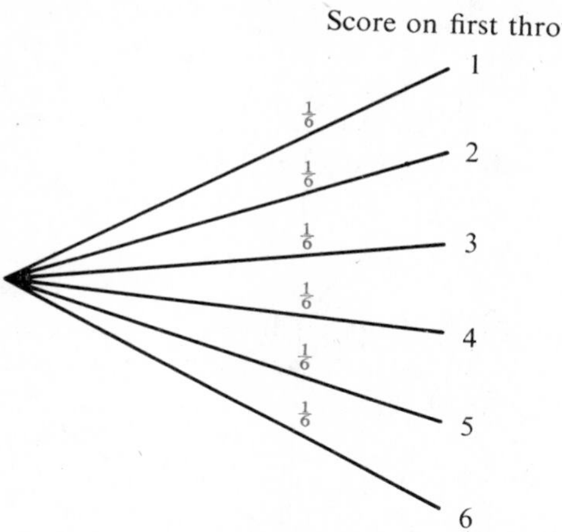

Fig. 4

The ends of the branches in Figure 4 give six growth points each with six branches corresponding to the six possible scores on the second throw.

However, it is tedious to draw 36 branches, and it is unnecessary since the diagram would give more information than we need. We are interested only in a '6' on the first throw and a number not greater than 2 on the second, we can therefore stack together some of the scores and their corresponding probabilities to obtain the simpler tree diagram shown in Figure 5. Notice that since throwing a '1', a '2', a '3', a '4' and a '5' are all mutually exclusive events, their probabilities can be added to give the probability of throwing 'not 6'.

Score on first throw	*Score on second throw*	*Outcome*	*Probability*
$\frac{1}{6}$ 6	$\frac{2}{6}$ 1 or 2	6, not greater than 2	$\frac{1}{6}\times\frac{2}{6}=\frac{2}{36}=\frac{1}{18}$
	$\frac{4}{6}$ 3, 4, 5 or 6	6, greater than 2	$\frac{1}{6}\times\frac{4}{6}=\frac{4}{36}=\frac{1}{9}$
$\frac{5}{6}$ not 6	$\frac{2}{6}$ 1 or 2	Not 6, not greater than 2	$\frac{5}{6}\times\frac{2}{6}=\frac{10}{36}=\frac{5}{18}$
	$\frac{4}{6}$ 3, 4, 5 or 6	Not 6, greater than 2	$\frac{5}{6}\times\frac{4}{6}=\frac{20}{36}=\frac{5}{9}$

Fig. 5

We are interested in a '6' followed by a number not greater than 2, that is, in the top line of the tree diagram. Suppose that we often throw a die twice in succession. We expect $\frac{1}{6}$ of our first throws to give a six and, after having achieved a six, we expect $\frac{2}{6}$ of our second throws to give a number not greater than 2. So we expect to obtain the desired result on $\frac{2}{6}\times\frac{1}{6}$ of our attempts. We therefore say that the probability of obtaining the desired result on any one attempt is $\frac{2}{6}\times\frac{1}{6}=\frac{1}{18}$. Thus all we need to do is multiply the probabilities from right to left, or from left to right, along the sequence of branches which lead to the desired outcome.

(*c*) Check that you agree with the probabilities which are given in Figure 5 for the other three possible outcomes. You will notice that the sum of the probabilities of all four possible outcomes is 1.

It is also true that the sum of the probabilities of all the branches from any growth point is 1. We can express this result in another way: if the probability of an event (or sequence of events) **E** is p then the probability of not-**E** is $1-p$.

(*d*) Drawing a tree diagram is unnecessary in dealing with simple problems like the one considered in (*a*) and (*b*) but can be very helpful in more complicated situations.

Example

A sub-committee of three is to be chosen at random from a committee of six boys and four girls. What is the probability that the members of the sub-committee will not all be of the same sex?

The complete tree diagram for this problem has eight branches, but we need draw only those branches which lead to three girls or three boys. See Figure 6.

Notice that if the first member chosen is a girl then there are nine members left to choose from and three of these are girls. So the probability that the second member is also a girl is $\frac{3}{9}$.

Make sure that you understand how to find the other probabilities which appear in Figure 6.

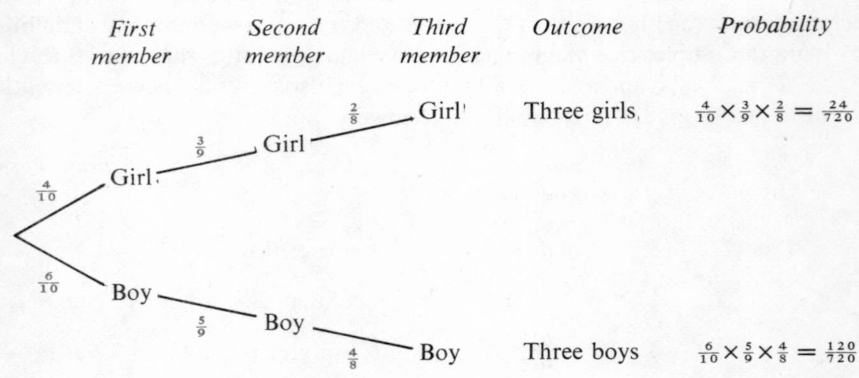

Fig. 6

Since choosing three girls and choosing three boys are mutually exclusive events, the probability that the members of the sub-committee will all be of the same sex is

$$\frac{24}{720}+\frac{120}{720}=\frac{144}{720}=\frac{1}{5}.$$

Hence the probability that they will not all be of the same sex is

$$1-\frac{1}{5}=\frac{4}{5}.$$

It is easier to subtract from 1 than to draw the complete tree diagram.

Exercise C

1 A die is thrown twice. List the set of possible outcomes which give a total score of 6 and represent this set on a diagram similar to Figure 3. What is the probability of scoring a total of 6 with two throws of a die? What is the probability of scoring a total other than 6?

2 Two dice are tossed together. What is the probability of throwing:

(i) at least one six;
(ii) a total of 8;
(iii) either a six or a total of 8 (or both);
(iv) both a six and a total of 8?

3 A card is drawn from a pack of 52. It is then replaced and a card is again drawn. Find the probability that both cards are diamonds (*a*) by considering the possible outcomes shown in Figure 7; (*b*) by drawing a tree diagram.

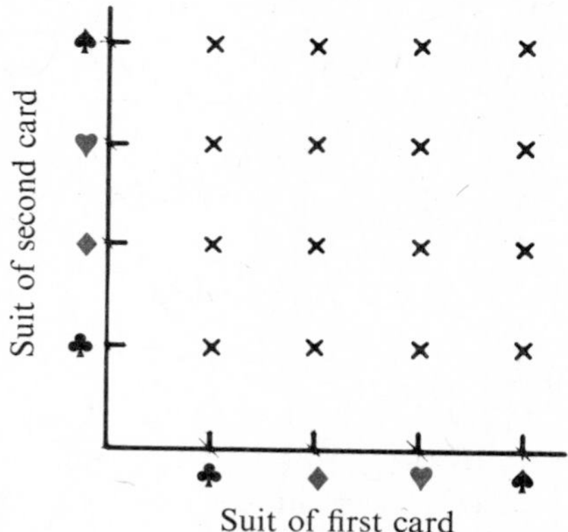

Fig. 7

4 Two cards are drawn one after the other from a pack of 52. What is the probability that they are both diamonds?

5 Two cards are drawn from a pack of 52. What is the probability that they are of the same suit (i) if the first card is replaced before the second is drawn, (ii) if the first card is not replaced before the second is drawn? [*Hint*: It does not matter which card is obtained on the first draw.]

6 Three cards are drawn one after the other from a pack of 52. What is the probability that each belongs to a different suit?

7 A housewife buys a dozen eggs of which two are bad. She decides to scramble three eggs. What is the probability that she chooses (i) at least one bad egg, (ii) exactly one bad egg?

11 Graphs

1 Graphs

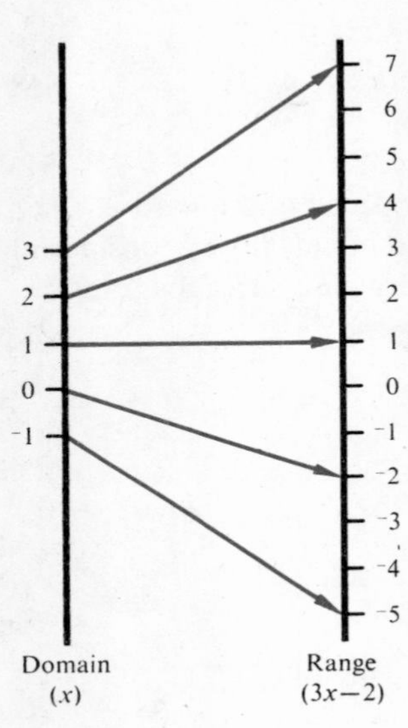

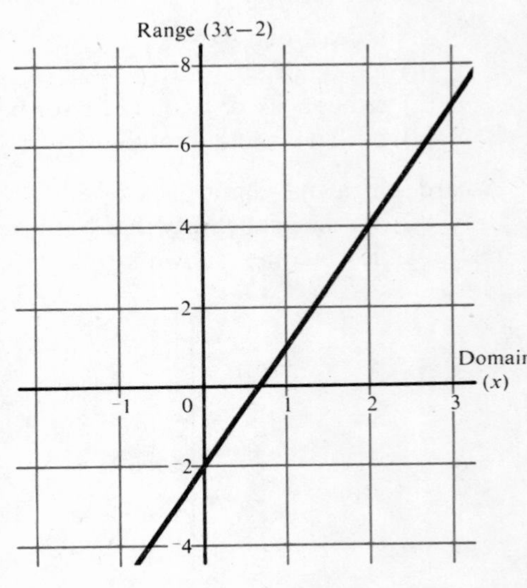

Fig. 1 (*a*) (*b*)

Figures 1(*a*) and 1(*b*) represent the relation $x \to 3x - 2$. Mapping diagrams like Figure 1(*b*) are called *graphs*.

To graph the relation $x \to x^2 - 4x + 3$

(i) choose a set of values for x which are elements of the domain (see Figure 2);

Fig. 2

(ii) calculate the corresponding elements of the range (y values);

Domain	x	$^-2$	$^-1$	0	1	2	3	4	5
	x^2	4	1	0	1	4	9	16	25
	^-4x	8	4	0	$^-4$	$^-8$	$^-12$	$^-16$	$^-20$
	$^+3$	3	3	3	3	3	3	3	3
Range	y	15	8	3	0	$^-1$	0	3	8

(iii) use these values to draw the graph (see Figure 3).

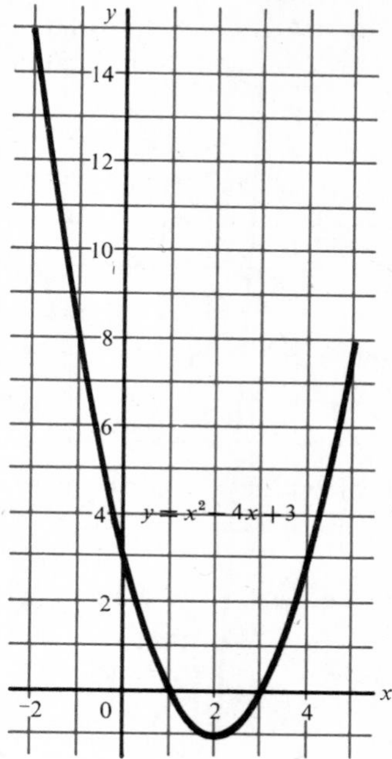

Fig. 3

Note that the *equation* of the curve is $y = x^2 - 4x + 3$.

2 Graphs of some common functions

(*a*) Linear functions

The expressions

$$x \to \tfrac{3}{2}x,$$
$$y = \tfrac{3}{2}x$$

are equivalent. When $y = kx$ for some constant k, then $y \propto x$ (y is directly proportional to x).

Figure 4 shows the graph of the relation $x \to \frac{3}{2}x$.

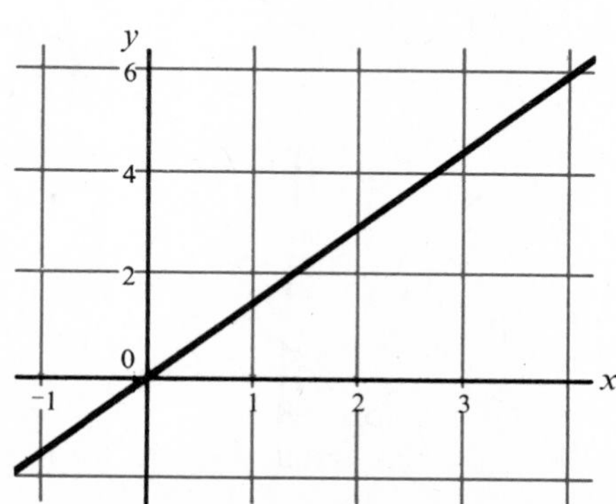

Fig. 4

The general equation of the straight line is $y = mx + c$ where m is the gradient of the line and c the intercept on the y-axis (see Figure 5).

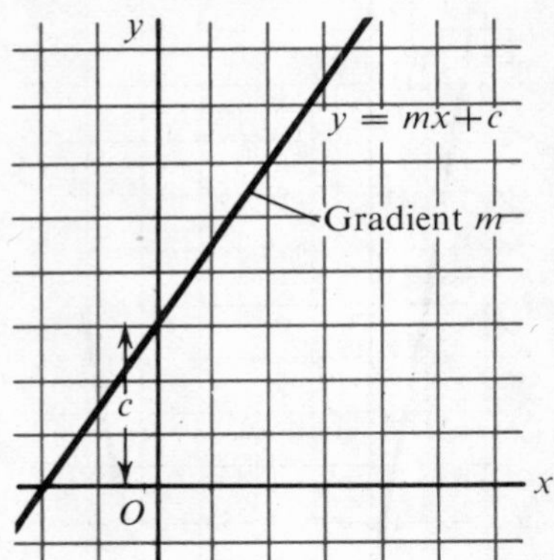

Fig. 5

Figure 6 shows the graphs of (*a*) $y = \frac{1}{2}x + 3$ and (*b*) $4y + 3x = 24$. Check that the gradients of the lines are $\frac{1}{2}$ and $^{-}\frac{3}{4}$ respectively, and that the intercepts on the y-axis are 3 and 6 respectively.

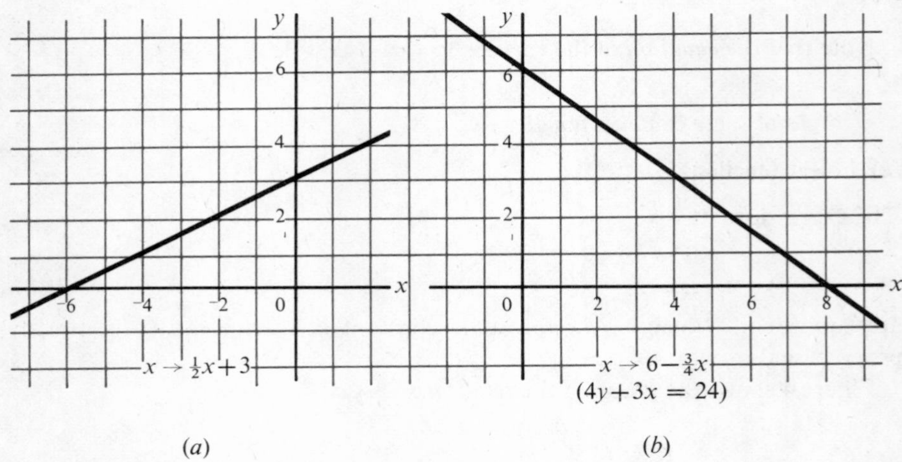

Fig. 6

(*b*) Quadratic functions

The expressions

$$y \propto x^2$$
$$y = kx^2$$

are equivalent. Functions of the form $x \to ax^2 + bx + c$ are called *quadratic* functions and their graphs are parabolas. Figure 7 shows some graphs of parabolas.

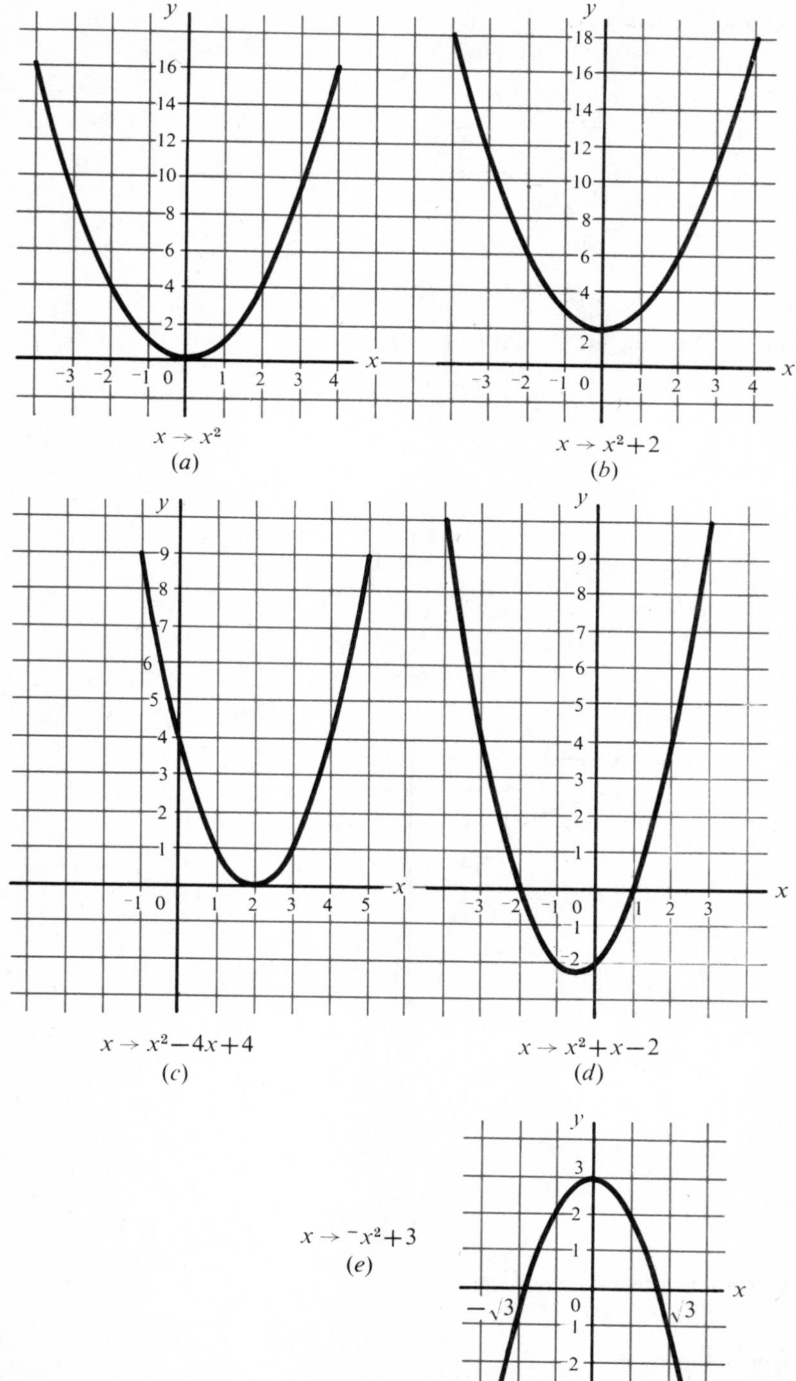

Fig. 7

A parabola always has a line of symmetry.

(*c*) Reciprocal function

The expressions

$$x \to \frac{k}{x},$$

y is proportional to the reciprocal of x,

$$y = \frac{k}{x}$$

can all be used to describe the 'reciprocal function'.

Figure 8 shows the graph of $x \to \frac{5}{x}$ $(xy = 5)$.

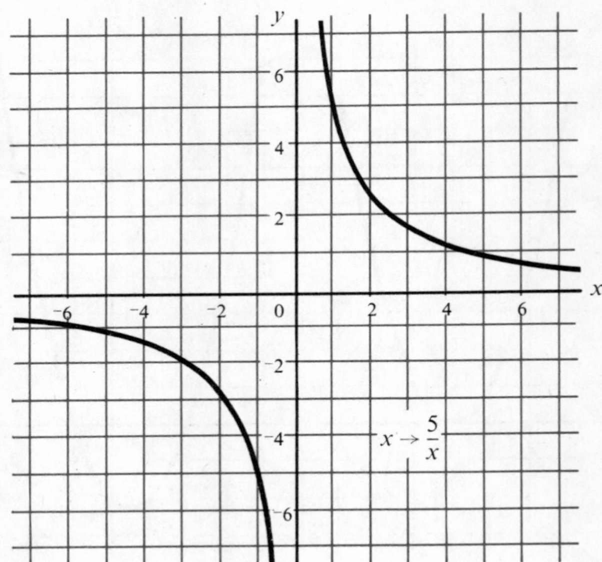

Fig. 8

The graph is a *rectangular hyperbola*.

(*d*) Growth functions

$x \to kA^x$ (see Figure 9).

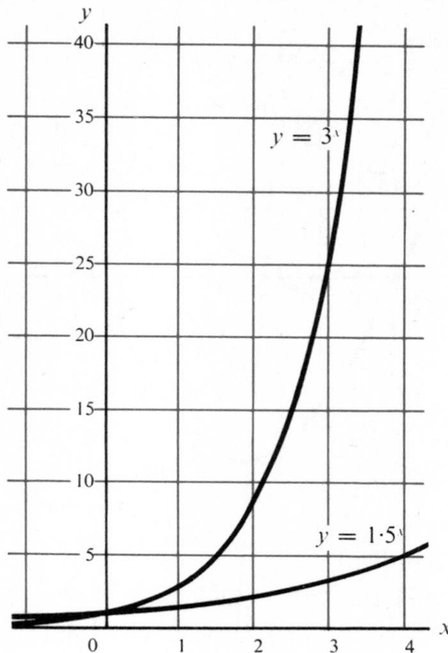

Fig. 9

(*e*) Sine and cosine functions

See Figures 10 and 11.

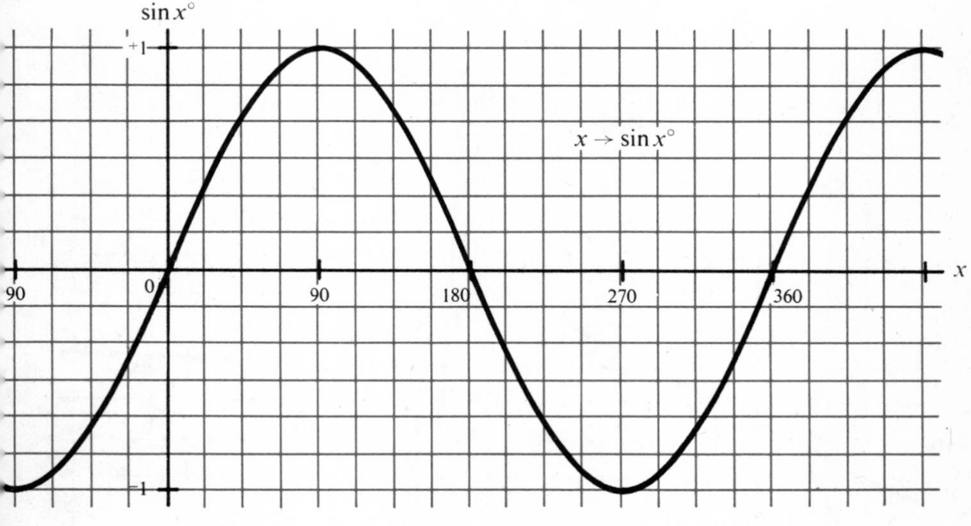

Fig. 10

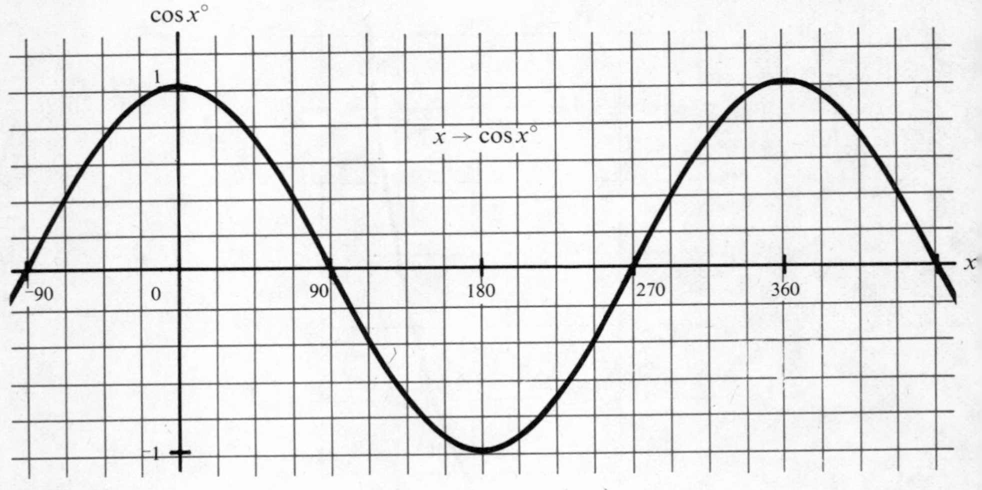

Fig. 11

3 Graphs of inverse functions

Figure 12 shows the graphs of the functions and their inverses listed in the following table. Notice that the graph of each function can be mapped onto the graph of its inverse by a reflection in $y = x$. If a graph is symmetrical about the line $y = x$, then the function is self-inverse (for example $x \to 5 - x$).

Function	Inverse
$x \to 2x + 1$	$x \to \dfrac{x-1}{2}$
$x \to 12^x$	$x \to \log_{12} x$
$x \to x^2$	$x \to \sqrt{x}$
$x \to 5 - x$	$x \to 5 - x$

(*a*)

(*b*)

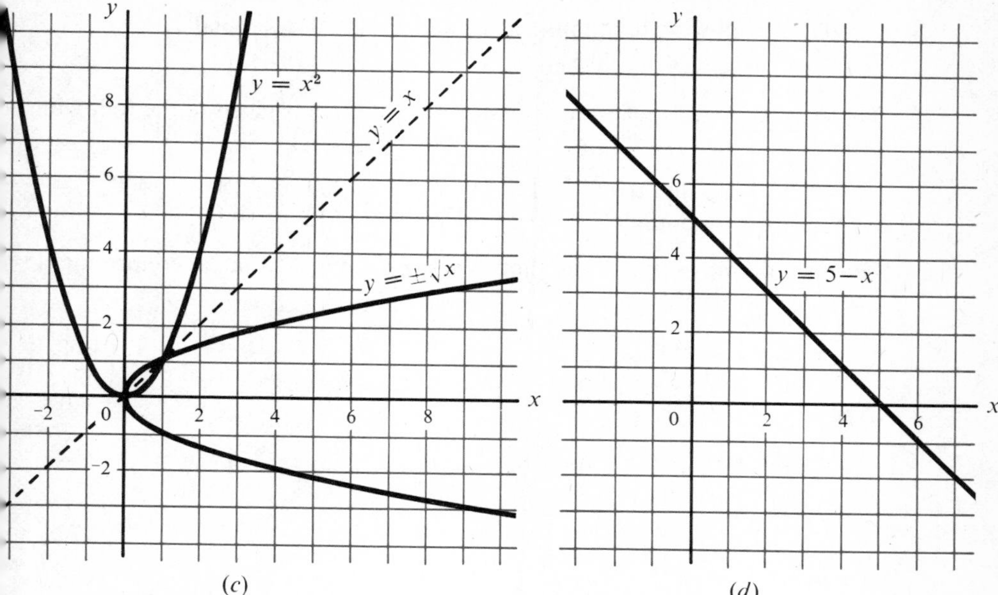

Fig. 12

Exercise A

1 Graph the following mappings for the domain $\{x: {}^-1 \leqslant x \leqslant 3\}$:

(a) $x \rightarrow x^2 - 2$; (b) $x \rightarrow (x-2)^2$;
(c) $x \rightarrow \sqrt{(3x)}$; (d) $x \rightarrow 2\sqrt{(x+1)}$;
(e) $x \rightarrow \frac{1}{2}x$; (f) $x \rightarrow \dfrac{1}{x-2}$;
(g) $x \rightarrow x^3$; (h) $x \rightarrow 3x$.

2 What are the inverse functions of the following:

(a) $x \rightarrow 5x+2$; (b) $x \rightarrow 2-3x$; (c) $x \rightarrow x^2-5$;
(d) $x \rightarrow 4x^2-4x-2$; (e) $x \rightarrow 2^x$; (f) $x \rightarrow 2\sqrt{x}$?

Draw the graphs of the functions above and of their inverses for the domain ${}^-2 \leqslant x \leqslant 3$.

3 For a given 'light value', the possible combinations of S, shutter speed (expressed as a fraction of a second, e.g. '60' means '$\frac{1}{60}$ s') and A, the aperture on a camera are as follows:

Aperture	4	5·6	8	11	16	22
Shutter-speed	60	30	15	8	4	2

Graph 'aperture/shutter speed' and find an approximate formula connecting S and A.

4 Graph the following functions for the domain $\{x: 0 \leqslant x \leqslant 360\}$:

$$x \to \sin 3x°; \qquad x \to 2 \sin x°; \qquad x \to 2 \cos \tfrac{1}{2}x°.$$

5 The cost, C pounds, of water butts to hold V litres is given in the following table:

Volume (in litres)	100	160	200	250	400
Cost (in pounds)	2·20	2·80	3·15	3·45	4·45

Graph 'volume → cost', and find an approximate formula connecting V and C.

4 Gradient

(*a*) All *straight* lines have constant gradients (see Figure 13).

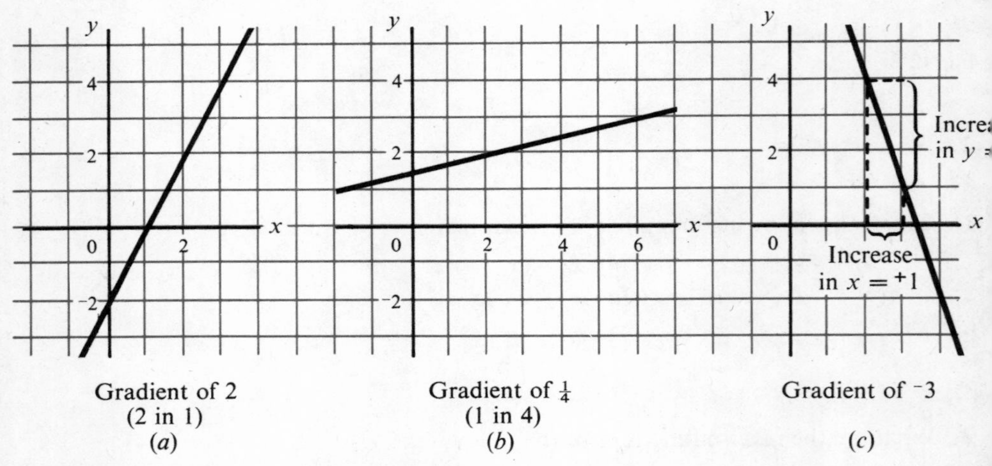

Gradient of 2 (2 in 1) (*a*)

Gradient of $\frac{1}{4}$ (1 in 4) (*b*)

Gradient of $^-3$ (*c*)

Fig. 13

(*b*) *Gradient at a point*

When a graph is a curve the function increases at *different* rates for different values of x. Consider the mapping $x \to 2x^2 + 1$. The gradient of the curve at, say, the point $A(1, 3)$ can be found by drawing the tangent to the curve at this point and calculating the value of PQ/QS (see Figure 14).

The gradient at $A(1, 3)$ is approximately 4.

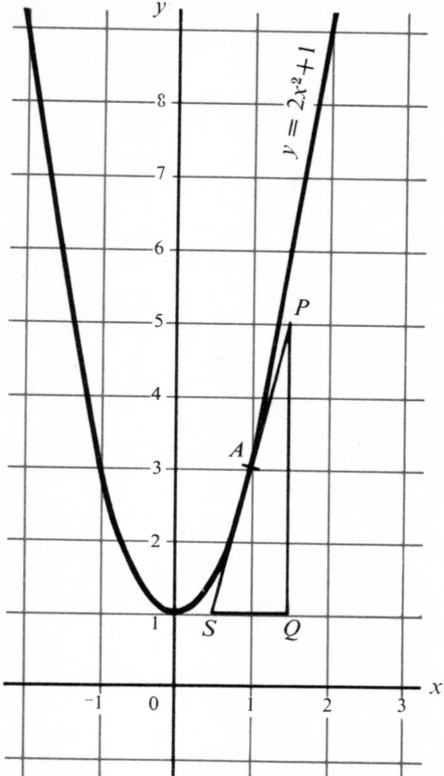

Fig. 14

5 Graphing other relations

The mapping $x \to 2x + 1$ involves all the number pairs of the type $(x, 2x + 1)$. The graph of the mapping is a straight line and all points on this line belong to the set $L = \{(x, y): y = 2x + 1\}$.

A point which is not on the line belongs either to the set P or to the set Q where:

$$P = \{(x, y): y > 2x + 1\},$$
$$Q = \{(x, y): y < 2x + 1\}.$$

For example:

$$(3, 8) \in P$$
$$(3, 6) \in Q \qquad \text{(see Figure 15).}$$

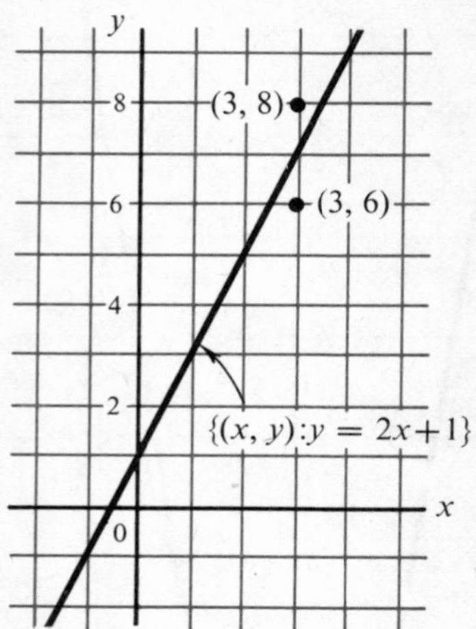

Fig. 15

The set P consists of all points on one side of the line and the set Q of all points on the other side of the line.

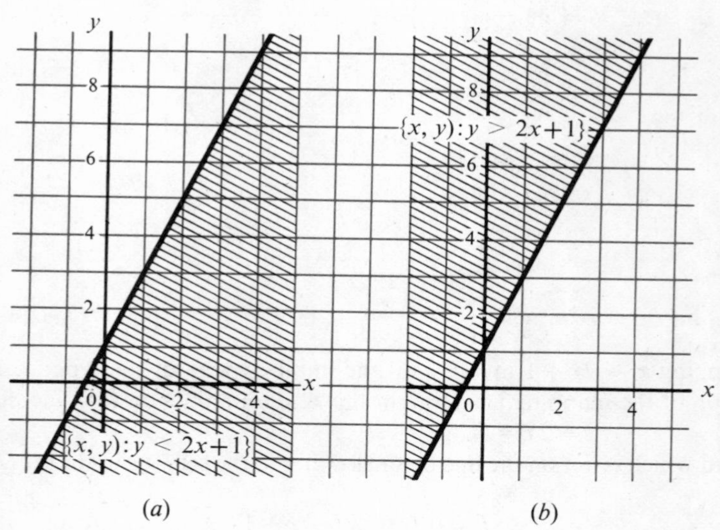

Fig. 16

Figure 16(a) shows the graph of Q and 16(b) the graph of P.

Exercise B

1 Graph the following mappings for the domain $\{x: {}^{-}2 \leqslant x \leqslant 3\}$:

(*a*) $x \rightarrow (2x)^2$; (*b*) $x \rightarrow x^2 + x - 2$;

(*c*) $x \rightarrow x^2 - 1$; (*d*) $x \rightarrow x^2 + \frac{x}{2} - 3$;

(*e*) $x \rightarrow \sqrt{(x-3)}$; (*f*) $x \rightarrow \frac{2}{x}$;

(*g*) $x \rightarrow \frac{1}{2x-1}$; (*h*) $x \rightarrow 2^x + 3$;

(*i*) $x \rightarrow 2^{x+1}$.

2 Draw a straight line through the points $({}^{-}5, 0)$ and $(0, 3)$. Shade the region in which lie all points whose coordinates (x, y) satisfy the inequality

$$5y - 3x < 15.$$

3 Shade on a diagram the set $A \cap B$ where

$$A = \{(x, y): x^2 \leqslant 4\}$$
$$B = \{(x, y): 1 \leqslant y \leqslant 3\}.$$

4 Graph the function $f: x \rightarrow x^3 - 6x^2 + 9x + 2$ for the domain $\{x: 0 \leqslant x \leqslant 4\tfrac{1}{2}\}$. (Plot the points whose x coordinates are 0, 1, 2, 3, 4, $4\tfrac{1}{2}$.)

(*a*) Estimate the gradients at $x = 0, 2, 4$.
(*b*) For what value or values of x is the gradient zero?

5 At noon the height of the tide on a certain sea wall was at M.S.L. (i.e. mean sea level: halfway between high and low tide). The height in metres (h) above this level t hours later is given approximately by the formula $h = 24\sin(30t)°$. Calculate

(i) the height of the tide above M.S.L. at 2.20 p.m.;
(ii) the first time after noon at which the tide is 2 m below M.S.L.;
(iii) the average rate (in metres per hour) at which the tide rose between 2.20 p.m. and 3 p.m.

Draw a graph showing t along the horizontal axis ($t = 0$ to $t = 24$) and h on the vertical axis. From the shape of your graph deduce the times at which the tide was rising most rapidly, giving reasons for your answer.

6 Graph the function $f: x \rightarrow x^4 - 6x^2 + 8x + 25$ for the domain $\{x: {}^{-}3 \leqslant x \leqslant 3\}$.

(*a*) Estimate the gradients when x is ${}^{-}2$, 0 and 2.
(*b*) For what values of x is the gradient zero?
(*c*) For what value of x, such that ${}^{-}3 \leqslant x \leqslant 3$, is the gradient least? Estimate the value of the gradient at this point.

12 Algebra

1 Elements and relations

(*a*) The bridge in the picture was built by combining together basic building materials such as metal girders and blocks of stone into a special kind of pattern or structure. In much the same way we can think of the elements of a set being combined under various operations and giving rise to an *algebraic structure*. The particular algebraic structure that emerges will depend on the 'building materials' (elements) we use.

Here are some of the more common examples of elements we have used in this course:

Integers: $^-2, ^-1, 0, 1, 2, \ldots$ Rationals: $\frac{1}{4}, \frac{7}{8}, \ldots$ Irrationals: $\pi, \sqrt{2}, \ldots$	In the algebra of numbers.
Matrices: $\begin{pmatrix} ^-1 & 0 \\ 2 & 3 \end{pmatrix}, \begin{pmatrix} 1 & 2 & 1 \\ 0 & 1 & 3 \end{pmatrix}, \ldots$	In the algebra of matrices.
Isometries, similarities, ...	In transformation algebra.
Sets: $A, B, \mathscr{E}, \varnothing, \ldots$	In the algebra of sets.
Functions: $f: x \rightarrow 2x, g: x \rightarrow 3x - 1, \ldots$	In the algebra of functions.

(*b*) To compare the elements of a set we use a *relation*. 'Is greater than' is an example of a relation and we write $2 > 1$ to mean '2 is greater than 1'.

Notice that a relation enables us to write a complete 'mathematical sentence', the relation acting rather like the verb in the English language.

Here are other examples of mathematical sentences:

(i) $A \subset B$ — 'A is a subset of B'.

(ii) $\begin{pmatrix} a \\ b \end{pmatrix} = \begin{pmatrix} 1 \\ 2 \end{pmatrix}$ — '$\begin{pmatrix} a \\ b \end{pmatrix}$ is the same translation as $\begin{pmatrix} 1 \\ 2 \end{pmatrix}$'.

(iii) $a \leqslant b$ — 'a is less than or equal to b'.

(*c*) Relations on a set, or between two sets of elements, can be represented by arrow diagrams. If $A = \{4, 12, 3\}$, $B = \{2, 3, 4\}$, the arrow diagram in Figure 1 represents the relation '>' connecting A and B:

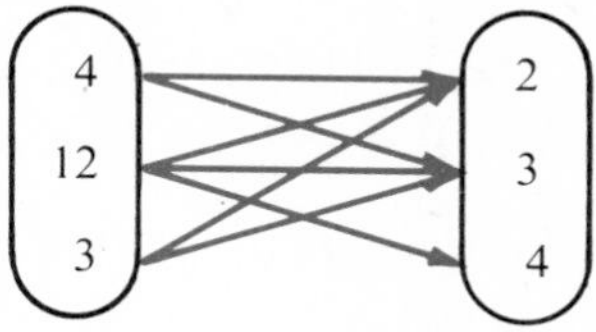

Fig. 1

The special kinds of relations which are either (i) one-to-one or (ii) many-to-one are called *mappings* or *functions*. Figure 2(i) represents the one-to-one function $f: x \rightarrow 2x - 1$, and Figure 2(ii) the many-to-one function $g: x \rightarrow x^2$ for the domain $\{^-2, ^-1, 0, 1, 2\}$ in each case:

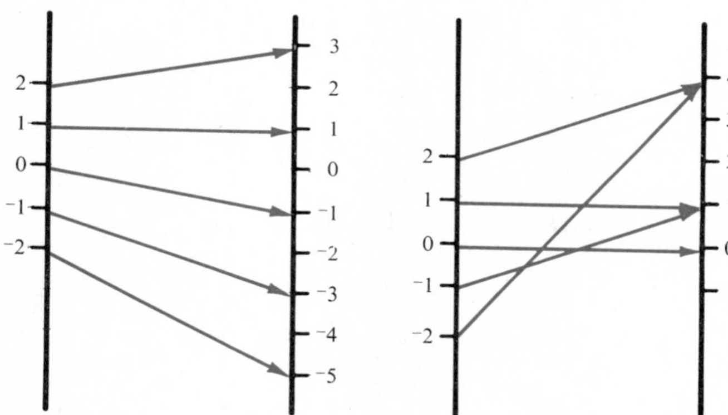

Fig. 2

2 Operations

(*a*) A binary operation combines two (hence *binary*) elements of a set to give a third element, not necessarily a member of the original set.

If we use the operation of 'addition' to combine 2 and 5 we obtain the result, 7, and we write

$$2 + 5 = 7.$$

Similarly, if we use the operation of 'intersection' to combine A and B (see Figure 3), for which $A \subset B$, we obtain A, and we write

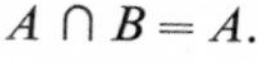

$$A \cap B = A.$$

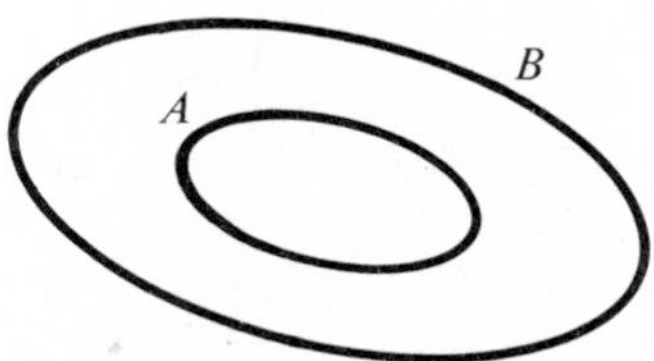

Fig. 3

The most common operations we have used are:

(i) $+, -, \times, \div$ in the algebra of numbers;

(ii) $\cap$, $\cup$ in set algebra;

(iii) matrix multiplication, addition and subtraction in matrix algebra;

(iv) 'follows' in the algebra of transformations.

(*b*) Often no particular symbol is used to denote an operation. When we write $\begin{pmatrix} 2 & 1 \\ 1 & 3 \end{pmatrix}\begin{pmatrix} 1 & 0 \\ 0 & 2 \end{pmatrix} = \begin{pmatrix} 2 & 2 \\ 1 & 6 \end{pmatrix}$ we imply that the operation is matrix multiplication.

In the same way **HT** in transformation algebra means **T** 'followed by' **H** (see Figure 4).

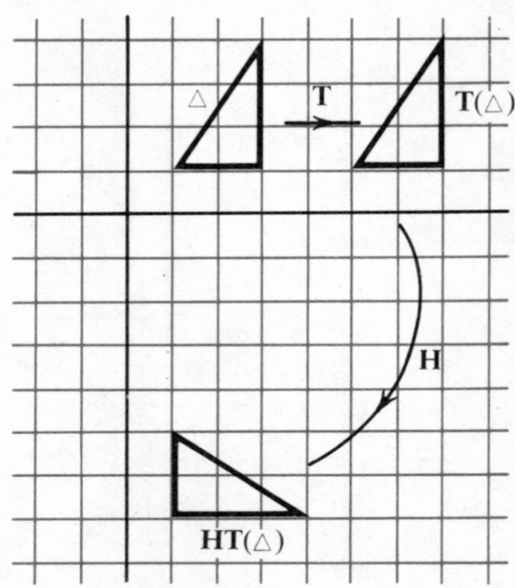

Fig. 4

(*c*) Some operations combine more than one element – for example, 'find the H.C.F. of'; and some texts define mappings such as

$x \to \sqrt{x}$	('find the square root of')
$A \to A'$	('find the complement of')
$T \to \mathbf{P}(T)$	('find the image under the transformation **P** of')

as *unary* operations (i.e. operations on one element).

(*d*) To help us compare and contrast sets of elements under various operations we can use operation tables. Figure 5 shows some examples of operation tables you will have seen before:

$1 \times 2 = 2$
$2 \times 2 = 4$
$3 \times 2 = 6$
$4 \times 2 = 8$
$5 \times 2 = 10$
⋮

The 'two times table'.

(*a*)

$\oplus$	0	1	2
0	0	1	2
1	1	2	0
2	2	0	1

'Clock arithmetic modulo 3 under addition'.

(*b*)

'*f*'	**I**	$\mathbf{H}_1$	$\mathbf{H}_2$
I	**I**	$\mathbf{H}_1$	$\mathbf{H}_2$
$\mathbf{H}_1$	$\mathbf{H}_1$	$\mathbf{H}_2$	**I**
$\mathbf{H}_2$	$\mathbf{H}_2$	**I**	$\mathbf{H}_1$

'Rotational symmetries of an equilateral triangle under followed by'.

(*c*)

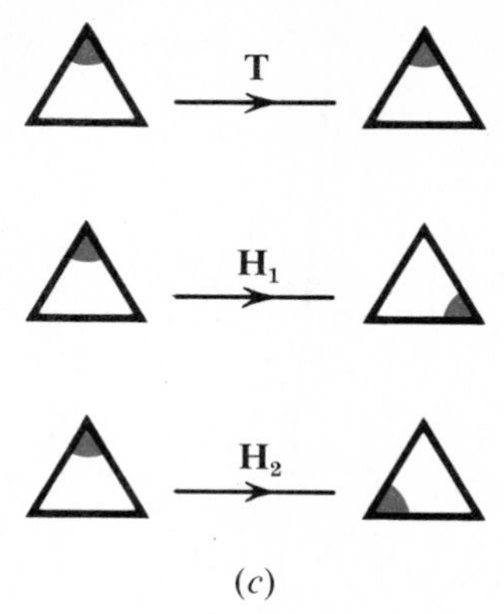

(c)

Fig. 5

Notice that tables (*b*) and (*c*) have a very similar structure.

Exercise A

1 Enter a suitable relation between each of the following pairs of elements to make a sensible mathematical statement:

(i) 2 3; (ii) 2 1; (iii) $\frac{1}{2}$ $\frac{1}{3}$; (iv) $\frac{7}{15}$ $\frac{8}{17}$; (v) $\frac{9}{18}$ $\frac{1}{2}$;
(vi) $\frac{7}{37}$ $\frac{21}{111}$; (vii) $\sqrt{2}$ π; (viii) $\{a, b, c\}$ $\{a, b\}$;
(ix) $\{1, 2, 3\}$ $\{1, 2, 3, 4\}$; (x) $\{2, 5, \frac{1}{2}\}$ $\{\frac{1}{2}, 5, 2\}$;
(xi) $\begin{pmatrix} 1 & 2 \\ 2 & 1 \end{pmatrix}$ $\begin{pmatrix} 1 & 2 \\ 2 & 1 \end{pmatrix}$; (xii) $\begin{pmatrix} 1 & 2 \\ 1 & 1 \end{pmatrix}$ $\begin{pmatrix} -1 & 2 \\ 1 & -1 \end{pmatrix}$;
(xiii) $\begin{pmatrix} 2 \\ 4 \end{pmatrix}$ $\begin{pmatrix} 2 \\ 4 \end{pmatrix}$; (xiv) $\begin{pmatrix} 2 \\ 4 \end{pmatrix}$ $\begin{pmatrix} -2 \\ -4 \end{pmatrix}$; (xv) $\begin{pmatrix} 3 \\ 4 \end{pmatrix}$ $\begin{pmatrix} -4 \\ -3 \end{pmatrix}$;
(xvi) 2·001 2; (xvii) {Factors of 12} {Factors of 24}.

2 Use Figure 6 to help you enter a suitable relation between each of the following pairs of elements:

(i) AC $A'C'$; (ii) AC PR;
(iii) $\triangle ABC$ $\triangle A'B'C'$; (iv) $\triangle ABC$ $\triangle PQR$;
(v) $\angle BAC$ $\angle RPQ$; (vi) $\angle B'A'C'$ $\angle A'C'B'$.

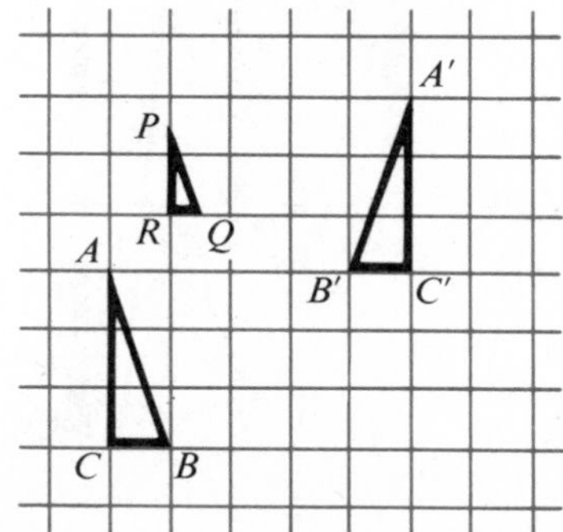

Fig. 6

3 A relation on a set is defined by the arrows in Figure 7. What is the relation in each case?

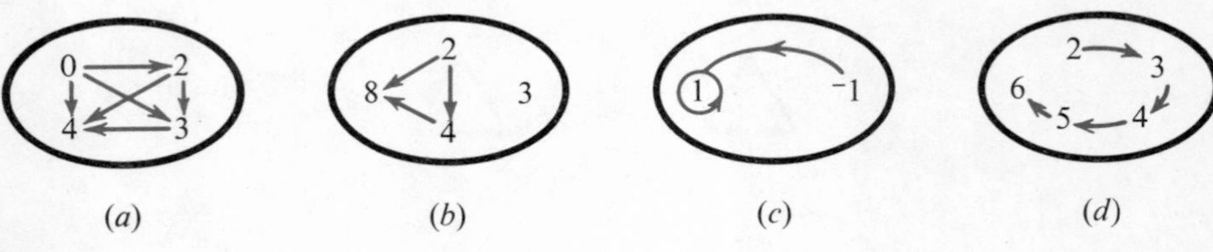

Fig. 7

4 A/B means 'the set of elements in A but not in B':

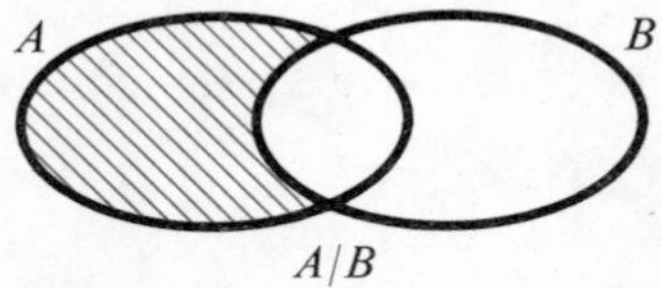

Fig. 8

(*a*) If $A=\{p, q, r\}$; $B=\{q, r, s\}$ what is (i) A/B; (ii) B/A?
(*b*) If $A=\{\text{Integers}\}$; $B=\{\text{Positive integers}\}$ what is (i) A/B; (ii) B/A?
(*c*) Draw Venn diagrams to represent the following statements:

(i) $A/B=A$; (ii) $A/B=\varnothing$; (iii) $B/A=B$; (iv) $B/A=\varnothing$.

5 The statement $\mathbf{MT}(P)=P'$ refers to Figure 9.

(*a*) (i) There is an operation implied in this statement. What is it?
(ii) Is P an element, a relation or an operation?
(iii) Is $\mathbf{M}$ an element, a relation or an operation?

(*b*) Answer (i) and (iii) for the statement

$$\mathbf{MT}=\mathbf{G}.$$

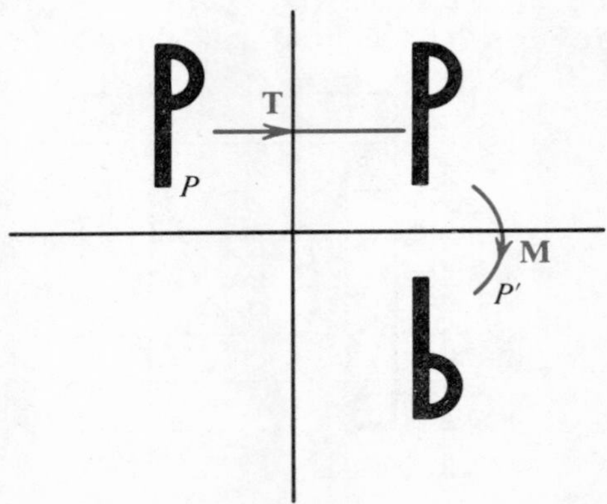

Fig. 9

6 Enter a suitable operation between the two elements on the left-hand side of each statement:

(i) $2 \quad 7=9$; (ii) $7 \quad 2=3\frac{1}{2}$; (iii) $8 \quad 2<9$;
(iv) $\{a, b\} \quad \{b\}=\{b\}$; (v) $\{a, b\} \quad \{b\}=\{a, b\}$;
(vi) $\{1, 3\} \quad \{2, 4\}=\varnothing$; (vii) $\{a, b\} \quad \{b, c\}=\{a, b, c\}$;
(viii) $\begin{pmatrix}1 & 2\\3 & 4\end{pmatrix} \quad \begin{pmatrix}3 & 4\\5 & 6\end{pmatrix}=\begin{pmatrix}^-2 & ^-2\\^-2 & ^-2\end{pmatrix}$; (ix) $\begin{pmatrix}1\\3\end{pmatrix} \quad \begin{pmatrix}^-1\\^-5\end{pmatrix}=\begin{pmatrix}2\\8\end{pmatrix}$;
(x) $3 \quad 7=0$ (modulo 10); (xi) $3 \quad 5=3$ (mod 6);
(xii) $\begin{pmatrix}1 & 2\\1 & 2\end{pmatrix} \quad \begin{pmatrix}3 & ^-1\\1 & 2\end{pmatrix}=\begin{pmatrix}5 & 3\\5 & 3\end{pmatrix}$; (xiii) $\mathscr{E} \quad \varnothing=\varnothing$;
(xiv) $101_2 \quad 11_2=1000_2$; (xv) $23_9 \quad 3_9=7_9$.

7 The symbol '$\Rightarrow$' means 'implies that'; $*$ means 'is a multiple of'; @ means 'is a factor of'; $\|$ means 'is parallel to'; $\perp$ means 'is perpendicular to'. Which of the following are always true statements (a and b being suitably defined for each operation)?

(i) $a * b \Rightarrow b * a$; (ii) $a \| b \Rightarrow b \| a$; (iii) $a * b \Leftrightarrow b$ @ a;
(iv) $a \perp b \Rightarrow b \perp a$; (v) $a \perp b$ and $b \perp c \Rightarrow a \| c$.

8 The operation $*$ means 'add twice the first number to a half the second'. For example, $2 * 1=4+\frac{1}{2}=4\frac{1}{2}$. Calculate:

(i) $1 * 2$; (ii) $3 * 4$; (iii) $4 * 3$; (iv) $0 * 2$; (v) $2 * 0$; (vi) $0 * 0$.

Is $a * b=b * a$ a true statement for all a and b?

9 (*a*) a @ b means a^2b. Find:

(i) 1 @ 2; (ii) 2 @ 1; (iii) 0 @ 4; (iv) 4 @ 0;
(v) (2 @ 1) @ 2; (vi) 2 @ (1 @ 2); (vii) $^-1$ @ 3;
(viii) 3 @ $^-1$; (ix) ($^-1$ @ 3) @ 2; (x) $^-1$ @ (3 @ 2).

Is a @ $b=b$ @ a? Is (a @ b) @ $c=a$ @ (b @ c)?
(*b*) Answer the same question if a @ b means a^2+b^2.

10 Give a meaning to each of the following combination tables by defining a suitable operation:

	1	2	3	4
1	2	3	4	0
2	3	4	0	1
3	4	0	1	2
4	0	1	2	3

(i)

	1	2	3	4
1	1	2	3	4
2	2	4	1	3
3	3	1	4	2
4	4	3	2	1

(ii)

	1	2	3	4
1	1	2	3	4
2	2	4	6	8
3	3	6	9	12
4	4	8	12	16

(iii)

11 $f: x \rightarrow 2x$; $g: x \rightarrow x+1$;

(*a*) Complete (i) $fg: x \rightarrow$; (ii) $gf: x \rightarrow$
(*b*) (i) What binary operation is implied in (*a*)?
(ii) What are the elements being used in this algebra?

3 Closure, identity and inverse

(*a*) If we combine two integers using addition, the result is another integer. That is, if a and b are integers then $a+b$ is an integer. The integers are said to be *closed* under addition. They are also closed under multiplication and subtraction.

However, since $2 \div 3 = \frac{2}{3}$ and $\frac{2}{3}$ is not an integer, then the set is not closed under division.

In the same way the irrationals are not closed under addition because, for example,

$$\sqrt{2} + (2 - \sqrt{2}) = 2, \text{ and 2 is not irrational.}$$

(*b*) The *identity element* of a set under an operation $*$ is an element e such that $e * a = a * e = a$, for any a.

1 is the identity element for real numbers under multiplication because $1 \times a = a \times 1 = a$, where a is any real number.

$\begin{pmatrix} 1 & 0 \\ 0 & 1 \end{pmatrix}$ is the identity element for 2×2 matrices under matrix multiplication because $\begin{pmatrix} 1 & 0 \\ 0 & 1 \end{pmatrix}\begin{pmatrix} a & b \\ c & d \end{pmatrix} = \begin{pmatrix} a & b \\ c & d \end{pmatrix}\begin{pmatrix} 1 & 0 \\ 0 & 1 \end{pmatrix} = \begin{pmatrix} a & b \\ c & d \end{pmatrix}$ for any 2×2 matrix $\begin{pmatrix} a & b \\ c & d \end{pmatrix}$.

$\mathscr{E}$ is the identity element for subsets of $\mathscr{E}$ under $\cap$ because $\mathscr{E} \cap A = A \cap \mathscr{E} = A$ for every subset A of $\mathscr{E}$.

Some sets do not have identity elements under certain operations. For example there is no integer e such that

$$4 - e = e - 4 = 4.$$

There is no identity element for the integers under subtraction.

If you study the combination tables in Figure 5 you will see that 1, 0 and **I** are the identity elements for these sets.

(*c*) In the algebra of functions the operation is 'follows' and the identity element is $h: x \to x$.

If $g: x \to 2x$ and $h: x \to x$ then $hg: x \to 2x$, $gh: x \to 2x$ and $hg = gh = g$.

(*d*) An inverse of x under the operation $*$ is an element y such that $x * y = y * x = e$, where e is the identity element. The inverse of x is written x^{-1}. Obviously if a set does not have an identity element then inverses cannot exist.

Since $2 \times \frac{1}{2} = \frac{1}{2} \times 2 = 1$, and 1 is the identity element for real numbers under multiplication, then $\frac{1}{2}$ is the inverse of 2 (and vice versa).

Similarly, since

$$\begin{pmatrix} 2 & 1 \\ 1 & 3 \end{pmatrix}\begin{pmatrix} \frac{3}{5} & -\frac{1}{5} \\ -\frac{1}{5} & \frac{2}{5} \end{pmatrix} = \begin{pmatrix} \frac{3}{5} & -\frac{1}{5} \\ -\frac{1}{5} & \frac{2}{5} \end{pmatrix}\begin{pmatrix} 2 & 1 \\ 1 & 3 \end{pmatrix} = \begin{pmatrix} 1 & 0 \\ 0 & 1 \end{pmatrix}$$

then $\begin{pmatrix} \frac{3}{5} & -\frac{1}{5} \\ -\frac{1}{5} & \frac{2}{5} \end{pmatrix}$ is the inverse of $\begin{pmatrix} 2 & 1 \\ 1 & 3 \end{pmatrix}$ under matrix multiplication.

Some elements do not have inverses. For example, there is no real number x such that

$$0 \times x = x \times 0 = 1,$$

so 0 has no inverse under multiplication.

(*e*) When two functions combine to form the identity function $i: x \to x$, then as we would expect each is the inverse of the other. If

$$g: x \to 3x + 1, \qquad \text{and} \quad h: x \to \frac{x-1}{3},$$

then $$gh = hg = i \qquad \text{(see Figure 10).}$$

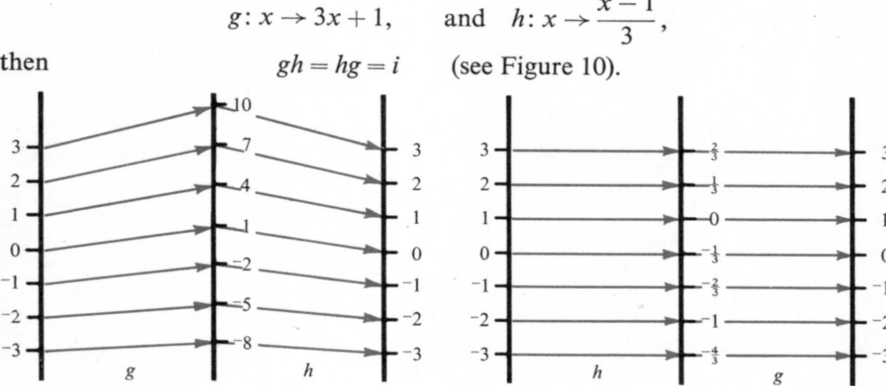

Fig. 10

The table below summarizes this section.

Elements	Operation	Identity	Are there inverses?	Inverse if it exists
Real numbers x	$+$	0	Yes	^{-}x
Real numbers x	$\times$	1	Yes if $x \neq 0$	$1/x$
Subsets of a set $\mathscr{E}$	$\cup$	$\varnothing$	No	
Subsets of a set $\mathscr{E}$	$\cap$	$\mathscr{E}$	No	
Two-dimensional vectors $\begin{pmatrix} x \\ y \end{pmatrix}$	$+$	$\begin{pmatrix} 0 \\ 0 \end{pmatrix}$	Yes	$\begin{pmatrix} ^{-}x \\ ^{-}y \end{pmatrix}$
Clock arithmetic on $\{0, 1, \ldots (n-1)\}$, modulo n	$+$	0	Yes	$n - x$
Clock arithmetic on $\{1, \ldots (n-1)\}$, modulo n	$\times$	1	Yes, if n is prime No, if n is not prime	No simple formula
Isometries of the plane	'Follows'	**I** ('stay put')	Yes	
2 by 2 matrices $\begin{pmatrix} a & b \\ c & d \end{pmatrix}$	$+$	$\begin{pmatrix} 0 & 0 \\ 0 & 0 \end{pmatrix}$	Yes	$\begin{pmatrix} ^{-}a & ^{-}b \\ ^{-}c & ^{-}d \end{pmatrix}$
2 by 2 matrices $\begin{pmatrix} a & b \\ c & d \end{pmatrix}$	Matrix multiplication	$\begin{pmatrix} 1 & 0 \\ 0 & 1 \end{pmatrix}$	Yes if $ad - bc \neq 0$	$\dfrac{1}{ad-bc}\begin{pmatrix} d & ^{-}b \\ ^{-}c & a \end{pmatrix}$

Exercise B

1 (*a*) Draw up a combination table for the set $\{0, 1, 3, 6, 9, 12\}$ under multiplication modulo 15.

(i) Is the set closed under the operation? (ii) What identity element, if any, is there? (iii) Which elements have an inverse? (iv) List the inverses.

(*b*) Answer the same question for $\{0, 1, 3, 6, 9, 12\}$ under addition modulo 15.

2 (*a*) Draw up combination tables for the following sets under multiplication modulo 12. If there is an identity find it, if not, say so.

(i) $\{4, 8\}$; (ii) $\{2, 4, 8\}$; (iii) $\{1, 2, 4, 8\}$; (iv) $\{1, 5, 7, 11\}$.

(*b*) In which of the tables does each element have an inverse?

(*c*) Write down the inverse of each element in table (iv).

3 (*a*) Write down the inverse under matrix multiplication of:

(i) $\begin{pmatrix} 1 & 1 \\ 1 & 2 \end{pmatrix}$; (ii) $\begin{pmatrix} 1 & 1 \\ 1 & 3 \end{pmatrix}$; (iii) $\begin{pmatrix} 1 & 1 \\ 1 & x \end{pmatrix}$.

(*b*) If the inverse of $\begin{pmatrix} 1 & 1 \\ 1 & x \end{pmatrix}$ is one composed of integers, what must x be?

(*c*) Write down the inverse matrix of $\begin{pmatrix} y & ^-2 \\ 2 & 1 \end{pmatrix}$ and find all the values of y which make the elements of the inverse integers.

(*d*) Find all the integer values of x such that the inverse of each of the following matrices has integer elements. If impossible, say so:

(i) $\begin{pmatrix} 5 & 2 \\ x & 3 \end{pmatrix}$; (ii) $\begin{pmatrix} x & ^-1 \\ 8 & 3 \end{pmatrix}$; (iii) $\begin{pmatrix} x & 4 \\ 5 & ^-2 \end{pmatrix}$; (iv) $\begin{pmatrix} x & ^-5 \\ 1 & x \end{pmatrix}$.

4 (*a*) Write down the inverse of each of the functions:

(i) $f: x \to 2x - 3$; (ii) $f: x \to \dfrac{x^2}{3}$; (iii) $f: x \to \dfrac{3}{2-x}$;

(iv) $f: x \to x^2 + 2x + 1$.

(*b*) Find functions g which combine with functions (i)–(iv) above to produce the function $gf: x \to 2x + 1$ (your answers to (*a*) will help you).

5 Copy and complete this combination table of functions.

	$x \to x$	$x \to 2x$	$x \to \dfrac{2}{x}$	$x \to \dfrac{x}{2}$
$x \to x$	$x \to x$			
$x \to 2x$			$x \to \dfrac{1}{x}$	
$x \to \dfrac{2}{x}$				
$x \to \dfrac{x}{2}$				$x \to \dfrac{x}{4}$

(*a*) What operation is implied by the table?
(*b*) Write down the identity element.
(*c*) Write down the inverse of each element (if one exists in the table). If no inverse exists, say so.
(*d*) Is the set closed?

6 If $\mathscr{E}=\{a, b, c, d\}$, $A=\{b, c, d\}$, $B=\{a, c, d\}$ and $C=\{a, b\}$, find

(i) $A \cap B \cap C$; (ii) $A' \cup B'$; (iii) $(A \cup B)'$.

What is the identity element under the operation of intersection for the set of sets $X=\{\mathscr{E}, \varnothing, A, B, C\}$?

Which of the elements of X have an inverse under the operation of intersection which is also a member of X?

7 The symmetry transformations of an equilateral triangle are **I**, **P**, **Q**, **R**, **S**, **T** where **I** represents identity, **P**, a rotation through 120°, **Q**, a rotation through 240°, and **R**, **S**, **T** mean reflection in the lines r, s, t respectively.

Draw up a combination table for these transformations under the operation 'follows'.

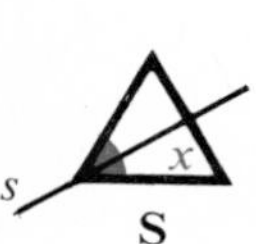

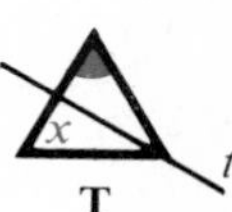

Fig. 11

Remember **XY** means **Y** followed by **X**, and that the mirror lines r, s, t are fixed to the page and do not move round with the triangle.

List each element with its inverse.

8 For each of the following combination tables

(i) state the identity element;
(ii) list the element alongside their inverses (if they exist);
(iii) state whether or not the set is closed under the operation $*$.

$*$	a	b	c
a	c	a	b
b	a	b	c
c	b	c	a

(*a*)

$*$	a	b	c
a	b	a	a
b	a	b	c
c	b	c	a

(*b*)

$*$	p	q	r
p	r	r	p
q	p	r	q
r	p	q	r

(*c*)

4 The laws of algebra

(*a*) If $a * b = b * a$ for every pair of elements a and b in a set (a and b might be the same element), then the operation $*$ obeys the *commutative* law.

For example, since $A \cap B = B \cap A$ for each pair of subsets A and B of a set $\mathscr{E}$, then $\cap$ is commutative.

Also (i) $+$ is commutative for numbers and for clock arithmetics, vectors and matrices.

(ii) $\times$ is commutative for numbers and for clock arithmetics but multiplication is generally not commutative for matrices.

(iii) $\cup$ and $\cap$ are commutative for sets.

(iv) $-$ and $\div$ are not commutative.

(v) 'followed by' is generally not commutative for transformations and functions.

(*b*) If $(a * b) * c = a * (b * c)$ for all a, b, c then the operation $*$ is *associative*.

For example, $(1 + 2) + 3 = 1 + (2 + 3)$, and this is true whichever numbers we use. Addition is associative for numbers.

However, $(1 - 2) - 3 \neq 1 - (2 - 3)$, and hence subtraction is not associative for numbers.

Generally speaking the following operations obey the associative law:

(i) those with 'addition' or 'multiplication' in their names;

(ii) 'follows' for transformations and functions;

(iii) $\cap$ and $\cup$ for sets.

(*c*) The figure below helps to show that $3(5 + 2) = 3.5 + 3.2$.

$$3\left\{\begin{array}{cc} \overbrace{\cdot\cdot\cdot\cdot\cdot}^{5} & \overbrace{\cdot\cdot}^{2} \\ \cdot\cdot\cdot\cdot\cdot & \cdot\cdot \\ \cdot\cdot\cdot\cdot\cdot & \cdot\cdot \end{array}\right. \qquad \begin{array}{ccc} \cdot\cdot\cdot\cdot\cdot & & \cdot\cdot \\ \cdot\cdot\cdot\cdot\cdot & + & \cdot\cdot \\ \cdot\cdot\cdot\cdot\cdot & & \cdot\cdot \end{array}$$

$$3(5 + 2) \quad = \quad (3 \times 5) + (3 \times 2)$$

$\times$ is *distributive* over $+$ for numbers.

Generally we have $a(b + c) = ab + ac$.

Obviously $+$ is not distributive over $\times$ because, for example

$$2 + (3 \times 5) \neq (2 + 3) \times (2 + 5).$$

The law can be used 'in the opposite direction' to simplify expressions:

$$x^2y + 3xy = xy(x + 3).$$

If you draw Venn diagrams to represent

(i) $A \cap (B \cup C)$ and (ii) $(A \cap B) \cup (A \cap C)$,

you will see that $A \cap (B \cup C) = (A \cap B) \cup (A \cap C)$, i.e. $\cap$ is distributive over $\cup$ for sets.

(*d*) We use the three laws when we are simplifying algebraic expressions. Consider some examples from the algebra of numbers:

(i) $$\frac{2a + 4b}{3a + 6b} = \frac{2(a + 2b)}{3(a + 2b)} = \frac{2}{3} \quad (\text{if } a \neq -2b).$$

Here the distributive law has been used before 'cancelling'.

(ii) It would be incorrect to write

$$\frac{a-2b}{2b-a} = \frac{a-2b}{a-2b} = 1,$$

because the commutative law does not hold for subtraction.

However, it is correct to write

$$\frac{a-2b}{2b-a} = \frac{^{-}1(2b-a)}{2b-a} = {}^{-}1,$$

using the distributive law, and 'cancelling'.

(iii) It is incorrect to write $\frac{1}{2}/3 = 1/\frac{2}{3}$ because $(1 \div 2) \div 3 \neq 1 \div (2 \div 3)$ (the associative law does not hold for $\div$).

(iv) We cannot write $\dfrac{2}{3-5} = 2 \div (3-5) = \dfrac{2}{3} - \dfrac{2}{5}$ because $\div$ is not distributive over $-$.

Hence, generally

$$\frac{a}{b-c} \neq \frac{a}{b} - \frac{a}{c} \quad \text{and} \quad \frac{a}{b+c} \neq \frac{a}{b} + \frac{a}{c}.$$

Exercise C

1 Which of the following are true statements?

(i) $1 \div 2 = 2 \div 1$; (ii) $\begin{pmatrix} 1 & 2 \\ 3 & 1 \end{pmatrix} - \begin{pmatrix} 2 & 0 \\ 1 & 3 \end{pmatrix} = \begin{pmatrix} 2 & 0 \\ 1 & 3 \end{pmatrix} - \begin{pmatrix} 1 & 2 \\ 3 & 1 \end{pmatrix}$;

(iii) $\begin{pmatrix} 1 \\ 2 \end{pmatrix} + \begin{pmatrix} ^{-}1 \\ 3 \end{pmatrix} = \begin{pmatrix} ^{-}1 \\ 3 \end{pmatrix} + \begin{pmatrix} 1 \\ 2 \end{pmatrix}$; (iv) $\begin{pmatrix} 2 \\ 1 \end{pmatrix} - \begin{pmatrix} 3 \\ 1 \end{pmatrix} = \begin{pmatrix} 3 \\ 1 \end{pmatrix} - \begin{pmatrix} 2 \\ 1 \end{pmatrix}$;

(v) $\frac{1}{2} - \frac{3}{7} = \frac{3}{7} - \frac{1}{2}$; (vi) $(2 + \frac{1}{2}) \div 3 = 2 + (\frac{1}{2} \div 3)$;

(vii) $\{p, q, r\} \cap \{q, r\} = \{q, r\} \cap \{p, q, r\}$; (viii) $2 \div (3+4) = \frac{2}{3} + \frac{2}{4}$;

(ix) $(2 \quad 1)\begin{pmatrix} 1 \\ 4 \end{pmatrix} = (1 \quad 4)\begin{pmatrix} 2 \\ 1 \end{pmatrix}$; (x) $(a \quad b)\begin{pmatrix} b \\ a \end{pmatrix} = \begin{pmatrix} b \\ a \end{pmatrix}(a \quad b)$;

(xi) $(p \quad q)\begin{pmatrix} r \\ s \end{pmatrix} = (r \quad s)\begin{pmatrix} p \\ q \end{pmatrix}$; (xii) $\frac{1}{2} + \frac{1}{3} = \dfrac{1}{2+3}$.

2 (*a*) $a * b$ means $a + 2b$, where a and b are real numbers. Work out:

(i) $1 * 2$; (ii) $2 * 1$; (iii) $(1 * 2) * 3$; (iv) $1 * (2 * 3)$.

(*b*) Is $*$ (i) commutative; (ii) associative?

(*c*) Work out (i) $p * (q * r)$; (ii) $(p * q) * r$. What does your result show?

(*d*) Show that if $c = 0$ then $(a * b) * c = a * (b * c)$.

(*e*) Write down a set of 2 elements upon which $*$ can be defined and is associative.

3 $\mathbf{p} * \mathbf{q} = \mathbf{p} + \frac{1}{2}\mathbf{q}$. Draw diagrams to show the vectors:

(i) $\begin{pmatrix} 1 \\ 2 \end{pmatrix} * \begin{pmatrix} 3 \\ 2 \end{pmatrix}$; (ii) $\begin{pmatrix} 3 \\ 2 \end{pmatrix} * \begin{pmatrix} 1 \\ 2 \end{pmatrix}$.

Is $*$ (*a*) commutative; (*b*) associative?

4 Copy Figure 12 four times. On separate diagrams shade the regions
(i) $A \cap (B \cup C)$; (ii) $(A \cap B) \cup (A \cap C)$; (iii) $A \cup (B \cap C)$;
(iv) $(A \cup B) \cap (A \cup C)$.

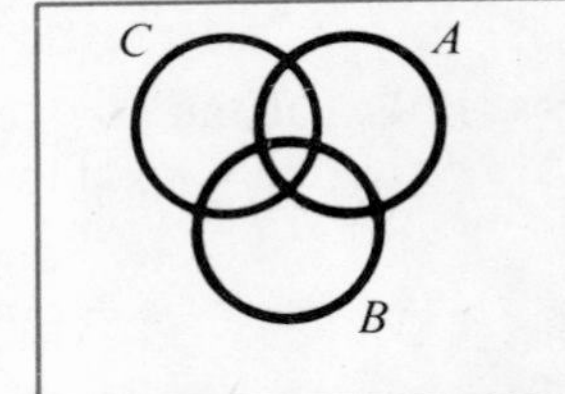

Is $\cap$ distributive over $\cup$?
Is $\cup$ distributive over $\cap$?

Fig. 12

5

	p	q	r
p	p	q	r
q	q	r	p
r	r	p	q

(*a*) Use the combination table to write down the element x so that
(i) $x = p * q$; (ii) $x = q * p$; (iii) $x = r * q$;
(iv) $x = q * r$; (v) $x = p * r$; (vi) $x = r * p$.

What do your results suggest about $*$?

(*b*) The 'leading diagonal' of the table contains p, r and q. Notice that the table is symmetrical about this diagonal. What does this tell you about $*$?

(*c*) Find x so that

(i) $x = p * (q * r)$; (ii) $x = (q * p) * q$; (iii) $x = (p * r) * r$.

(*d*) From your results in (*c*) (and without using the table again), write down x so that

(i) $x = p * (r * q)$; (ii) $x = q * (q * p)$; (iii) $x = r * (r * p)$.

Explain why it is possible to write the answers down without using the table.

6 $\mathbf{M}_1$ is a reflection in the lines $x = 1$.
$\mathbf{R}_\theta$ is a rotation through an angle $\theta°$ about (0, 0).
$\mathbf{E}_h$ is an enlargement (centre (0, 0)), scale-factor h.
$\mathbf{S}_t$ is a shear through t units parallel to the x axis leaving the x axis invariant. (i.e. $(0, 1) \to (t, 1)$.)

(*a*) Draw diagrams to show the images of the flag F in Figure 13 under the transformations

(i) $\mathbf{M}_0$; (ii) $\mathbf{R}_{90}$; (iii) $\mathbf{E}_2$; (iv) $\mathbf{S}_2$;
(v) $\mathbf{M}_0\mathbf{M}_2$; (vi) $\mathbf{R}_{90}\mathbf{R}_{180}$.

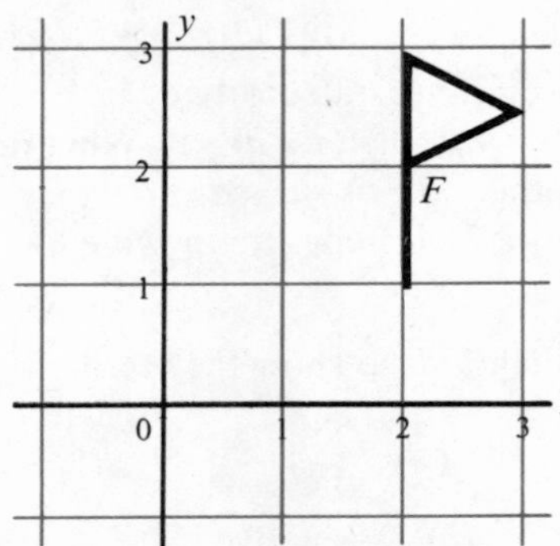

Fig. 13

(*b*) Is (i) $\mathbf{M}_p\mathbf{M}_q = \mathbf{M}_q\mathbf{M}_p$; (ii) $\mathbf{R}_\theta\mathbf{R}_\alpha = \mathbf{R}_\alpha\mathbf{R}_\theta$;
(iii) $\mathbf{E}_l\mathbf{E}_t = \mathbf{E}_t\mathbf{E}_l$; (iv) $\mathbf{S}_t\mathbf{S}_q = \mathbf{S}_q\mathbf{S}_t$?

(*c*) Does the operation 'follows' obey the commutative law when defined on the sets:

(i) {reflections in lines parallel to the y axis};
(ii) {rotations, centre (0, 0)};
(iii) {enlargements, centre (0, 0)};
(iv) {shears parallel to the x axis leaving the x axis invariant}?

7 $\mathbf{G}_1$, $\mathbf{G}_2$ are two glide reflections. When is it true that $\mathbf{G}_1\mathbf{G}_2 = \mathbf{G}_2\mathbf{G}_1$?

8 Which of the following are necessarily true statements:

(i) $a + b = b + a$ (in number algebra);
(ii) $\mathbf{MP} = \mathbf{PM}$ (in matrix algebra);
(iii) $\mathbf{MP} = \mathbf{PM}$ (in the algebra of transformations);
(iv) $a > b$ and $b > c \Rightarrow a > c$ (in number algebra);
(v) $A \cap B = B \cap A$ (in set algebra)?

For the statements which are not necessarily true, give an example to support your answer.

9 a @ b means $(\frac{1}{2}a).b$; $a \gamma b$ means $\frac{1}{2}a + b$.
(*a*) Calculate:

(i) 2 @ 3; (ii) 3 @ 2; (iii) $2 \gamma 3$; (iv) $3 \gamma 2$.

(*b*) State whether or not the following are true statements. Justify your 'not true' answers with an example:

(i) @ is commutative; (ii) @ is associative;
(iii) γ is commutative; (iv) γ is associative;
(v) @ is distributive over γ; (vi) γ is distributive over @.

5 Equations and inequalities

(*a*) We saw in *Book X* that equations such as

$$\tfrac{2}{3}(x - 1) = 2x$$

could be solved by 'doing the same thing to both sides'. That is, the equality still holds if (i) the same element is added to both sides;
(ii) the same element is subtracted from both sides;
(iii) both sides are multiplied by the same element;
(iv) both sides are divided by the same element.

Thus the equation above can be solved in the following way:

Multiply by $\frac{3}{2}$	$\frac{3}{2}.\frac{2}{3}(x - 1) = \frac{3}{2}.2x$
i.e.	$x - 1 = 3x$
Subtract x	$^{-}1 = 2x$
Divide by 2	$x = {}^{-}\frac{1}{2}$.

(*b*) The same method can be used in matrix algebra. The equation

$$\begin{pmatrix}1\\2\end{pmatrix}=\begin{pmatrix}2\\3\end{pmatrix}+\begin{pmatrix}2&1\\1&0\end{pmatrix}\begin{pmatrix}x\\y\end{pmatrix}$$

can be solved as follows:
Add $\begin{pmatrix}^-2\\^-3\end{pmatrix}$ to both sides

$$\begin{pmatrix}^-1\\^-1\end{pmatrix}=\begin{pmatrix}2&1\\1&0\end{pmatrix}\begin{pmatrix}x\\y\end{pmatrix}.$$

We need to isolate $\begin{pmatrix}x\\y\end{pmatrix}$, but we cannot divide by $\begin{pmatrix}2&1\\1&0\end{pmatrix}$ because division by a matrix is meaningless. However, we can premultiply both sides by $\begin{pmatrix}0&1\\1&^-2\end{pmatrix}$, the multiplicative inverse of $\begin{pmatrix}2&1\\1&0\end{pmatrix}$:

i.e.

$$\begin{pmatrix}0&1\\1&^-2\end{pmatrix}\begin{pmatrix}^-1\\^-1\end{pmatrix}=\begin{pmatrix}0&1\\1&^-2\end{pmatrix}\left\{\begin{pmatrix}2&1\\1&0\end{pmatrix}\begin{pmatrix}x\\y\end{pmatrix}\right\}.$$

Hence $$\begin{pmatrix}^-1\\1\end{pmatrix}=\begin{pmatrix}1&0\\0&1\end{pmatrix}\begin{pmatrix}x\\y\end{pmatrix}$$

i.e. $$\begin{pmatrix}^-1\\1\end{pmatrix}=\begin{pmatrix}x\\y\end{pmatrix}.$$

(note that we have used the associative law, and also that both sides *must* be *premultiplied* by $\begin{pmatrix}0&1\\1&^-2\end{pmatrix}$ because matrix multiplication is not commutative).

(*c*) 'Inequalities' such as $2(1-3x)<3+x$ can be solved in a similar way but we must remember to 'change the sign round' when multiplying or dividing by a negative number:

$$2(1-3x)<3+x.$$

Using the distributive law

$$2-6x<3+x$$

Subtract 2 $\quad -6x<1+x$

Subtract x $\quad -7x<1$

Divide by $^-7$ $\quad x>{^-\tfrac{1}{7}}$ ($<$ has been changed to $>$).

(*d*) If we attempt to solve the equation

$$2x+3=2x+4,$$

we obtain, by subtracting $2x$ from each side,

$$0.x+3=4$$

and hence $\quad 0.x=1.$

The equation obviously has no solution.

In solving equations we need to use

(i) the identity elements for multiplication and addition [in (*b*) we added $\begin{pmatrix}^-2\\ ^-3\end{pmatrix}$ to obtain $\begin{pmatrix}0\\0\end{pmatrix}+\begin{pmatrix}2&1\\1&0\end{pmatrix}\begin{pmatrix}x\\y\end{pmatrix}=\begin{pmatrix}2&1\\1&0\end{pmatrix}\begin{pmatrix}x\\y\end{pmatrix}$];

(ii) inverses under multiplication and addition (in (*a*) we multiplied by the inverse of $\frac{2}{3}$ to obtain $1.(x-1)=1.3x$, i.e. $(x-1)=3x$);

(iii) the three laws of algebra.

The equation above proved impossible to solve because 0 has no inverse under multiplication. Hence we could not multiply both sides of $0.x=1$ by a number to obtain $x=\ldots$.

Exercise D

1 Solve the following equations:

(i) $2x+1=3$; (ii) $3x-4=5$; (iii) $1-x=1$;
(iv) $\dfrac{1}{x}=3$; (v) $\dfrac{x}{4}=7$; (vi) $\dfrac{1}{x}-3=2$;
(vii) $1-3x={}^-2$; (viii) $\dfrac{3}{x}-\dfrac{4}{x}=4$; (ix) $2(3-x)=x$;
(x) $3(x-7)=2(x-4)$; (xi) $\dfrac{3}{x+1}=\dfrac{2}{x-1}$; (xii) $\dfrac{1}{x}+\dfrac{1}{4}=\dfrac{1}{2x}$.

2 Solve the following:

(i) $x+2\geqslant 3$; (ii) $3x-4<5$; (iii) $1-x\geqslant 1$;
(iv) $1-\dfrac{x}{2}>3$; (v) $\dfrac{x}{4}<7$; (vi) $2-x\leqslant x-2$;
(vii) $2(x-3)\leqslant 3(x+1)$; (viii) $x^2>4$;
(ix) $x^3<1$; (x) $3<x+2<4$; (xi) $x<2x-1\leqslant x+3$;
(xii) $x^2-4\geqslant 2$.

3 Solve:

(i) $\begin{pmatrix}1&0\\3&1\end{pmatrix}-\begin{pmatrix}a&b\\c&d\end{pmatrix}=\begin{pmatrix}2&3\\1&0\end{pmatrix}$; (ii) $\begin{pmatrix}3&4\\1&2\end{pmatrix}+\begin{pmatrix}a&b\\c&d\end{pmatrix}=\begin{pmatrix}^-1&0\\3&1\end{pmatrix}$;
(iii) $\begin{pmatrix}1&2\\1&3\end{pmatrix}\begin{pmatrix}a&b\\c&d\end{pmatrix}=\begin{pmatrix}2&3\\-1&-1\end{pmatrix}$; (iv) $\begin{pmatrix}1&2\\3&4\end{pmatrix}\begin{pmatrix}x\\y\end{pmatrix}=\begin{pmatrix}4\\2\end{pmatrix}$;
(v) $\begin{pmatrix}2&3\\2&3\end{pmatrix}\begin{pmatrix}x\\y\end{pmatrix}=\begin{pmatrix}3\\1\end{pmatrix}$.

4 Multiply $(p\quad q)\begin{pmatrix}2\\1\end{pmatrix}$.

(*a*) Find positive integral values of p and q so that $(p\quad q)\begin{pmatrix}2\\1\end{pmatrix}=2$.

(*b*) If p and q are integers give three possible solutions to the equation in (*a*).

5 If $A=\{a, b, c, d\}$ and $\mathscr{E}$ is composed of all the subsets of A, i.e.

$$\mathscr{E}=\{\varnothing, A, \{a\}, \{b\}, \ldots \{a, b\}, \ldots\}$$

find all possible solutions from $\mathscr{E}$ for each of the following:

(i) $A \cap B=\{b\}$;
(ii) $\{a, b, c\} \cup C=A$;
(iii) $\{a, b\} \cap D=\varnothing$;
(iv) $[\{a, b, d\} \cap C] \cup \{b, c\}=\{b, c, d\}$;
(v) $\{a, b, c\}' \cap E=\{a\}$.

6 Solve the following equations:

(i) $\begin{pmatrix}1\\2\end{pmatrix}=\begin{pmatrix}3\\4\end{pmatrix}+\begin{pmatrix}2 & 1\\1 & 0\end{pmatrix}\begin{pmatrix}x\\y\end{pmatrix}$;

(ii) $\begin{pmatrix}1 & ^-1\\0 & 1\end{pmatrix}\begin{pmatrix}x\\y\end{pmatrix}+\begin{pmatrix}2\\1\end{pmatrix}=\begin{pmatrix}3\\^-1\end{pmatrix}$;

(iii) $\begin{pmatrix}1 & 0\\0 & 2\end{pmatrix}\left[\begin{pmatrix}x\\y\end{pmatrix}+\begin{pmatrix}2\\1\end{pmatrix}\right]=\begin{pmatrix}2\\1\end{pmatrix}$.

7 State whether the following equations have (i) a finite number of solutions; (ii) an infinite number of solutions; (iii) no solutions, in the set of real numbers. For those which have a finite number, write down the solution(s).

(i) $x+3=x+4$;
(ii) $2(x-1)=2x-2$;
(iii) $x^2+4=3$;
(iv) $\dfrac{3}{x}=\dfrac{4}{x}$;
(v) $x+y=2$;
(vi) $x(1-x)=3-x^2$;
(vii) $x(1-x)=x^2-x$;
(viii) $x^2-9=(x-3)(x+3)$.

13 Computation 1

1 Introduction

Here are some questions designed to help you check your *understanding* of various operations with numbers. None of them requires any complicated calculation, so you should not need a slide-rule or other computational aid to help find the answers.

Exercise A

1 What is the value of:

(*a*) $\sqrt{5} \times \sqrt{5}$; (*b*) $\sqrt{2} \times \sqrt{8}$; (*c*) $\sqrt{0{\cdot}04}$?

2 Write down the value of:

(*a*) $\log_{10} 1000$; (*b*) $\log_{10} 0{\cdot}001$.

3 How do you recognize multiples of five when numbers are written in base five?

4 How do you recognize multiples of four when numbers are written in binary form?

5 $M_3 = \{\text{multiples of three}\}$; $M_6 = \{\text{multiples of six}\}$. What is:

(*a*) $M_3 \cap M_6$; (*b*) $M_3 \cup M_6$?

6 Given that $21 \times 29 = 609$, *write down* the values of:

(*a*) 22×29; (*b*) 21×30; (*c*) 21×58;
(*d*) $2{\cdot}1 \times 2{\cdot}9$; (*e*) $0{\cdot}21 \times 0{\cdot}029$; (*f*) $6{\cdot}09 \div 2{\cdot}9$;
(*g*) $0{\cdot}609 \div 29$.

7 Which of the following statements are true?

(*a*) If two numbers in a subtraction are both increased by 10 the answer is unchanged.
(*b*) If two numbers in a multiplication are both made 10 times bigger, the answer is ten times bigger.
(*c*) If two numbers in a division are both made 10 times bigger, the answer is unchanged.
(*d*) If two numbers in an addition are both increased by 10 the answer is also increased by 10.

8 What is the value of:

(*a*) $\dfrac{1}{1/5}$; (*b*) $\dfrac{1}{0{\cdot}01}$; (*c*) $(0{\cdot}2)^3$;
(*d*) $(^-2)^2$; (*e*) $(^-2)^3$; (*f*) $(^-1)^{17}$?

2 Number bases

In base ten, or denary notation, 3742 is short for

$$(3 \times 10^3) + (7 \times 10^2) + (4 \times 10) + (2 \times 1).$$

Here are some examples of numbers written in full in other number bases:

In base six: $42_6 = (4 \times 6) + (2 \times 1)$.
In base four: $231_4 = (2 \times 4^2) + (3 \times 4) + (1 \times 1)$.
In base two (binary):

$$101{\cdot}11_2 = (1 \times 2^2) + (0 \times 2) + (1 \times 1) + (1 \times \tfrac{1}{2}) + \left(1 \times \frac{1}{2^2}\right).$$

It is important to remember that in base five, for example, only the symbols 0, 1, 2, 3, 4 are used. In general the largest symbol employed is one less than the base number.

In base seven 0, 1, 2, 3, 4, 5, 6 are used.
In base two 0, 1 are used.
In base twelve (duodecimal) 0, 1, 2, 3, 4, 5, 6, 7, 8, 9, T, E are used.

The methods which you use for adding, subtracting, multiplying and dividing numbers in denary notation can easily be adapted for use with numbers in other bases.

Comparisons between numbers written in denary notation and other bases are interesting and helpful. For example:

In base ten 53 $\xrightarrow{\text{multiply by ten}}$ 530.

In base six $34_6 \xrightarrow{\text{multiply by six}} 340_6$.

In base two $1101_2 \xrightarrow{\text{divide by two}} 110{\cdot}1_2$.

Exercise B

1 Express:

(*a*) 234_5 as a numeral to base ten;
(*b*) 37_8 in binary notation.

2 Multiply 145_6 by six, giving your answer:

(*a*) as a base six number;
(*b*) as a base ten number.

3 Find the sum and difference of 111010_2 and 10011_2 and give your answers in base two.

4 Express in binary notation:

(*a*) 17; (*b*) 17·5; (*c*) 17·25.

5 Find the sum and difference of 325_6 and 144_6 giving your answers in the scale of six.

6 Assuming $n > 3$, write $3n^2 + n + 2$ as a numeral to base n.

7 Find n if $154_n = 88_{10}$.

8 Solve the equation $11x = 1001$, the numbers being in binary form, giving your answer also in binary notation.

3 Patterns of numbers

Here are some 'useful' patterns:

(*a*)
$$10^2 = 100$$
$$10^1 = 10$$
$$10^0 = 1$$
$$10^{-1} = \frac{1}{10}$$
$$10^{-2} = \frac{1}{100}$$

(*b*)

$3 \times 2 = 6$	$3 \times {}^-2 = {}^-6$
$3 \times 1 = 3$	$2 \times {}^-2 = {}^-4$
$3 \times 0 = 0$	$1 \times {}^-2 = {}^-2$
$3 \times {}^-1 = {}^-3$	$0 \times {}^-2 = 0$
$3 \times {}^-2 = {}^-6$	${}^-1 \times {}^-2 = 2$
$3 \times {}^-3 = {}^-9$	${}^-2 \times {}^-2 = 4$

(*c*)
$$\sqrt{0{\cdot}49} = 0{\cdot}7$$
$$\sqrt{4{\cdot}9} \approx 2{\cdot}21$$
$$\sqrt{49} = 7$$
$$\sqrt{490} \approx 22{\cdot}1$$
$$\sqrt{4900} = 70$$

(*d*)
$$\frac{1}{0{\cdot}08} = 12{\cdot}5$$
$$\frac{1}{0{\cdot}8} = 12{\cdot}5$$
$$\frac{1}{8} = 0{\cdot}125$$
$$\frac{1}{80} = 0{\cdot}0125$$
$$\frac{1}{800} = 0{\cdot}00125$$

Other common number patterns include:

(*e*) Fibonacci:

1, 1, 2, 3, 5, 8, 13, 21 . . .

(each number is the sum of the two before it).

(*f*) Pascal's triangle:

```
        1
      1   1
    1   2   1
  1   3   3   1
1   4   6   4   1
.   .   .   .   .   .
```

(*g*) Square numbers: 1, 4, 9, 16, 25 . . .
Triangular numbers: 1, 3, 6, 10, 15 . . .
Cubes: 1, 8, 27, 64, 125 . . .

The prime numbers: 2, 3, 5, 7, 11, 13, 17 . . . do not follow any obvious pattern.

Exercise C

1 Copy and add three more lines to the pattern:

$$\begin{aligned} 1^3 &= 1^2 \\ 1^3 + 2^3 &= 3^2 \\ 1^3 + 2^3 + 3^3 &= 6^2 \end{aligned}$$

2 Find the values of x and y if $3^2 + 4^2 = x^2$ and $3^3 + 4^3 + 5^3 = y^3$.
Write down the next statement suggested to you by these two results.
Check carefully to see if it is correct.

3 Given that $\sqrt{7 \cdot 32} \approx 2 \cdot 71$ and $\dfrac{1}{6 \cdot 18} \approx 0 \cdot 162$, find the value of:

(*a*) $\sqrt{732}$; (*b*) $\dfrac{1}{0 \cdot 618}$; (*c*) $\sqrt{0 \cdot 0732}$; (*d*) $\dfrac{1}{618}$.

4 The nth number in a sequence of positive numbers is $3n + 2$.
How many members of the sequence are less than 30?

5 Consider the sequence 2, 3, 5, 8, 12 . . . .
(*a*) Write down the next two terms.
(*b*) Write down the first seven terms of the sequence whose nth term is $4 + n(n - 1)$.
(*c*) State a relationship between the two sequences.
(*d*) Deduce the 25th term of the first sequence.

6 (*a*) What is the value of $1^2 + 3^2 + 3^2 + 1^2$?
(*b*) Where does your answer to (*a*) appear in Pascal's triangle?
(*c*) Work out $1^2 + 4^2 + 6^2 + 4^2 + 1^2$.
(*d*) Where does your answer to (*c*) appear in Pascal's triangle?
(*e*) Suggest a general rule about adding the squares of numbers in a row of Pascal's triangle.
(*f*) Check you rule in detail for two more cases.

7 Write the numbers 11^0, 11^1, 11^2, 11^3, 11^4 without using indices. How are they related to Pascal's triangle? Does 11^5 follow the pattern?

4 Indices

3^7 is a convenient shorthand notation for $3 \times 3 \times 3 \times 3 \times 3 \times 3 \times 3$.
7^3 is a convenient shorthand notation for $7 \times 7 \times 7$.
Some important results about indices are shown by the following examples:

(*a*) $2^4 \times 2^3 = (2 \times 2 \times 2 \times 2) \times (2 \times 2 \times 2) = 2^7$,
so $2^4 \times 3^3 = 2^{4+3} = 2^7$, and *in general* $a^m \times a^n = a^{m+n}$.

(*b*) $6^5 \div 6^2 = \dfrac{6 \times 6 \times 6 \times 6 \times 6}{6 \times 6} = 6^3$,
so $6^5 \div 6^2 = 6^{5-2} = 6^3$, and *in general* $a^m \div a^n = a^{m-n}$.

(*c*) $(10^2)^3 = 10^2 \times 10^2 \times 10^2 = 10^6$,
so $(10^2)^3 = 10^{2\times 3} = 10^6$, and *in general* $(a^m)^n = a^{m\times n}$.
(*d*) $3^4 \div 3^4 = 1$ (a number divided by itself gives the answer one);
but $3^4 \div 3^4 = 3^{4-4} = 3^0$.
So $3^0 = 1$, and *in general* $a^0 = 1$.
(*e*) $10^2 \div 10^5 = 10^{2-5} = 10^{-3}$;

$$\text{but } 10^2 \div 10^5 = \frac{10 \times 10}{10 \times 10 \times 10 \times 10 \times 10} = \frac{1}{10^3}.$$

So $10^{-3} = \dfrac{1}{10^3} = \dfrac{1}{1000}$ and *in general* $a^{-n} = \dfrac{1}{a^n}$.

(*f*) $16^{\frac{1}{2}} \times 16^{\frac{1}{2}} = 16^{\frac{1}{2}+\frac{1}{2}} = 16^1$,
so $16^{\frac{1}{2}} = 4 = \sqrt{16}$, and *in general*, $a^{\frac{1}{2}} = \sqrt{a}$.

Exercise D

1 Given that $3^{10} = 59\,049$ and $5^7 = 78\,125$, find the value of:
(*a*) 3^{11}; (*b*) 5^8; (*c*) 5^6; (*d*) 3^8.

2 Solve:
(*a*) $9^x = 729$; (*b*) $10^x = 0{\cdot}001$; (*c*) $7^x = 1$;
(*d*) $17^x = 289$; (*e*) $289^x = 17$.

3 Write in the form a^n:
(*a*) $5^7 \times 5^3$; (*b*) $5^7 \times 5^{-3}$; (*c*) $3^6 \div 3^2$; (*d*) $\dfrac{3^2}{3^6}$;
(*e*) $3^6 \div 3^{-2}$; (*f*) $3^{-6} \div 3^{-2}$; (*g*) $2^5 \times 2^{-2} \times 2^3$.

4 Find the value of $\dfrac{ab}{c}$ where $a = 10^6$, $b = 10^7$, $c = 10^{-3}$.

5 Write in *ascending* order of size: $6^3 - 5^3$, 3^4, 4^3, $2^3 \times 3^2$, $29^2 \times 10^{-1}$.

6 Say which of the statements are true and which are false:
(*a*) $\sqrt{2^8} = 2^4$; (*b*) $3^2 \times 2^2 = 6^2$; (*c*) $3^2 + 2^2 = 5^2$;
(*d*) $\dfrac{10^3}{5^3} = 2^3$; (*e*) $10^3 - 4^3 = 6^3$.

5 Standard index form

Number	Standard index form
7280	$7{\cdot}28 \times 10^3$
728	$7{\cdot}28 \times 10^2$
72·8	$7{\cdot}28 \times 10^1$
7·28	$7{\cdot}28 \times 10^0$
0·728	$7{\cdot}28 \times 10^{-1}$
0·0728	$7{\cdot}28 \times 10^{-2}$
0·00728	$7{\cdot}28 \times 10^{-3}$

A number in standard index form consists of:

(i) a number between 1 and 10

and (ii) a power of 10,

multiplied together.

You may find it useful to make up rules for conversion to standard index form but it can always be carried out by common sense. For example:

$$537 = 53{\cdot}7 \times 10^1 = 5{\cdot}37 \times 10^2$$

(notice that as the first number decreases, the second increases but the product remains constant).

$$0{\cdot}00851 = 0{\cdot}0851 \times 10^{-1} = 0{\cdot}851 \times 10^{-2} = 8{\cdot}51 \times 10^{-3}$$

(the first number increases, the second decreases and the product remains constant).

Exercise E

1 Given that $0{\cdot}00073 = 7{\cdot}3 \times 10^p$ and $892\,000 = 8{\cdot}92 \times 10^q$ find the values of p and q.

2 Express the following numbers in standard index form:

(*a*) 630; (*b*) 0·038; (*c*) $0{\cdot}0604 \times 10^6$; (*d*) $0{\cdot}71 \times 10^{-2}$;

(*e*) 2 million.

3 Multiply 0·04 by 0·0012 giving your answer in standard index form.

4 Write in the form $a \times 10^n$, where $1 \leqslant a < 10$ and n is an integer:

(*a*) $\dfrac{4{\cdot}5 \times 10^6}{1{\cdot}5 \times 10^2}$; (*b*) $\dfrac{7{\cdot}6 \times 10^5}{1{\cdot}9 \times 10^{-2}}$; (*c*) $\dfrac{1{\cdot}4 \times 10^3}{2{\cdot}8 \times 10^5}$.

5 Find the product of $2{\cdot}4 \times 10^7$ and 5×10^{-3} and express the answer in standard index form.

6 The speed of light is $3{\cdot}00 \times 10^5$ km/s. If the sun is approximately $1{\cdot}53 \times 10^8$ km from the earth, how long does its light take to reach us?

6 Fractions

$$A = \{\tfrac{3}{4}, \tfrac{6}{8}, \tfrac{9}{12}, \tfrac{12}{16}, \tfrac{15}{20} \ldots\}.$$
$$B = \{\tfrac{5}{6}, \tfrac{10}{12}, \tfrac{15}{18}, \tfrac{20}{24}, \tfrac{25}{30} \ldots\}.$$

Set A contains fractions which are equivalent to *three-quarters*.

Set B contains fractions which are equivalent to *five-sixths*.

We can use equivalent fractions to compare size, add and subtract. For example:

(i) Since $\frac{10}{12} > \frac{9}{12}$, we know that $\frac{5}{6} > \frac{3}{4}$.

(ii) $\frac{3}{4}+\frac{5}{6}=\frac{9}{12}+\frac{10}{12}$
$=\frac{19}{12}$
$=1\frac{7}{12}$.

(iii) $3\frac{3}{4}-1\frac{5}{6}=3\frac{9}{12}-1\frac{10}{12}$
$=2\frac{21}{12}-1\frac{10}{12}$
$=1\frac{11}{12}$.

In multiplying fractions, we make use of the rule:

$$\frac{a}{b}\times\frac{c}{d}=\frac{a\times c}{b\times d}.$$

Here are some examples:

(i) $\frac{2}{3}\times\frac{1}{5}=\frac{2}{15}$;

(ii) $\frac{4}{9}\times\frac{3}{4}=\frac{12}{36}$
$=\frac{1}{3}$;

(iii) $4\frac{1}{2}\times 1\frac{1}{4}=\frac{9}{2}\times\frac{5}{4}$
$=\frac{45}{8}$
$=5\frac{5}{8}$.

Since $\frac{4}{3}\times\frac{3}{4}=1$, we say that $\frac{4}{3}$ is the inverse, under multiplication, of $\frac{3}{4}$. We make use of inverses in dividing one fraction by another. For example:

$\frac{2}{5}\div\frac{3}{4}=\frac{2}{5}\times\frac{4}{3}$ (division by $\frac{3}{4}$ has the same effect as multiplication by $\frac{4}{3}$)
$=\frac{8}{15}$.

Here are two examples of division:

(i) $\frac{1}{6}\div\frac{3}{8}=\frac{1}{6}\times\frac{8}{3}$
$=\frac{8}{18}$
$=\frac{4}{9}$;

(ii) $2\frac{1}{4}\div 2\frac{1}{10}=\frac{9}{4}\div\frac{21}{10}$
$=\frac{9}{4}\times\frac{10}{21}$
$=\frac{90}{84}$
$=1\frac{6}{84}$
$=1\frac{1}{14}$.

When a fraction is converted to decimal form by a division there are two possibilities:

(i) the division 'stops' and the answer is a *terminating* decimal. For example:

$$8\,\underline{|3{\cdot}000}\quad 0{\cdot}375;\qquad \tfrac{3}{8}=0{\cdot}375.$$

(ii) the division 'goes on for ever' and the answer is a *recurring* decimal. For example:

$$6\,\underline{|5{\cdot}000}\quad 0{\cdot}833\ldots;\qquad \tfrac{5}{6}=0{\cdot}8\dot{3}.$$

The following method can be adapted to convert any recurring decimal to its fractional form:

$$0{\cdot}317\;317\;317\;\ldots=0{\cdot}\dot{3}1\dot{7}$$

Let $x=0{\cdot}\dot{3}1\dot{7}$

then $1000x=317{\cdot}\dot{3}1\dot{7}$.

Subtracting $999x=317$,

and so $x=\frac{317}{999}$.

Exercise F

1 Calculate:

(*a*) $\frac{1}{5}+\frac{1}{15}$; (*b*) $5\frac{1}{4}-3\frac{5}{8}$; (*c*) $\frac{1}{2}-\frac{1}{3}+\frac{1}{4}$; (*d*) $\frac{3}{4}\times\frac{2}{9}$;

(*e*) $3\frac{1}{2}\times 5\frac{1}{3}$; (*f*) $\frac{2}{7}\div\frac{3}{5}$; (*g*) $\dfrac{1\frac{1}{2}}{2\frac{1}{4}}$.

2 Express as decimals:

(*a*) $\frac{16}{25}$; (*b*) $\frac{11}{12}$; (*c*) $\frac{1}{11}$; (*d*) $\frac{5}{7}$.

3 If $\frac{4}{5}<\frac{n}{60}<\frac{5}{6}$, find the value of n, where n is an integer.

4 When a certain number is added to both the top and bottom numbers of the fraction $\frac{32}{81}$ it becomes equivalent to $\frac{1}{2}$. What is the number?

5 Write as fractions:

(*a*) 0·3; (*b*) $0{\cdot}\dot{3}$; (*c*) $0{\cdot}\dot{2}\dot{3}$; (*d*) $0{\cdot}2\dot{3}$.

6 Which of the following are true and which are false?

(*a*) $\frac{91}{143}$ and $\frac{7}{11}$ are equivalent fractions; (*b*) $(1\frac{1}{2})^2=2\frac{1}{4}$;
(*c*) $(2\frac{1}{8})^2=4\frac{1}{64}$; (*d*) $\sqrt{\frac{9}{16}}<\frac{9}{16}$.

7 Evaluate $(1\frac{1}{2}+1\frac{1}{3})\div 1\frac{3}{4}$.

7 Percentages

15 % means $\frac{15}{100}$, 15 % of 200 means $\frac{15}{100}\times 200$. So 15 % $=\frac{3}{20}$ and 15 % of 200 = 30.

In general, expressing a fraction as a percentage or vice versa can be carried out as follows:

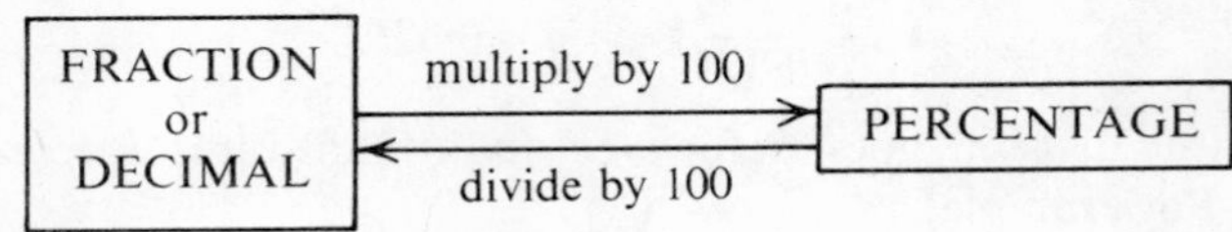

If the price, £X, of a transistor radio is increased by 20 % giving a new price of £30, then

$$\tfrac{120}{100}\times X=30$$

and
$$X=\tfrac{100}{120}\times 30.$$

If the mass, Y grams, of a bar of chocolate is reduced by 5 % giving a new mass of 80 g, then

$$\tfrac{95}{100}\times Y=80$$

and
$$Y=\tfrac{100}{95}\times 80.$$

Exercise G

1 Express as fractions in their simplest forms:

(*a*) 9 %; (*b*) 40 %; (*c*) 64 %; (*d*) $12\frac{1}{2}$ %; (*e*) $87\frac{1}{2}$ %; (*f*) $8\frac{1}{3}$ %.

2 (*a*) What is 5 % of £76? (*b*) Increase £72 by 15 %.

3 In each case, express the first number as a percentage of the second:

(*a*) 1, 5; (*b*) 2, 25; (*c*) 11, 12; (*d*) 7, 16; (*e*) 62, 50.

4 Find the number of which 15 is 30 %.

5 If a price is increased by 20 % to £96, the original price was:

(*a*) £76; (*b*) £116; (*c*) £80; (*d*) none of these.

6 40 % of the population of an island is male. 25 % of the male population and 30 % of the female population are under 21 years of age. What percentage of the total population is under 21?

7 A shop marks an article so as to make a profit of 30 % on the cost price. In a sale, a discount of 10 % was allowed off the marked price. If the article was sold in the sale, state the actual percentage profit made by the shop.

8 Limits of accuracy

5·74 is nearer 5·7 than 5·8. We say that 5·74 = 5·7 to 2 S.F.

5·78 is nearer 5·8 than 5·7. We say that 5·78 = 5·8 to 2 S.F.

5·75 is exactly half way between 5·7 and 5·8 and by convention we 'round up' and say that 5·75 = 5·8 to 2 S.F.

If a number is given as 2·4 correct to 2 S.F. we know that

$$2{\cdot}35 \leqslant \text{actual number} < 2{\cdot}45.$$

5·2958 = 5·30 to 3 S.F.	7·849 = 7·8 to 1 D.P.
= 5·296 to 3 D.P.	7·850 = 7·9 to 1 D.P.
0·0543 = 0·05 to 2 D.P.	49 348 = 49 000 to 2 S.F.
= 0·054 to 2 S.F.	30 062 = 30 100 to 3 S.F.

Percentage error is always obtained from the fraction $\dfrac{\text{error}}{\text{true value}}$.

For example, the percentage error which results from taking 2·34 as 2·3 to 2 S.F. is

$$\frac{0{\cdot}04}{2{\cdot}34} \times 100 \approx 1{\cdot}7\ \%.$$

Whenever you do a calculation, you should consider carefully the accuracy of the answer which you state. It is not possible to give rules which apply to all the situations you are likely to encounter. However, here are two points to bear in mind:

(*a*) The problem which gives rise to the calculation may itself indicate what degree of accuracy is desirable. For example, if grass seed is to be sown on an area of 143 m^2 at the rate of 60 g/m^2, it might be realistic to give the amount of grass seed required as 9 kg (since $150 \times 60 = 9000$).

(*b*) As a general rule, never give an answer which contains more significant figures than the least accurate of the numbers you use. This is particularly important to bear in mind if you are using a computational aid and also if the numbers you are using have been obtained as readings in an experiment.

Exercise H

1 Give the following (i) correct to 2 S.F. and (ii) correct to 2 D.P.

(*a*) 7·428; (*b*) 456·055; (*c*) 0·0964.

2 The dimensions of a rectangular sheet of metal are given as length 9·8 cm, breadth 0·76 cm, each being correct to 2 S.F. What two numbers must be multiplied together to give the greatest possible area of the sheet in cm^2?

3 Write the number 4·997 correct to 3 S.F.

4 The three sides of a triangle are measured correct to the nearest millimetre. The lengths obtained are 6·4 cm, 7·0 cm and 5·8 cm. State the limits between which the perimeter must lie.

5 What are the approximate percentage errors in reading:

(*a*) a temperature difference of about 4 degrees Celsius correct to 0·1°;
(*b*) an increase in mass of about 17 grams correct to 0·01 g?

6 In a doll's house the height of each of the 15 steps in the stairs is 1·3 cm, correct to the nearest 0·1 cm. If the height between the floors is h cm, state the limits between which h must lie.

9 Sets of numbers

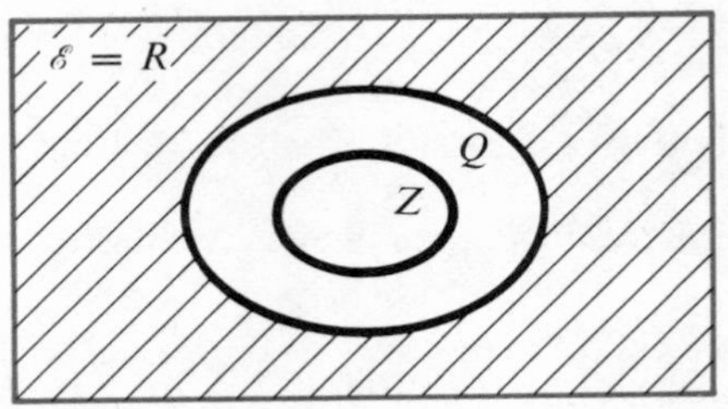

Fig. 1

$N = \{\text{counting numbers}\} = \{1, 2, 3, 4, \ldots\}$
$Z = \{\text{integers}\} = \{\ldots\ ^-3, ^-2, ^-1, 0, ^+1, ^+2, ^+3, \ldots\}$

$Q = \{\text{rational numbers}\} = \left\{\frac{p}{q} : p \in Z, q \in Z, p \text{ and } q \text{ have no common divisor and } q \neq 0\right\}$. The shaded region represents Q' the set of irrational numbers which cannot be expressed exactly as fractions. For example:

$$\pi, \quad \sqrt{2}, \quad \log_{10} 50$$

are all irrational numbers. The universal set, in this case, is R the real numbers.

Exercise I

1 A circle of radius 3·5 cm has a circumference of 7π cm. Which of the following sets does the number 7π belong to:

(*a*) R; (*b*) Q; (*c*) Q'; (*d*) Z?

2 Which of the following numbers are irrational:

(*a*) $\sqrt{6400}$; (*b*) $\sqrt{640}$; (*c*) $\frac{22}{7}$; (*d*) $2 + \sqrt{3}$; (*e*) $\log_{10} 100$?

3 Give an example of two irrational numbers whose sum is rational.

4 Which of the statements are true and which are false?

(*a*) $Z \subset Q$; (*b*) $Q \cup Q' = R$; (*c*) $\sqrt{\frac{1}{2}} \in Q$;
(*d*) $\sqrt{\frac{1}{4}} \in Q$; (*e*) $2^0 \in N$; (*f*) $0{\cdot}\dot{3} \in Q$.

14 Computation II

1 Aids to computation

When you are using a slide rule or a table of logarithms to carry out a computation it is most important that you should make a rough estimate to check that your answer is reasonable.

Logarithms

The mapping $x \to \log_{10} x$ is the inverse of the mapping $x \to 10^x$.

Using logarithms converts multiplication to addition and division to subtraction.

(*a*) $3{\cdot}25 \times 4{\cdot}19 \approx 13{\cdot}6$
(rough estimate
$3 \times 4 = 12$)

x	$\to \log_{10} x$
3·25	$\to$ 0·512
4·19	$\to$ 0·622 +
13·6	$\leftarrow$ 1·134
10^x	$\leftarrow$ x

(*b*) $53{\cdot}2 \div 19{\cdot}7 \approx 2{\cdot}71$
(rough estimate
$60 \div 20 = 3$)

x	$\to \log_{10} x$
53·2	$\to$ 1·726
19·7	$\to$ 1·294 −
2·71	$\leftarrow$ 0·432
10^x	$\leftarrow$ x

When using logarithms for computation involving large and small numbers it is advisable to write the numbers in standard index form first.

The slide rule

Your slide rule is an addition rule with a logarithmic scale. For example the numbers 1, 2, 4, 8 are equally spaced on both *A* and *D* scales. Here are the slide-rule settings for some typical calculations:

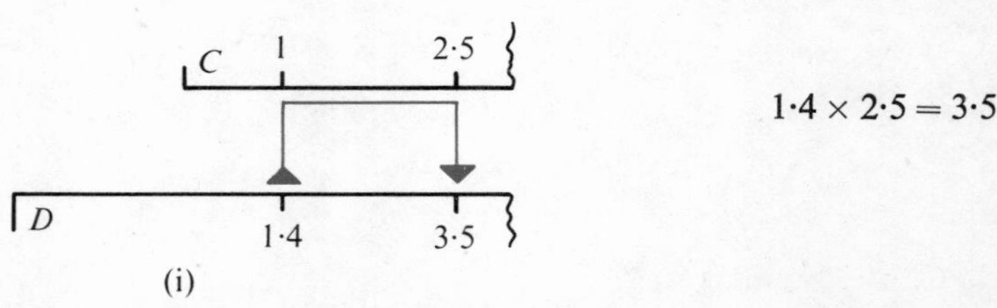

(i)

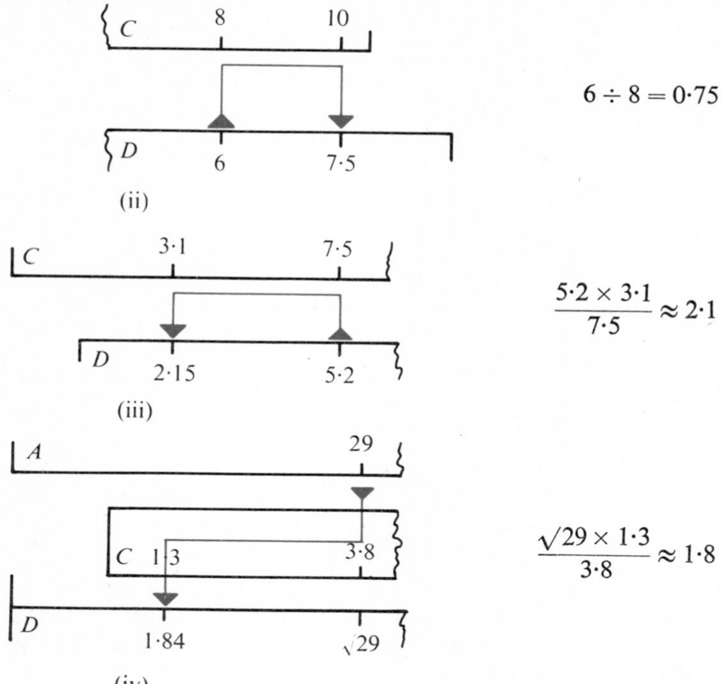

Fig. 1

Exercise A

1 Evaluate, using slide rule and/or tables:

(a) $\dfrac{2{\cdot}1 \times \sqrt{14{\cdot}7}}{1{\cdot}85}$; (b) $\dfrac{\pi(3{\cdot}8)^2}{7{\cdot}1}$; (c) $\sqrt{(2{\cdot}5 - 0{\cdot}76)}$.

2 Express 97 marks out of 136 as a percentage.

3 If n is an integer such that $n < 7\sqrt{0{\cdot}8} < n+1$, find the value of n.

4 Find $\sqrt{50}$ using tables. Which of the following can be written down without further use of tables:

(a) $\sqrt{50\,000}$; (b) $\sqrt{0{\cdot}5}$; (c) $\sqrt{500}$; (d) $\sqrt{0{\cdot}005}$?

5 Work out $2{\cdot}73 \times 10^4 \times 8{\cdot}45 \times 10^{-7}$ giving your answer in standard index form.

6 Evaluate:

(a) $7{\cdot}92^3$; (b) $\dfrac{1}{\sqrt{800}}$; (c) $\sqrt{(\sqrt{29} + \sqrt{69})}$.

2 Area

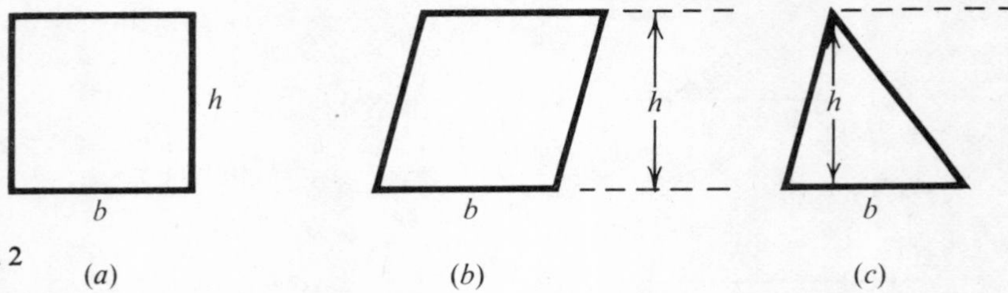

Fig. 2 (a) (b) (c)

Area of rectangle = bh.
Area of parallelogram = bh (area invariant under shear).
Area of triangle = $\frac{1}{2}$ × area of parallelogram = $\frac{1}{2}bh$.

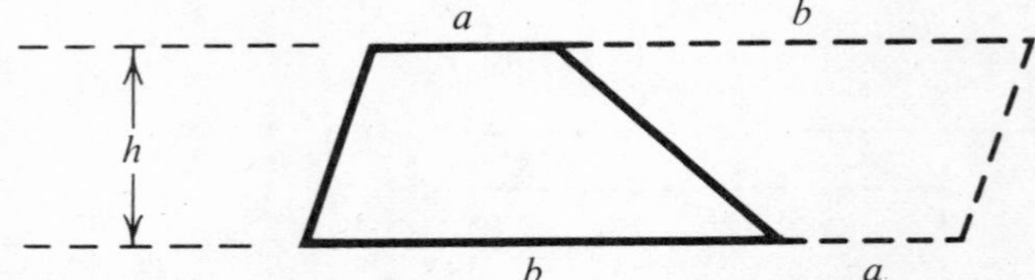

Fig. 3

Area of trapezium (two of them make a parallelogram) = $\frac{1}{2}(a+b)h$.

The area under a graph can be found approximately by dividing it into trapeziums. Figure 4 shows this method used to find the area under the graph of $y = 1/x$ between $x = 1$ and $x = 4$.

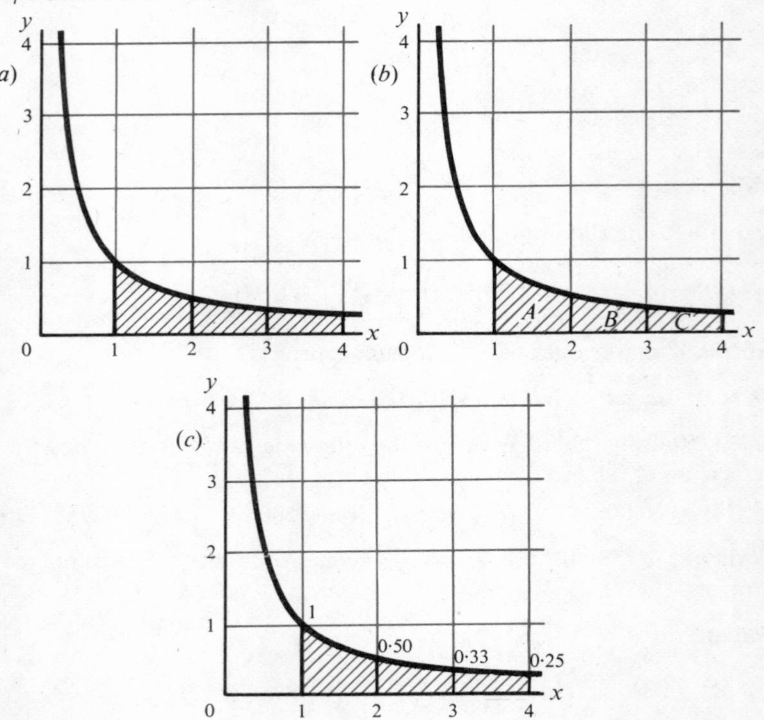

Fig. 4

Area of trapezium $A = \frac{1}{2}(1 + 0{\cdot}50) \times 1 = 0{\cdot}75\ \text{cm}^2$.
Area of trapezium $B = \frac{1}{2}(0{\cdot}50 + 0{\cdot}33) \times 1 = 0{\cdot}42\ \text{cm}^2$.
Area of trapezium $C = \frac{1}{2}(0{\cdot}33 + 0{\cdot}25) \times 1 = 0{\cdot}29\ \text{cm}^2$.

The area of the region shaded red is $1{\cdot}46\ \text{cm}^2$ and this provides an estimate of the area shaded black. We could obtain a better estimate by using, say, six trapeziums each of width 0·5 cm.

Exercise B

1 $ABCD$ is a rectangle in which $AB = 7$ cm and $AD = 10{\cdot}5$ cm. P and Q lie on AB and AD respectively, with $AP = 4$ cm and $AQ = 6$ cm. Calculate the area of the triangle CPQ.

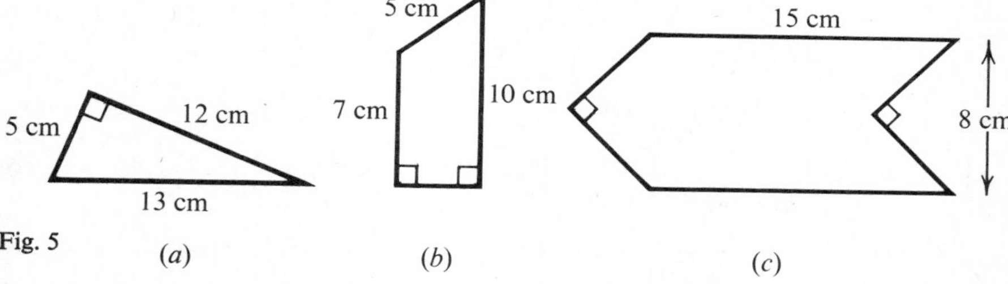

Fig. 5

2 Find the areas of the shapes in Figure 5.

3 A parallelogram with sides 12 cm and 6 cm long has one pair of opposite sides 8 cm apart. What is the distance between the other pair of opposite sides?

4 Find the area of the triangle with its vertices at the points with coordinates $(^-1, 1)$, $(5, 1)$ and $(8, 8)$.

5 A rhombus has diagonals 15 cm and 10 cm long respectively. What is its area?

6 Calculate approximately the shaded area under the graph of $y = x^2$ in Figure 6. Use four trapeziums and say whether your estimate is too large or too small.

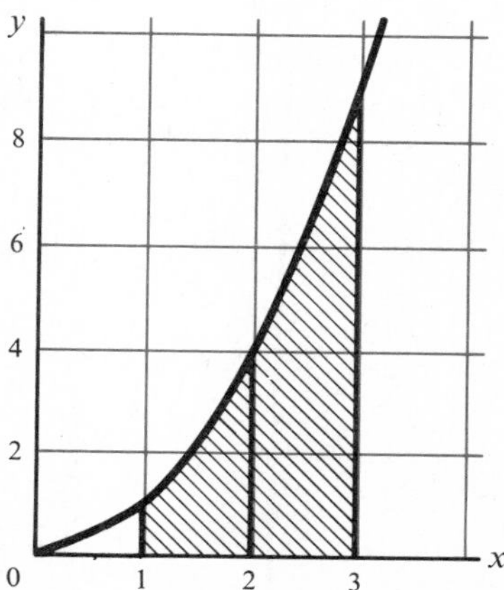

Fig. 6

3 The circle

For a circle of radius R,

$$\text{circumference} = 2\pi R \text{ and area} = \pi R^2,$$

(π is a number without units which cannot be written down exactly as either a terminating or recurring decimal. Its value correct to 4 D.P. is 3·1416).

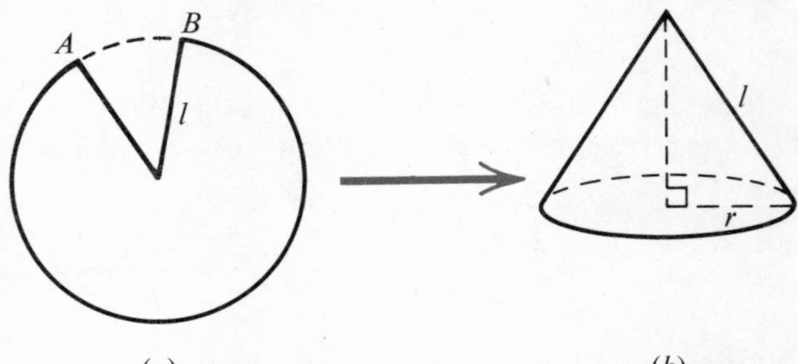

Fig. 7 (a) (b)

If the corners A, B of the sector in Figure 7(a) are brought together the curved surface of a cone is produced (Figure 7(b)).

$$\begin{aligned}\text{Area of curved surface} &= \pi \times \text{base radius} \times \text{slant height of cone}\\ &= \pi rl.\\ \textit{Total} \text{ surface area of a solid cone} &= \pi rl + \pi r^2\\ &= \pi r(l+r).\end{aligned}$$

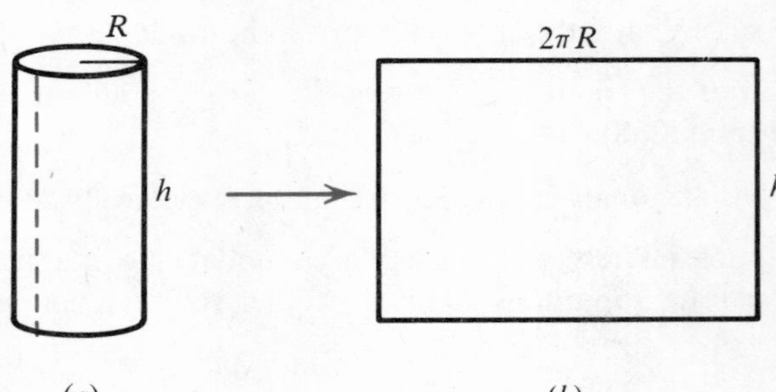

Fig. 8 (a) (b)

If the surface of the cylinder in Figure 8(a) is cut along the dotted line and flattened, the rectangle of Figure 8(b) is produced.

$$\begin{aligned}\text{Area of curved surface of cylinder} &= 2\pi Rh.\\ \textit{Total} \text{ surface area (including ends)} &= 2\pi Rh + 2\pi R^2\\ &= 2\pi R(h+R).\end{aligned}$$

The surface area of a sphere of radius R (see Figure 9) $= 4\pi R^2$.

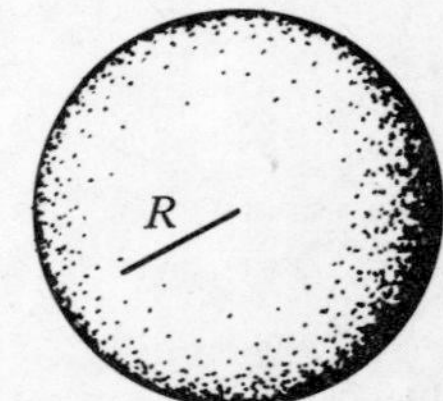

Fig. 9

Exercise C

1 If the circumference of a circle is increased from 10π cm to 12π cm, by how much is the radius increased?

2 The area between two concentric circles of radii 5 cm and 4·5 cm is $p\pi$ cm^2. Find the value of p.

3 A 400 m running track is to have two parallel straights each 80 m long and two semi-circular ends. What should the radius of each semi-circle be? Find the area enclosed by the track.

4 A 120° sector of a circle of radius 24 cm is folded over to form a circular cone so that its arc becomes the circumference of the base. What is the radius of the base circle?

5 In Figure 10, $ABCD$ is a square of side 10 cm. The boundary of the shaded shape is formed by two quarter circles and a semi-circle all of radius 5 cm. Find (i) the perimeter and (ii) the area of the shaded portion.

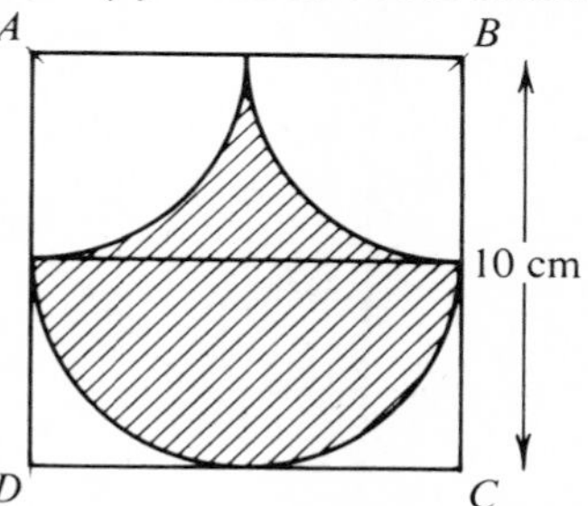

Fig. 10

6 Find the total surface area of a piece of magnesium in the shape of a cylinder 7·4 cm long with a diameter of 5 mm.

7 The curved surface area of a cylinder of height 10 cm is equal to its base area. Find the radius of the cylinder.

8 (*a*) What is the total surface area of a solid hemisphere of radius 5 cm?
(*b*) What is the radius of a sphere which has a surface area of 100 cm^2?

4 Volume

The volume of any prism = cross-section area × length

$$= Al.$$

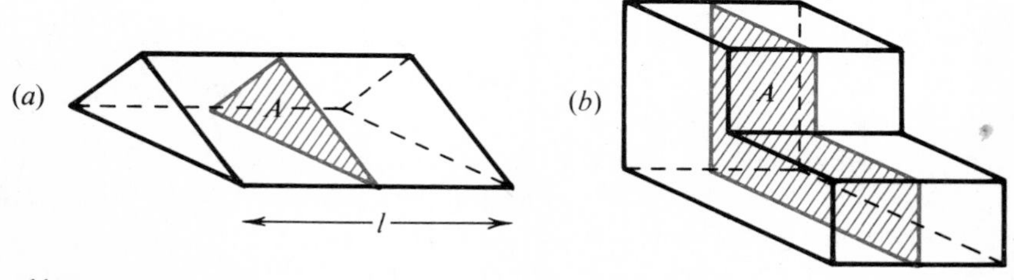

Fig. 11

The cuboid and cylinder are special cases.

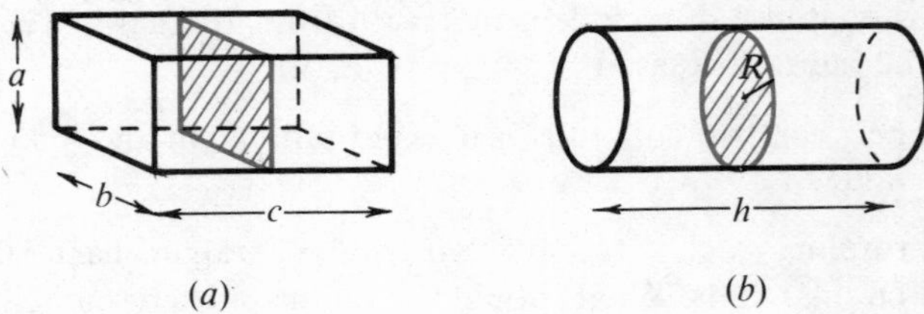

Fig. 12

Volume of cuboid $= a \times b \times c$. Volume of cylinder $= \pi R^2 h$.

$$\text{The volume of any pyramid} = \tfrac{1}{3}(\text{base area} \times \text{height}) = \tfrac{1}{3} \times B \times h.$$

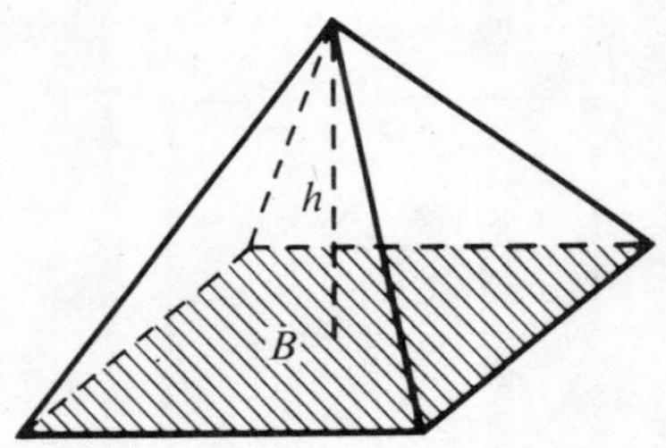

Fig. 13

The cone is a special case.

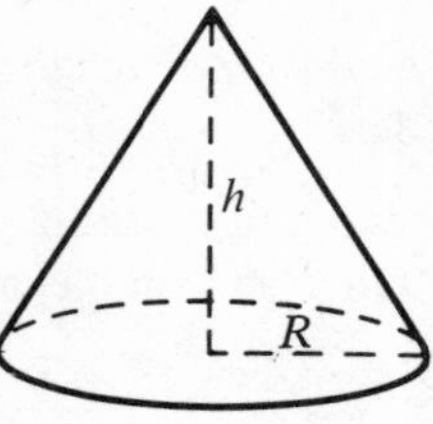

Fig. 14

Volume of cone $= \frac{1}{3}\pi R^2 h$.
Volume of sphere of radius $R = \frac{4}{3}\pi R^3$.

Exercise D

1 The base of an open wooden box has external measurements 52 cm by 40 cm and it is 25 cm high. The bottom and sides are 1 cm thick. Find the capacity of the box and the volume of wood used to make it.

2 A cube has a volume of 167 cm³. Find, correct to 2 S.F., the length of its edge.

3 Figure 15 is a sketch of the vertical cross section of a swimming bath. If the bath is 11 m wide, what is the maximum amount of water it will hold?

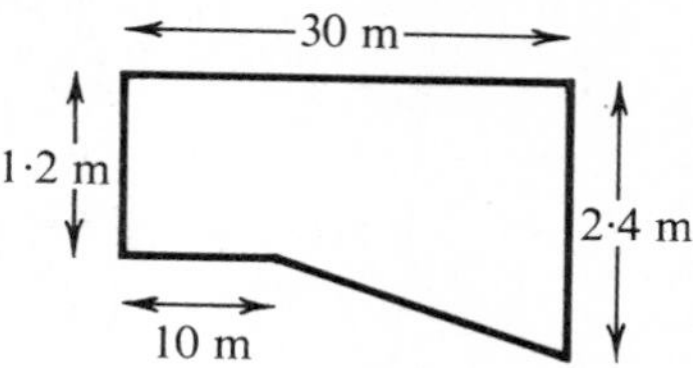

Fig. 15

4 A certain type of biscuit is made by cutting cylinders of radius 4 cm from dough 0·5 cm thick. How many complete biscuits would you expect to make from a lump of dough of volume 50 cm^3?

5 Figure 16 shows the net of a pyramid with a square base. Calculate the volume of the completed solid shape?

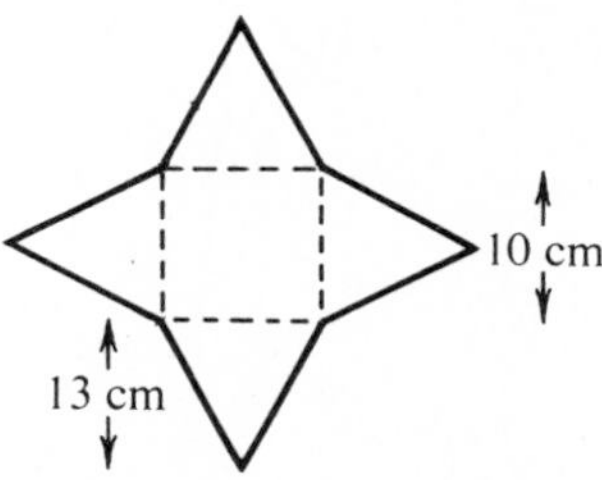

Fig. 16

6 Find the volume of:

(*a*) a cone of radius 2·5 cm and height 5·4 cm;
(*b*) a sphere of radius 3·5 cm.

7 A cone has a volume of 600 cm^3 and a height of 15 cm. Find:

(*a*) the area of its base;
(*b*) the radius of its base.

5 Ratio and proportion

Ratio compares quantities measured in the same units and has no units itself. For example, the ratio of £8 to £0.50 is:

(*a*) (working in £) $8:\frac{1}{2} = 16:1$;
(*b*) (working in p) $800:50 = 16:1$ also.

When a ratio is expressed in the form $1:n$, n is called the *scale factor*. In the above example $n = \frac{1}{16}$.

With each ratio there is an associated fraction. If the ratio is 2:3, the associated fraction is $\frac{2}{3}$.

Areas and volumes of similar shapes. If the *linear* scale factor is n, then the *area* scale factor is n^2. For example:

$$\text{ratio of dimensions} = 1:3$$
$$\text{ratio of areas} = 4:36$$
$$= 1:9 \text{ (or } 1:3^2\text{)}.$$

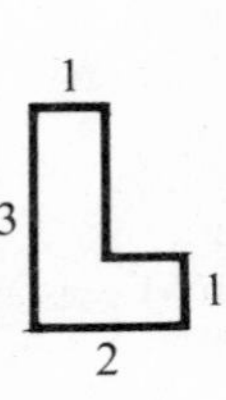

Area = 4 cm²

Area = 36 cm²

Fig. 17

If the *linear* scale factor is n, then the *volume* scale factor is n^3. For example:

$$\text{ratio of dimensions} = 1:2$$
$$\text{ratio of volumes} = 3:24$$
$$= 1:8 \text{ (or } 1:2^3\text{)}.$$

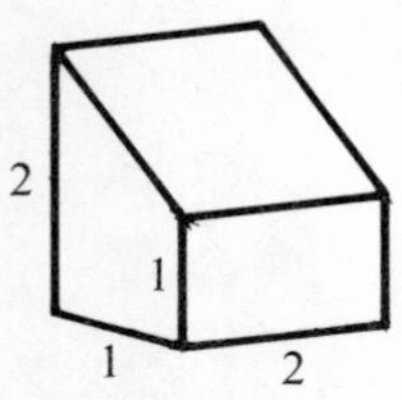

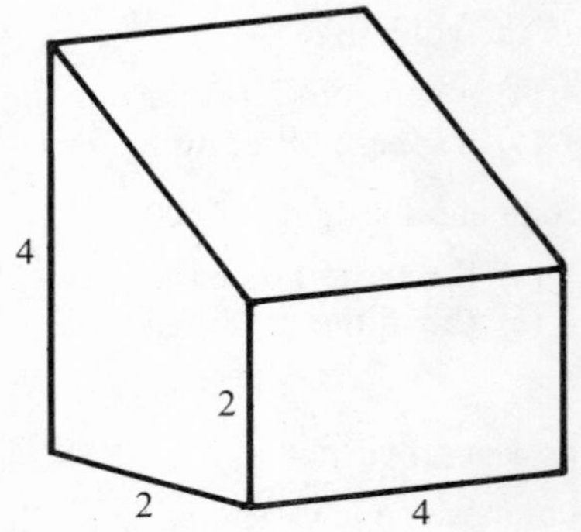

Volume = 3 cm³

Volume = 24 cm³

Fig. 18

Proportion. Since $\frac{2}{6} = \frac{3}{9} = \frac{5}{15} = \frac{8}{24}$, the ordered sets,

$$A = \{2, 3, 5, 8\} \quad \text{and} \quad B = \{6, 9, 15, 24\}$$

are proportional with scale factor 3.

Proportional sets can be displayed like this:

$$\begin{array}{ccccccccc} A=\{ & 2 & \xrightarrow{\times\frac{3}{2}} & 3 & \xrightarrow{\times\frac{5}{3}} & 5 & \xrightarrow{\times\frac{8}{5}} & 8 & \} \\ & \downarrow\times 3 & & \downarrow\times 3 & & \downarrow\times 3 & & \downarrow\times 3 & \\ B=\{ & 6 & \xrightarrow{\times\frac{3}{2}} & 9 & \xrightarrow{\times\frac{5}{3}} & 15 & \xrightarrow{\times\frac{8}{5}} & 24 & \} \end{array}$$

The *multipliers* $\frac{3}{2}$, $\frac{5}{3}$, $\frac{8}{5}$ are the same for both sets.

We write $M_A = \{\frac{3}{2}, \frac{5}{3}, \frac{8}{5}\}$ and $M_B = \{\frac{3}{2}, \frac{5}{3}, \frac{8}{5}\}$.

Exercise E

1 Express the ratio 64 cm:5 m in the form $1:x$ giving x correct to 2 S.F.

2 In a pie chart showing a family's weekly expenditure, the angle of the sector representing food is 108°. If the total expenditure was £30, how much was spent on food?

3 In a scale model of a building, the ground area is $\frac{1}{100}$th of the actual ground area. What is the ratio of the volume of the building to the volume of the model?

4 If $5x = 11y$, state the ratio of x to y.

5 On a map whose scale is 1:100 000, a lake is represented by a blue region of area 3 cm^2. What is the corresponding area on a map whose scale is 1:50 000?

6 In a will, £600 is left to be shared by three nieces, Anne, Barbara and Claire, in the ratio 4:5:6. How much should each receive?

7 A solid sphere has a surface area of 50 cm^2. What is the surface area of another solid sphere with a radius three times as big? If the first sphere has a mass of 70 g, what is the mass of the second assuming they are made of the same material?

8 A and B are the ordered sets,

$$A = \{1{\cdot}0, 1{\cdot}5, 2{\cdot}1, 3{\cdot}2\} \quad \text{and} \quad B = \{4{\cdot}0, 9{\cdot}0, 17{\cdot}6, 41{\cdot}0\}.$$

(*a*) List the elements of M_A and M_B in decimal form.
(*b*) Square each element of M_A and compare with M_B.
(*c*) What type of rule relates corresponding pairs of elements from A and B?

6 Pythagoras' rule

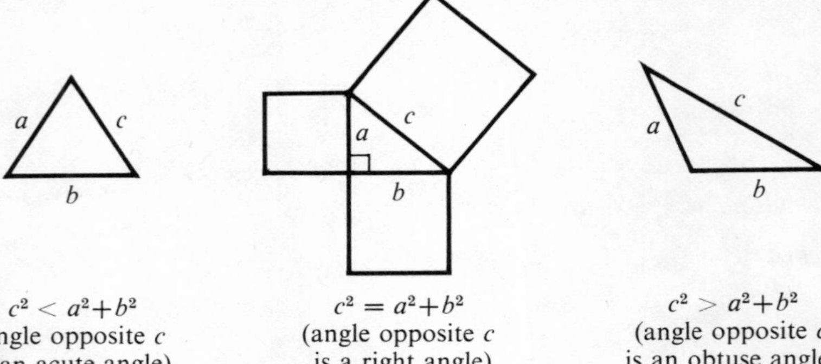

Fig. 19

Some right-angled triangles with sides of whole number length are:

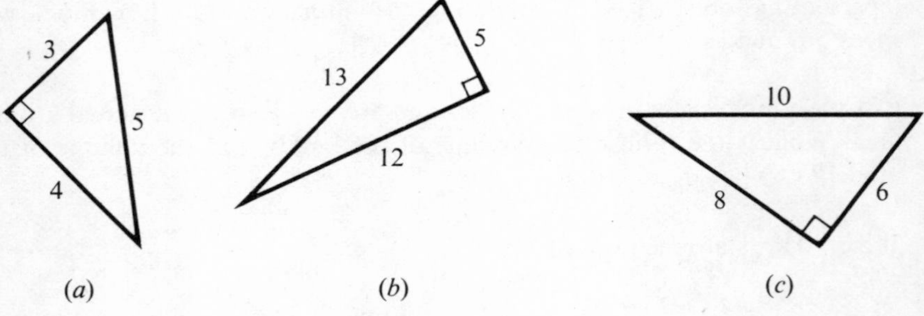

Fig. 20

Notice that (*a*) and (*c*) are similar triangles.

In two dimensions, we can use Pythagoras' rule to find the length of a diagonal of a rectangle.

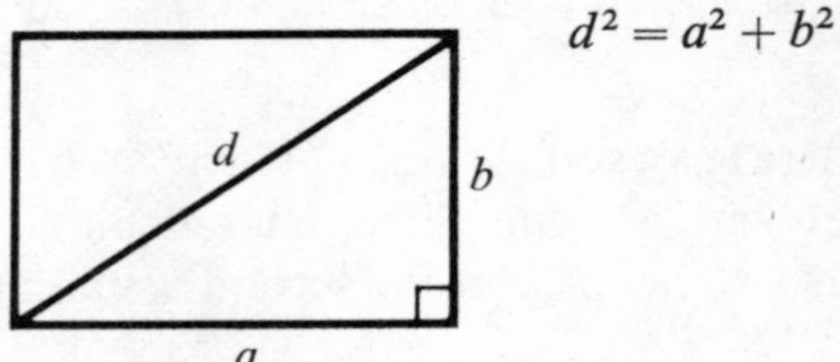

$$d^2 = a^2 + b^2$$

Fig. 21

In three dimensions, we can use Pythagoras' rule to find the length of the diagonal of a cuboid.

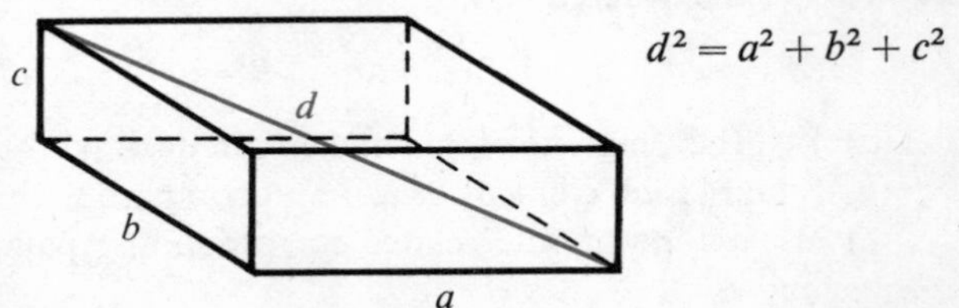

$$d^2 = a^2 + b^2 + c^2$$

Fig. 22

Pythagoras' rule can also be used for calculating the distance between two points on a graph:

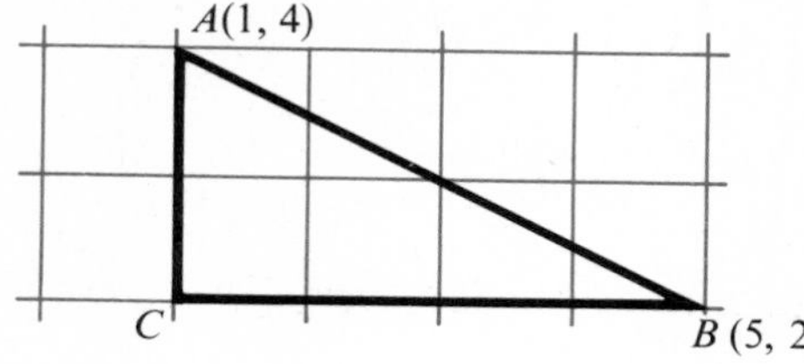

$$AB^2 = AC^2 + CB^2$$
$$= (4-2)^2 + (5-1)^2$$
$$= 2^2 + 4^2$$
$$= 20.$$

So $AB \approx 4{\cdot}47$.

Fig. 23

Exercise F

1 What are the lengths of the sides of the triangle formed by joining points with coordinates $(3, {}^-2)$, $(0, 1)$ and $({}^-1, {}^-1)$? Is the triangle obtuse-angled?

2 A chord of a circle is 30 cm long and 8 cm from the centre. What is the radius of the circle?

3 Find the length of the diagonal of a rectangular box whose dimensions are 4 cm by 3 cm by 2 cm.

4 A circular cone has a base of diameter 15·0 cm and a slant height of 18·0 cm. Calculate its vertical height.

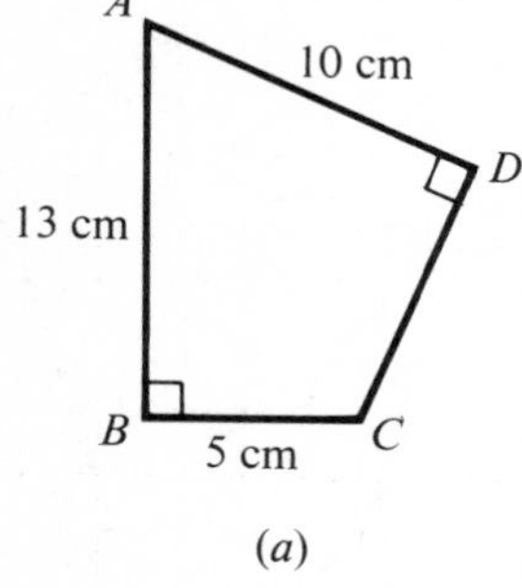

(*a*)

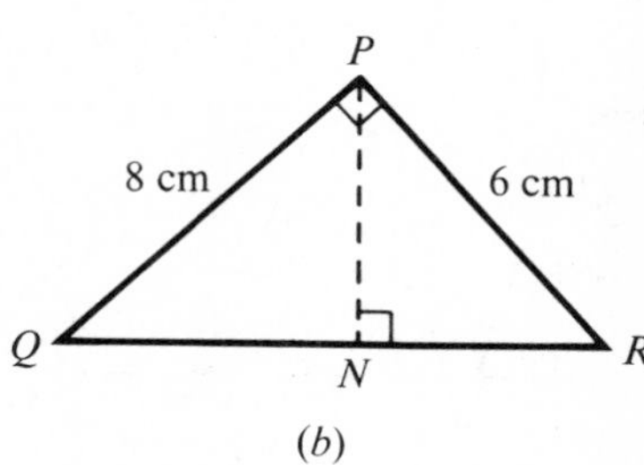

(*b*)

Fig. 24

5 Find the length of CD in Figure 24(*a*).

6 Find the length of PN in Figure 24(*b*).

7 A 20 cm straw is placed in a tumbler 13 cm high and of uniform diameter 7 cm. What is the least possible length of straw which projects above the rim of the tumbler?

8 Find the distance between the points:

(*a*) $(9, {}^-1, 7)$ and $(10, 1, 9)$;
(*b*) $(1, {}^-4, {}^-10)$ and $(6, 2, 20)$.

7 Trigonometry

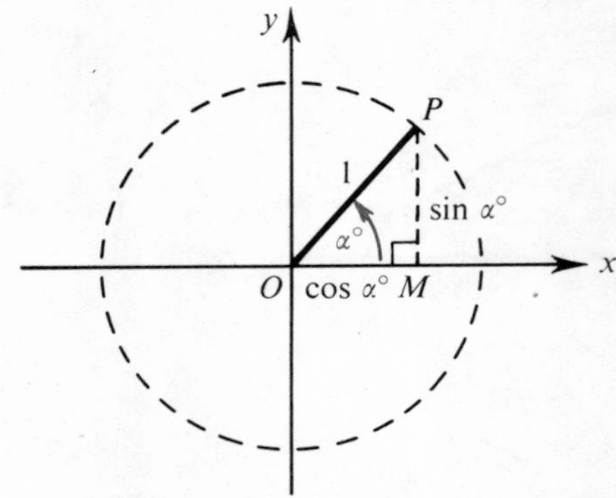

Fig. 25

In Figure 25, OP has unit length and $\cos \alpha°$, $\sin \alpha°$ are defined as follows:

the x-coordinate of P is $\cos \alpha°$;
the y-coordinate of P is $\sin \alpha°$.

Applying Pythagoras' rule to triangle OPM we obtain:

$$\sin^2 \alpha° + \cos^2 \alpha° = 1.$$

(Remember that $\sin^2 \alpha°$ is short for $\sin \alpha° \times \sin \alpha°$.)

With these definitions we can find the sine and cosine of *any* angle.

Figure 26 shows how $\sin \alpha°$ and $\cos \alpha°$ change as α varies.

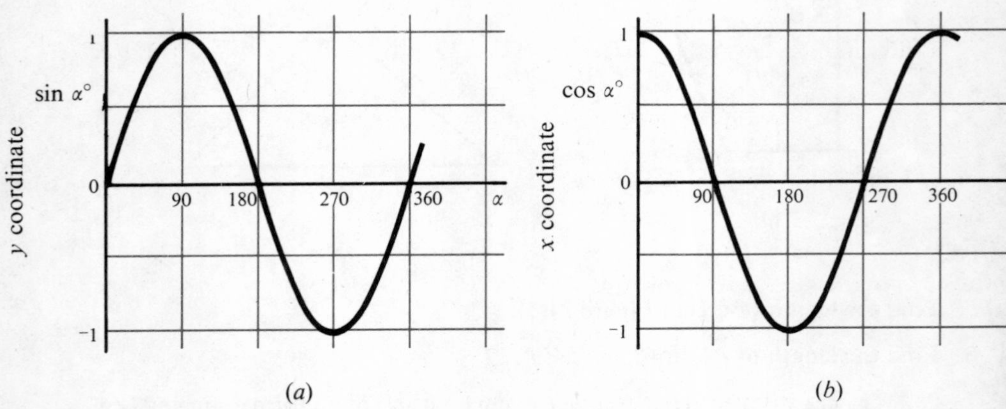

Fig. 26

The tangent of an angle $\alpha°$ is defined as follows:

$$\tan \alpha° = \frac{\sin \alpha°}{\cos \alpha°}.$$

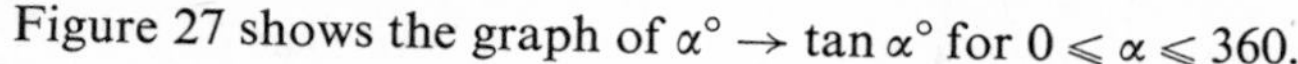

Figure 27 shows the graph of $\alpha° \to \tan \alpha°$ for $0 \leqslant \alpha \leqslant 360$.

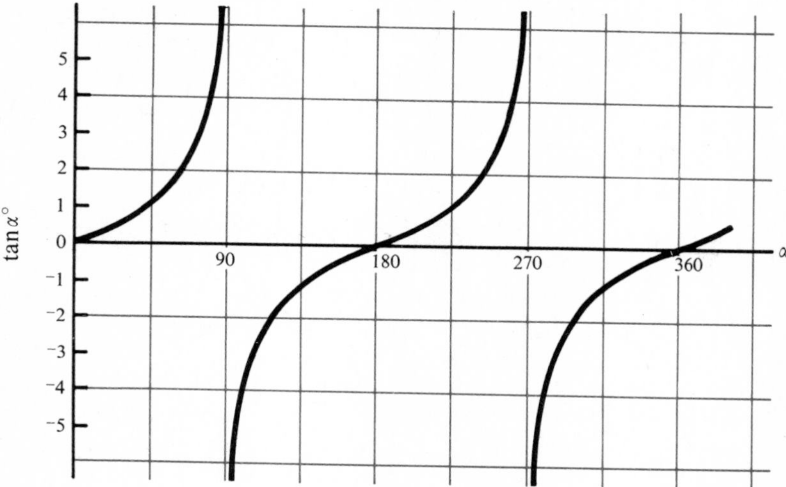

Fig. 27

Useful forms for $\sin \alpha°$, $\cos \alpha°$, $\tan \alpha°$ can be obtained from Figure 28.

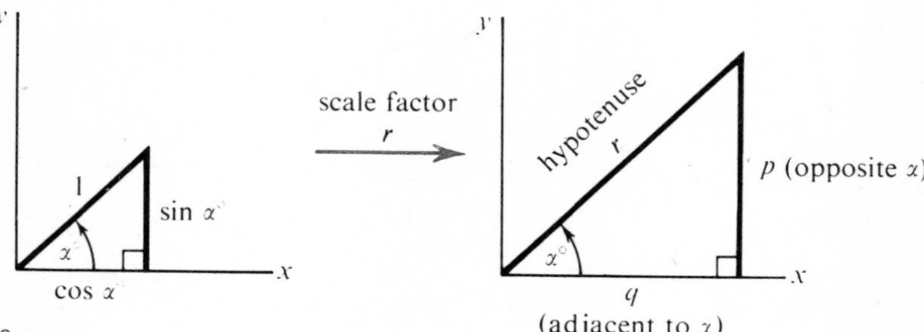

Fig. 28

$$p = r \sin \alpha° \quad \text{or} \quad \frac{p}{r} = \sin \alpha°$$

$$q = r \cos \alpha° \quad \text{or} \quad \frac{q}{r} = \cos \alpha°$$

$$\frac{p}{q} = \frac{r \sin \alpha°}{r \cos \alpha°} \quad \text{or} \quad \frac{p}{q} = \tan \alpha°$$

The following examples make use of these trigonometrical ideas.

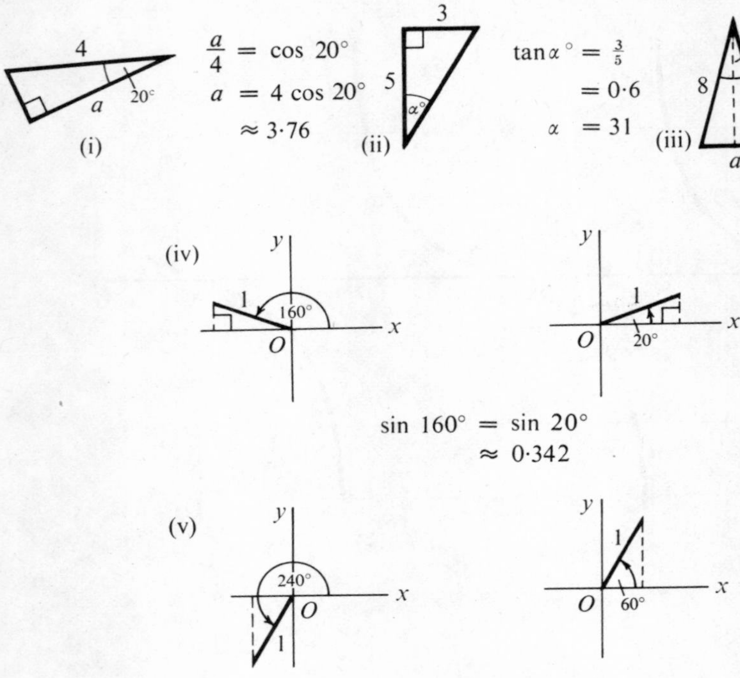

Fig. 29

Exercise G

1 *P* is a point 8·0 cm from the origin *O* and *OP* makes an angle of 30° with the *x*-axis. Calculate the coordinates of *P* correct to 2 S.F.

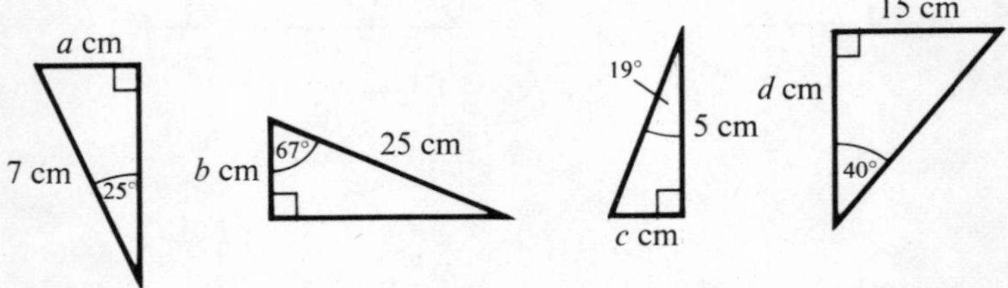

Fig. 30

2 Calculate the lengths indicated by letters in Figure 30.

3 Oakham is 29 km due north of Kettering and 18 km due west of Stamford. Find the bearing of Stamford from Kettering.

4 A building 26 m high casts a shadow 30 m long. What is the angle of elevation of the sun?

5 A rocket ascending vertically is seen by an observer who is 2 km from the launching pad. If the observer estimates its angle of elevation to be 70° at a certain moment, what is the rocket's height at that moment?

6 Use tables to find:

(*a*) sin 210°; (*b*) cos 300°.

7 In triangle ABC, $AB = AC = 6{\cdot}0$ cm and angle $ABC = 50°$. Calculate the length of BC.

8 A parallelogram has sides 12 cm and 8 cm long and one of its angles is 35°. Find the area of the parallelogram.

9 ABC is an equilateral triangle with sides 10 cm long. It rests with the edge AB in contact with the table top, the plane of the triangle making an angle of 55° with the plane of the table. Calculate the height of C above the table.

Revision Exercises

Matrices

1 $\mathbf{P} = \begin{pmatrix} 1 & 2 \\ 0 & ^-3 \\ 3 & 1 \end{pmatrix}$ and $\mathbf{Q} = \begin{pmatrix} 1 & ^-1 & 5 \\ 7 & 4 & ^-9 \end{pmatrix}$.

(*a*) Calculate **PQ**.
(*b*) Calculate **QP**.
(*c*) What do your answers to (*a*) and (*b*) tell you about matrix multiplication?

2 Find values of *a* and *b* so that $\begin{pmatrix} a \\ 13 \end{pmatrix} = \begin{pmatrix} 2 & 0 \\ 1 & b \end{pmatrix}\begin{pmatrix} 4 \\ 3 \end{pmatrix}$.

3 What translation would map (2, 7) onto (5, 3)? Onto what point would the translation map ($^-2$, 1)?

4 Calculate: $(3 \quad 4 \quad ^-2)\begin{pmatrix} 7 & ^-1 & 0 \\ 3 & 2 & ^-1 \\ 4 & 5 & 2 \end{pmatrix}\begin{pmatrix} ^-4 \\ 1 \\ 3 \end{pmatrix}$.

5 A shear is represented by the matrix $\begin{pmatrix} 1 & 0 \\ 2 & 1 \end{pmatrix}$.

(*a*) State the equation of the line of invariant points for this shear.
(*b*) Write down the matrix which represents the inverse transformation.

6 A matrix **A** has *p* rows and *q* columns and can be premultiplied by a matrix **B** with *r* rows and *s* columns.

(*a*) Which of the letters *p*, *q*, *r*, *s* must represent the same number?
(*b*) How many rows and columns does the resulting matrix have?

7 This one-stage route matrix represents a network of roads:

$$\text{From}\ \begin{matrix} A \\ B \\ C \\ D \end{matrix}\overset{\text{To}}{\overset{\begin{matrix} A & B & C & D \end{matrix}}{\begin{pmatrix} 0 & 2 & 1 & 1 \\ 1 & 0 & 1 & 0 \\ 0 & 1 & 0 & 0 \\ 2 & 1 & 1 & 0 \end{pmatrix}}}.$$

(*a*) Draw the network.
(*b*) Is it possible to reach *D* from *C*? If so, describe the route. If not, give a reason.

8 (*a*) Write down the inverse of $\begin{pmatrix} 3 & 2 \\ ^-4 & 1 \end{pmatrix}$.

(*b*) Solve simultaneously the equations $3x + 2y = 6$ and $^-4x + y = 14$.

9 Mr and Mrs Brown have four children – three boys, Arthur, Benjamin and Charles, and one girl Zena. Each child is represented by the initial letter of his or her name.

(*a*) Describe the relation illustrated in Figure 1 in the form '*x* is . . . *y*', where *x* and *y* belong to the set $\{A, B, C, Z\}$.

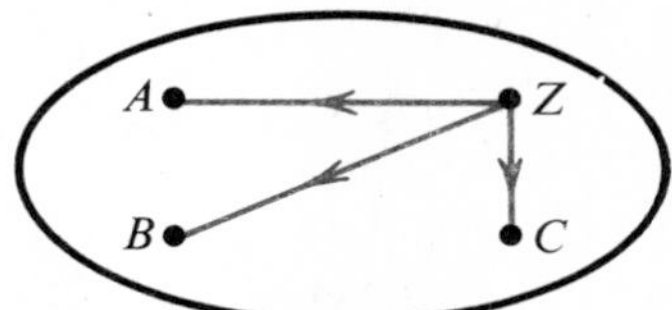

Fig. 1

(*b*) Draw a mapping diagram to show the relation '$x \to$ brother of x' on $\{A, B, C, Z\}$.

(*c*) Write down the matrix **R** which represents the relation of part (*b*).

10 $\mathbf{P} = \begin{pmatrix} 1 & 0 \\ 0 & 4 \end{pmatrix}$, $\mathbf{Q} = \begin{pmatrix} ^-1 & 2 \\ 0 & ^-3 \end{pmatrix}$ and $\mathbf{R} = \begin{pmatrix} 2 & ^-5 \\ ^-1 & 0 \end{pmatrix}$.

(*a*) Evaluate **P**(**QR**).

(*b*) Evaluate (**PQ**)**R**.

(*c*) Which property of the multiplication of matrices is illustrated by the answers to (*a*) and (*b*)?

11 A 2×2 matrix transforms the vector

$$\begin{pmatrix} 1 \\ 0 \end{pmatrix} \text{ to } \begin{pmatrix} 5 \\ 2 \end{pmatrix}$$

and the vector

$$\begin{pmatrix} 0 \\ 1 \end{pmatrix} \text{ to } \begin{pmatrix} ^-1 \\ 4 \end{pmatrix}.$$

(*a*) State the matrix for the transformation.

(*b*) Give the image of $\begin{pmatrix} ^-2 \\ 3 \end{pmatrix}$ under this transformation.

(*c*) Find the vector which is mapped to $\begin{pmatrix} 7 \\ 5 \end{pmatrix}$ by the transformation.

12 **A** and **B** are 2×2 matrices and **0** is the zero matrix $\begin{pmatrix} 0 & 0 \\ 0 & 0 \end{pmatrix}$.

Which of the following statements are true and which are false:

(*a*) $\mathbf{AB} = \mathbf{0} \Leftrightarrow \mathbf{A} = \mathbf{0}$ and $\mathbf{B} = \mathbf{0}$;

(*b*) $\mathbf{A} = \mathbf{0} \Rightarrow \mathbf{AB} = \mathbf{0}$;

(*c*) $\mathbf{AB} = \mathbf{0} \Leftrightarrow \mathbf{A} = \mathbf{0}$ or $\mathbf{B} = \mathbf{0}$;

(*d*) $\mathbf{A}^2 = \mathbf{0} \Rightarrow \mathbf{A} = \mathbf{0}$?

13 Draw a diagram to show the image of the unit square *OIAJ* under a rotation about the origin through an angle of $\theta°$. Find, in terms of $\sin\theta°$ and $\cos\theta°$, the coordinates of the images of *I* and *J* and hence write down the matrix for this transformation.

14 Three lorries *A*, *B* and *C* carry crates of beer, cider, lemonade and orangeade.

Lorry *A* carries 60 crates of beer.

Lorry *B* carries 20 crates of beer, 25 crates of cider and 20 crates of lemonade.

Lorry *C* carries 30 crates of cider, 10 crates of lemonade and 25 crates of orangeade.

(*a*) Write this information as a 3×4 matrix.

(*b*) Each crate of beer weighs 12 kg, each crate of cider weighs 10 kg and each crate of lemonade and orangeade weighs 9 kg. Express these weights as a 4×1 matrix.

(*c*) Use your matrices to find the weight in kilogrammes carried by each of the lorries.

15 $\mathbf{M}_1 = \begin{pmatrix} 1 & 0 \\ 0 & 2 \end{pmatrix}$, $\mathbf{M}_2 = \begin{pmatrix} 2 & 2 \\ 1 & 4 \end{pmatrix}$ and the unit square is denoted by *U*. U_1, U_2 are the images of *U* under the transformations represented by $\mathbf{M}_1$, $\mathbf{M}_2$.

(*a*) Show U_1, U_2 on separate sketches, making clear the coordinates of each vertex.

(*b*) Calculate the matrix for the transformation which maps U_1 onto U_2.

16 For simplicity, assume that every day can be described as either fine or wet. If it is fine today, the probability that it will be fine tomorrow is 0·7. If it is wet today, the probability that it will be wet tomorrow is 0·6.

(*a*) Copy and complete the following table, showing the probabilities of tomorrow's weather, given today's:

		Today: Fine	Today: Wet
Tomorrow	Fine	0·7	
	Wet		

(*b*) Calculate the probability that, if Monday is fine, (i) both Tuesday and Wednesday will be fine; (ii) Wednesday will be fine, regardless of Tuesday's weather.

(*c*) **M** is the matrix formed by the entries in the table; **N** is the corresponding matrix for the probabilities of the day-after-tomorrow's weather, given today's. Express **N** in terms of **M**.

17 (*a*) Show that the shear represented by

$$\begin{pmatrix} 1 & 0 \\ ^-2 & 1 \end{pmatrix}$$

maps the point $P(2, 4)$ onto a point P' on the x-axis. If it also maps $Q(^-3, ^-5)$ onto Q', find the coordinates of Q'. Draw a diagram showing the positions of P, Q, P' and Q'.

(*b*) Find the value of b for which the shear represented by

$$\begin{pmatrix} 1 & b \\ 0 & 1 \end{pmatrix}$$

maps $(^-3, 1)$ onto a point M on the y-axis. If it also maps P' onto L, find the coordinates of L and M.

(*c*) State the area of triangle OLM, where O is the origin, and name any other triangles in the diagram with the same area, giving your reasons.

18 Figure 2 shows part of the London underground system. Trains run in both directions round the inner circle and along the Northern Line.

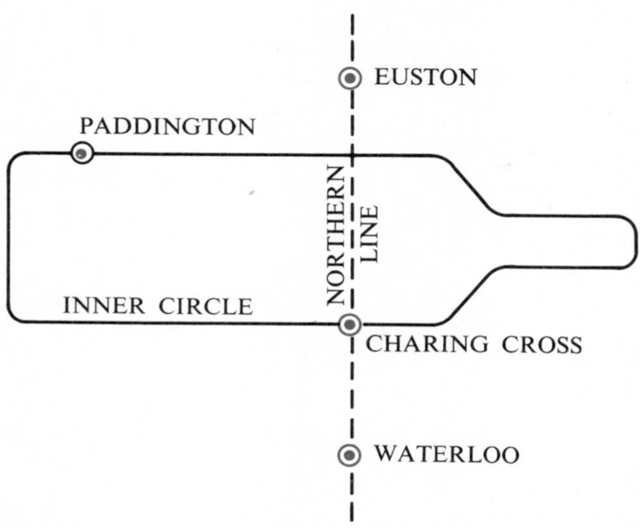

Fig. 2

(*a*) Copy and complete the following table which shows the number of ways of travelling from one of these stations to another without changing trains:

	P	E	CC	W
P	2	0		
E	0	0		
CC	2	1		
W	0			

(*b*) **M** is the matrix formed by the entries in the table. Calculate $\mathbf{M}^2$.

(*c*) (i) List the ways in which one can travel from Waterloo, change trains once (at one of the four named stations), and get out at Charing Cross.

(ii) State the total number of ways in (i) and explain where this number appears in your answer to (*b*).

(*d*) State the number of different ways which start from Charing Cross and end at Charing Cross, with one change of trains.

19 The average number of passengers per day travelling by rail from London to three towns *A*, *B* and *C*, is given by the matrix **P**.

$$\begin{matrix} & A & B & C & \\ \mathbf{P} = & \begin{pmatrix} 170 & 120 & 200 \\ 880 & 480 & 700 \end{pmatrix} & & & \begin{matrix} \text{1st Class} \\ \text{2nd Class} \end{matrix} \end{matrix}$$

The distances are respectively 200 km, 300 km and 400 km.

(*a*) Express these distances as a suitable vector **V** and calculate **PV**.

(*b*) First-class fares are calculated at the rate of 1·4 pence per km, second class at 0·9 pence per km. Express these rates as a suitable vector **F**, calculate **FPV** and say what information is conveyed by the product.

(*c*) It is proposed to abolish the two classes and charge all passengers at the same rate. Find this rate, in pence per km, if the total revenue is to remain the same as before.

20 (*a*) Copy and complete the 8×8 matrix **M** which represents the relation 'is a factor of' on the set of numbers {1, 2, 3, 4, 5, 6, 7, 8}.

$$\mathbf{M} = \begin{matrix} 1 \\ 2 \\ 3 \\ 4 \\ 5 \\ 6 \\ 7 \\ 8 \end{matrix} \overset{\begin{matrix} 1 & 2 & 3 & 4 & 5 & 6 & 7 & 8 \end{matrix}}{\begin{pmatrix} 1 & 1 & 1 & 1 & 1 & 1 & 1 & 1 \\ 0 & 1 & 0 & 1 & 0 & & & \\ 0 & 0 & 1 & 0 & 0 & & & \\ & & & & & & & \\ & & & & & & & \\ & & & & & & & \\ & & & & & & & \\ & & & & & & & \end{pmatrix}}$$

(*b*) What relation does **M**′ represent?

(*c*) Calculate $(1 \quad 1 \quad 1 \quad 1 \quad 1 \quad 1 \quad 1 \quad 1)\,\mathbf{M}$.

(*d*) Explain the meaning of each of the entries in your answer to (*c*).

21 If

$$\mathbf{A} = \begin{pmatrix} 4 & 4 \\ -2 & -1 \end{pmatrix} \quad \text{and} \quad \mathbf{B} = \begin{pmatrix} 2 & 1 \\ -2 & 1 \end{pmatrix},$$

calculate (i) $\mathbf{A}^2 - \mathbf{B}^2$; (ii) $(\mathbf{A} + \mathbf{B})(\mathbf{A} - \mathbf{B})$.

Explain why $\mathbf{A}^2 - \mathbf{B}^2 \neq (\mathbf{A} + \mathbf{B})(\mathbf{A} - \mathbf{B})$.

22 Each of two flasks A and B may contain either of two liquids H and K or a mixture of both. The matrix

$$\begin{pmatrix} p & q \\ r & s \end{pmatrix}$$

represents the state of affairs when A contains p cm³ of H and q cm³ of K, while B contains r cm³ of H and s cm³ of K.

(*a*) For a certain experiment, A contains 120 cm³ of H and B contains 160 cm³ of K. Write down the corresponding matrix **U**.

(*b*) One-third of the contents of A are now poured into B. Write down the matrix **V** representing the new situation and find the matrix **M** such that **MU** = **V**.

(*c*) Interpret $\begin{pmatrix} 1 & \frac{1}{4} \\ 0 & \frac{3}{4} \end{pmatrix}$ as an instruction to perform some similar operation with the flasks.

(*d*) Find the single matrix that gives instructions to perform in succession the operations referred to in (*b*) and (*c*).

23 Describe fully the transformation whose matrix is $\mathbf{R} = \begin{pmatrix} 0{\cdot}6 & {}^-0{\cdot}8 \\ 0{\cdot}8 & 0{\cdot}6 \end{pmatrix}$.

A matrix **X** is given by the relation **XM** = **R**, where $\mathbf{M} = \begin{pmatrix} 1 & 0 \\ 0 & {}^-1 \end{pmatrix}$.

Describe fully the transformations represented by **M** and **X** and evaluate **X**.

Geometry

1 If P is the point (3, 7) write down:

(i) the coordinates of the point obtained by reflecting P in the line $y = x$;

(ii) the coordinates of the point obtained by rotating P through 90° in the positive direction about the origin.

2

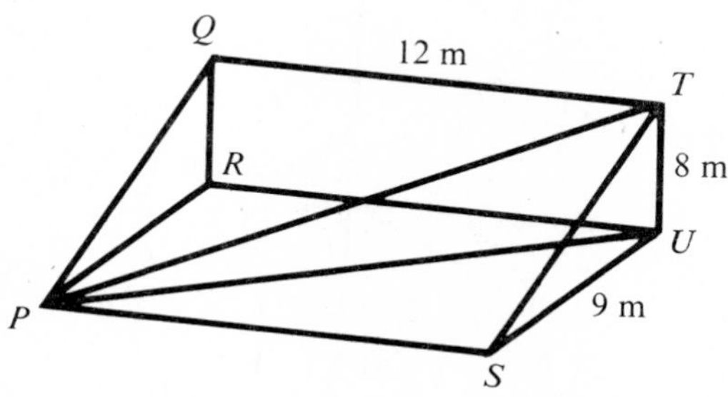

Fig. 3

In Figure 3, $PRUS$ is a rectangle in a horizontal plane and $QTUR$ is a rectangle in a vertical plane. Calculate (i) PU; (ii) PT.

3 Name a solid which is symmetrical about all the planes through a given line.

4 The statement, 'Each external angle of a given regular polygon is 25°', cannot be true. Explain why.

5 In Figure 4, shade in the set of all points, within the triangle ABC, which are nearer to B than to A and whose distance from C is less than CB.

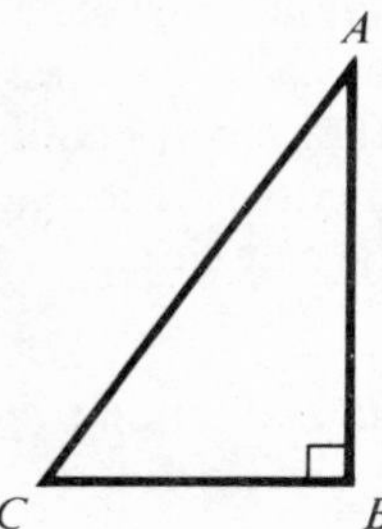

Fig. 4

6 Under a translation **T** the point $P(4, 5)$ maps onto the point $Q(1, 2)$. Under **T** the point Q maps onto the point R. State the coordinates of R.

7 Q is a set of quadrilaterals each of which has p lines of symmetry in its plane.

(i) State all possible values of p.

(ii) Name the type of quadrilateral for which $p > 3$.

8 A regular polygon has angles of 150°. How many sides has it?

9 The top and base of the solid in Figure 5 are parallel squares. The four sloping faces are congruent, and the four sloping edges are equal.

(i) State the number of planes of symmetry.

(ii) Describe, or indicate clearly on the diagram the position of one of these.

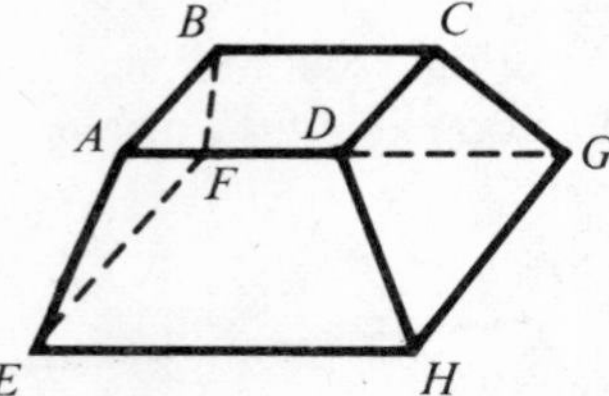

Fig. 5

10 O is the centre of the circle in Figure 6. Find x.

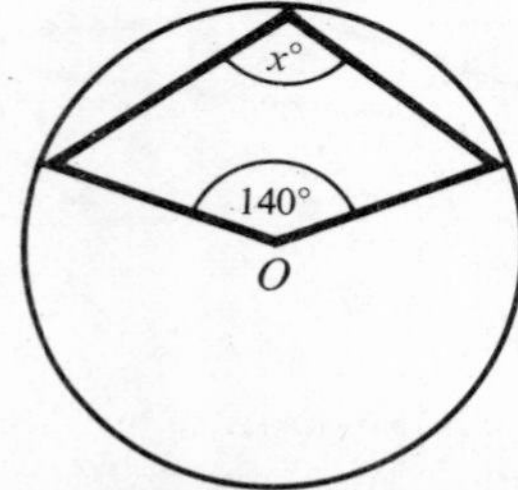

Fig. 6

11 In Figure 7, a square and a regular hexagon are placed together as shown.

(*a*) What is the size of $\angle ABH$?
(*b*) What is the size of $\angle ACH$?
(*c*) How many sides has the regular figure that has $\angle ABH$ as one of its angles?

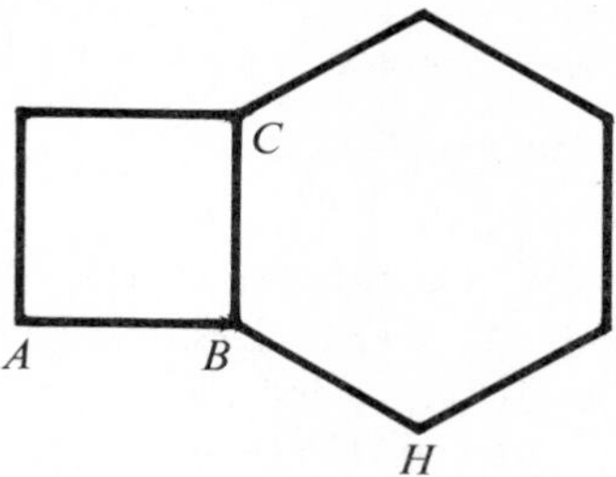

Fig. 7

12 In Figure 8, **P** denotes a positive quarter turn in the plane, about O; **Q** denotes the reflection in the axis Ox.

(i) Sketch the image of the square correctly lettered, under the transformation **QP** (i.e. P followed by Q).
(ii) State a single transformation which is equivalent to **QP**.

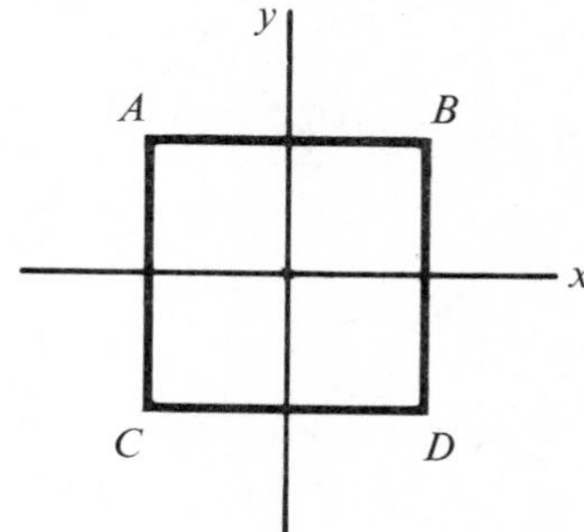

Fig. 8

13 In Figure 9, ABC is an equilateral triangle. P is a bead attached to A by a string of length 3 cm. P moves in the plane of the triangle so that the string is always taut.

If the string cannot cross the triangle, sketch the locus of P.

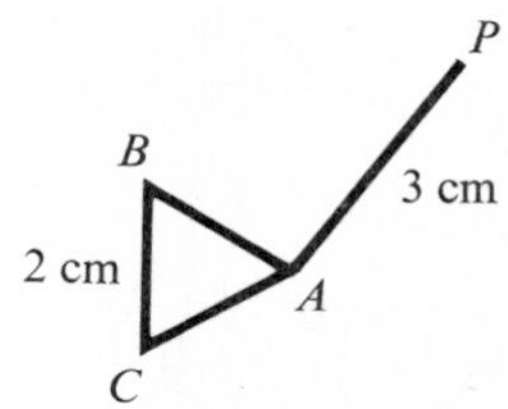

Fig. 9

14 **T** is the translation $\begin{pmatrix} ^-3 \\ 2 \end{pmatrix}$; **R** is a half turn about the origin; **S** is reflection in the x-axis.

(i) Give the image of ($^-3$, 1) under the transformation **RT**.
(ii) Give the image of (2, $^-1$) under **STR**.
(iii) State which pair out of the three transformations given is commutative. If no pair is commutative, say so.

15 A, B, C are three points on a circle, centre O (Figure 10). R is the image of O under reflection in AC.

(i) Explain why $AOCR$ is a rhombus.
(ii) State a relation between the sizes of angles OAC and OAR.
(iii) Q is the image of O under reflection in AB. State a similar relation between angles OAB and OAQ, and hence obtain a relation between angles BAC and QAR.
(iv) If **T** is the translation which maps A onto O name, with reasons, the images of R and Q under **T**.
(v) Deduce a relation between angles BAC and BOC.

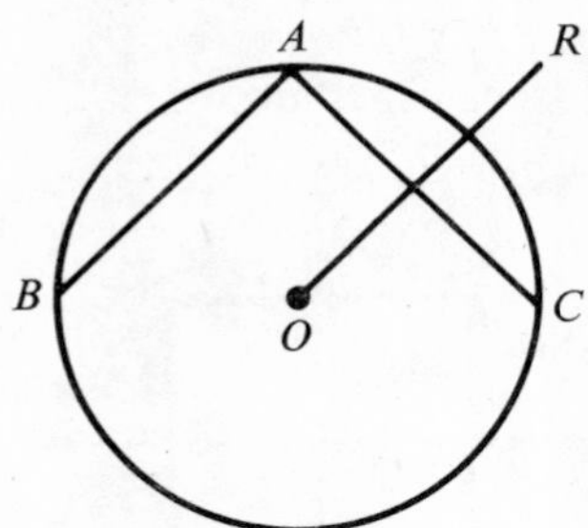

Fig. 10

16 (i) A given translation maps X onto X' and Y onto Y'. Copy and complete the following statements:

XX' is (a) equal to . . .;
(b) parallel to . . . .
XY is (c) . . . to . . .;
(d) . . . to . . . .

(ii) $ABCD$ is a trapezium in which AB is parallel to DC and shorter than DC. The translation which maps A onto D also maps B onto P. Say which of the results (a)–(d) lead you to make each of the following statements:

I $BP = AD$;
II P is on CD;
III $PC = DC - AB$.

(iii) Explain briefly, with a free-hand diagram, how you would use the results given in (ii) to construct such a trapezium given that $AB = 3$ cm, $BC = 2{\cdot}5$ cm, $CD = 5$ cm and $DA = 3{\cdot}5$ cm.

17 PQ is a diameter of a circle. C is the mid-point of one of the semi-circular arcs into which PQ divides the circumference and X is any point on the other semi-circular arc. **R** is the rotation with centre C which maps P onto Q, and $\mathbf{R}(X) = X'$.

(i) State the angle of the rotation **R**.

(ii) Explain why (*a*) PX is perpendicular to QX', (*b*) PX is perpendicular to XQ.

What do these facts show about X, Q, X'?

(iii) Show that $(X'X)^2 = 2(CX)^2$.

18 (i) Taking 2 cm as unit, draw on squared paper axes showing values of x from $^-5$ to $^+2$ and y from $^-1$ to $^+3$. Mark in the points $O(0, 0)$, $A(1, 0)$, $B(1, 1)$, $C(0, 1)$, $P(1, 2)$ and $Q(0, 2)$.

(ii) **S** is the shear with invariant line $y = 0$ which maps $(1, 1)$ onto $(^-1, 1)$. Mark the images $\mathbf{S}(O)$, $\mathbf{S}(A)$, etc., of all the six given points under the shear **S** using the letters O', A', etc. Show the form of the image of the rectangle $OAPQ$ by lightly shading within its outline.

(iii) **T** is the shear with invariant line $y = 1$ which maps $(0, 0)$ onto $(^-2, 0)$. Mark the images $\mathbf{T}(O')$, $\mathbf{T}(A')$, $\mathbf{T}(P')$, $\mathbf{T}(Q')$ of the points O', A', P', Q' under the shear **T** by the letters O'', A'', P'', Q'' and show clearly the outline of the figure which they form.

(iv) State what conclusions you reach about the nature of the transformation **TS**, giving your reasons.

19 (i) Using the theorem of Pythagoras obtain (do not merely state) a formula giving the distance from the origin to the point (x, y, z) in space.

(ii) C is the surface of the sphere whose centre is the origin and whose radius is 7 units. Regarding C as a set of points express it in the form $\{(x, y, z): \quad \}$.

(iii) Identify the locus $D = \{(x, y, 4)\}$.

(iv) Express $C \cap D$ in the form $\{(x, y, 4): \quad \}$ and interpret the result geometrically.

20 In Figure 11, ABE, ACF are the tangents to the circle at B, C and BD is parallel to AF.

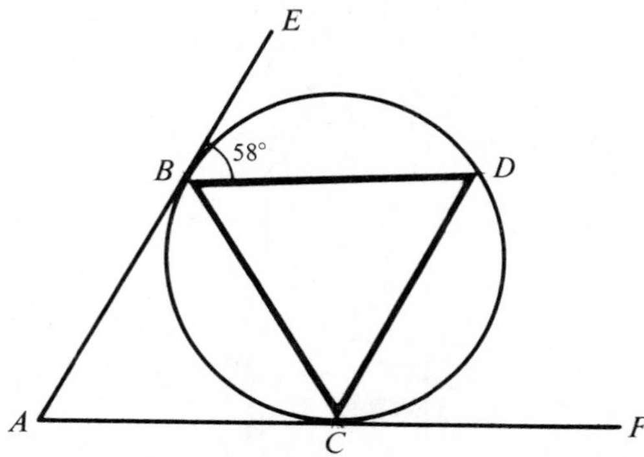

Fig. 11

(i) Use the symmetry properties of the circle to describe (with reasons) the symmetries of triangles ABC, BCD.

(ii) Give the reasons for each step in the argument

$$\begin{aligned}\angle ABC &= \angle ACB\\ &= \angle CBD\\ &= \angle CDB\end{aligned}$$

(iii) It is given that $\angle EBD = 58°$. Name (with reasons) two other angles of 58° in the diagram.

Statistics

1 Of 100 pupils 25 prefer French, 35 prefer German, 40 prefer Spanish. Illustrate with a pie chart and state the sector angles.

2 What number together with 7, 13, 18 and 30 will reduce their mean by 2?

3 If there are 10 million cars, 1 million commercial vehicles and 1 million motor cycles, and if this information is represented in a pie chart, what would be the angle of the sector representing motor cycles?

4 A batsman's average for five matches was 20. His scores in the first four matches were 63, 7, 10, 15 (out each time). What was his score in the last match?

5 The numbers 5, 2, 12, 9 have a mean m. The numbers 5, 2, 12, 9, p have a mean $m - 1$ and median q. Find p and q.

6 Group the following numbers into classes of five, where the first class is from 3 to 7:

26	17	42	28	23	17	39	3	38	10
42	29	40	37	18	6	22	15	15	32
27	30	30	12	40	45	4	26	24	22
44	25	4	10	33	11	38	31	38	23
32	15	12	37	38	37	12	29	12	18
33	17	8	3	7	42	39	6	41	20

7 For the following grouped frequency tables estimate the mean, mode, and median:

(*a*)

x	f
3–5	3
6–8	7
9–11	8
12–14	6
15–17	6

(*b*)

x	f
1–10	6
11–20	9
21–30	13
31–40	12
41–50	10

(*c*)

x	f
0–10	6
10–20	9
20–30	13
30–40	12
40–50	10

(*d*)

x	f
1–15	7
16–30	10
31–45	15
46–60	12
61–90	6

8 Write out ordered sets with the following properties:

	Number of elements	Mean	Median	Mode
(*a*)	5	3	–	–
(*b*)	5	3	4	–
(*c*)	5	–	2	3
(*d*)	5	7	–	5
(*e*)	4	–	6	–
(*f*)	7	4	5	6
(*g*)	6	5·2	–	–
(*h*)	5	3·7	2·6	–

9 A class of seven boys sat for a test paper in which 25 was full marks. They obtained the following marks: 15, 13, 21, 11, 19 ,7, 12. The pass mark was 14. Indicate which of the following statements are true and which are false.

(*a*) The median mark is 14.
(*b*) The mean mark is 14.
(*c*) The probability of a boy, chosen at random, passing the test is 3/7.
(*d*) The probability of a boy, chosen at random, scoring more than 18 is 2/5.

10 From a grouped frequency table of boys heights a bar chart is drawn and is found to be symmetrical.
Which of the following are true, which are false?

(*a*) The mean must be the same as the mode.
(*b*) The mode could be the same as the median.
(*c*) The median must be the same as the mean.
(*d*) The mode could be the same as the mean.

11 The table shows the numbers of goals scored in 43 Football League matches on the same day.

Total number of goals in a match	0	1	2	3	4	5	6	7
Number of matches with this total score	4	10	5	9	7	5	1	2

(i) State the mode of this distribution.
(ii) State the median.
(iii) Calculate the mean number of goals scored per match. (State the answer to the nearest 0·1.)

12 The following table gives the diameters (in mm) of the pearls in three necklaces each of 100 pearls.

Necklace	Mean	Median	Inter-quartile range
A	7·1	7·3	1·6
B	7·4	7·5	1·4
C	7·5	7·3	1·7

On the basis of these figures, which necklace would you choose

(*a*) to have the pearls most evenly matched for size;
(*b*) to have the longest necklace;
(*c*) to be sure of having at least 50 pearls more than 7·4 mm in diameter?

Give the reasons for your choices.

13 The scores of eleven marksmen firing at the same target but with two different rifles were as follows (maximum score, 36):

Rifle *A*	27	24	25	32	29	30	29	27	23	28	26
Rifle *B*	25	35	36	18	14	28	19	32	36	34	21

Find the mean score and the interquartile range for each rifle. Which do you think is the better rifle?

14 In a game, a player can score any number of points between zero and nine, inclusive. In a set of 100 games the score in points and frequency are given in the following table:

Score	0	1	2	3	4	5	6	7	8	9
Frequency	0	1	2	5	10	20	18	25	14	5

Calculate the mean and interquartile range for this set of games.

15 The table shows the distribution of masses of 115 fifteen-year-old boys (the masses being measured to the nearest kilogram):

Mass (kg)	50–54	55–59	60–64	65–69	70–74	75–79	80–84	85–89
Frequency	7	16	20	17	21	19	10	5

(i) What is the modal class?
(ii) Estimate the median as accurately as you can, giving some indication of your method.

16 A man firing at a target, obtains the following scores:

Score	0	1	2	3	4	5
Number of shots	2	8	11	12	6	1

With this score calculate (*a*) the mode, (*b*) the median, (*c*) the mean.

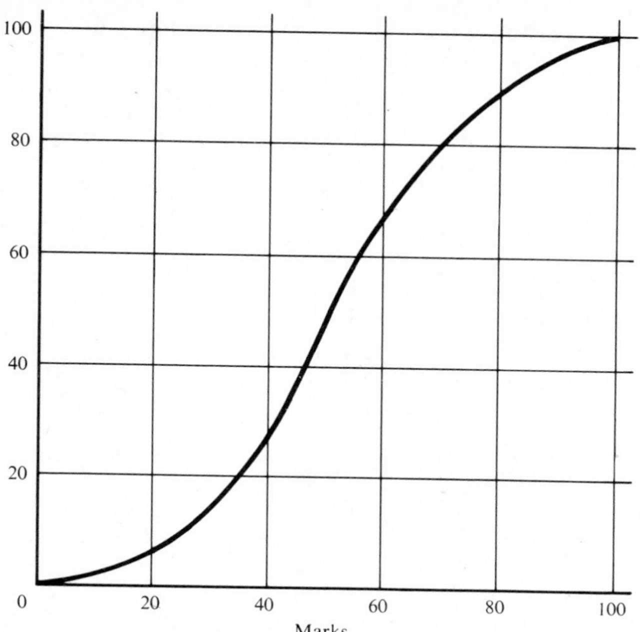

Fig. 12

17 Figure 12 shows a cumulative frequency graph for 100 candidates in an examination.

(i) State how many candidates would fail if the pass mark was 40.

(ii) Give the approximate range of marks in which an increase of 5 in the pass mark would make most difference in the number of passes.

(iii) If 60 candidates pass and they are to be divided into three grades with an equal number of candidates in each grade, state the range of marks which will include the candidates in the middle of these three grades.

18 The surnames of the 800 boys on a school roll vary in length from 3 letters to 11 letters as follows:

No. of letters:	3	4	5	6	7	8	9	10	11
No. of boys:	14	98	173	206	144	97	49	15	4

(*a*) State the mode and the median length of the boys' surnames.

(*b*) Calculate the mean length of the boys' surnames.

(*c*) State, with a reason, what you consider to be the most suitable way of representing these data graphically.

(*d*) If a boy is selected at random from those in this school, what is the probability that his name would contain less than five letters?

19 The following table is a grouped frequency distribution of percentage marks obtained by 350 candidates in an examination:

Mark	Frequency	Mark	Frequency
1–10	17	51–60	55
11–20	31	61–70	46
21–30	35	71–80	39
31–40	43	81–90	20
41–50	51	91–100	13

Draw a cumulative frequency diagram and use it to estimate:

(*a*) the median mark;
(*b*) the upper and lower quartiles;
(*c*) the pass mark in the examination if 60 % of the candidates passed.

20 In an orchard containing 100 young apple trees, half of the trees were treated with a special chemical JXS. The yields from these treated trees are summarized in the frequency table (Table 1).

TABLE 1

Yield (apples per tree)	0–4	5–9	10–14	15–19	20–24	25–29	30–34	35–39	40–44	45–49
Frequency (number of trees)	0	0	0	1	2	7	21	15	3	1

The other 50 trees received no special treatment, and their individual yields were:

27	18	30	13	17	36	24	42	15	22
26	28	43	21	2	11	28	25	34	14
8	30	19	30	23	25	32	16	23	26
20	33	24	17	13	38	6	20	19	39
13	19	27	16	32	25	48	10	24	29

(i) Calculate from Table 1 the mean yield of the treated trees.
(ii) Compile a Table 2, with the same classes as Table 1, for the frequency of yields from untreated trees.
(iii) Illustrate the data of Tables 1 and 2 by frequency diagrams.
(iv) Comment briefly on ways in which JXS appears to have had an effect. (The mean for Table 2 is 23·5 apples per tree.)

Probability

1 State the probability of throwing a number greater than 4 with one throw of an ordinary die.

2 A card is selected at random from a pack of 52 cards. State the probability:

(*a*) that the card is a queen;
(*b*) that the card is a heart;
(*c*) that the card is either a heart or a queen (or both).

3 You have four marbles in your pocket: three red and one blue. You take out three marbles. What is the probability that the marble left in your pocket is red?

4 A box contains 32 chocolates, some with hard centres and some with soft. If a chocolate is taken at random from the box, the probability that it has a hard centre is $\frac{1}{4}$. How many of the chocolates have soft centres?

5 A bag contains beads, some red, some white and some blue. If a bead is drawn at random from the bag, the probability that it is red is $\frac{1}{3}$ and the probability that it is white is $\frac{1}{4}$. What is the probability that it is blue?

6 I have four pieces of paper marked 3, 4, 5 and 9, respectively.

(*a*) How many different pairs of numbers can I select from these four numbers? List these pairs.

(*b*) I choose two of the four pieces at random. What is the probability that the numbers on these pieces will add to more than eight?

7 Figure 13 shows the numbers of netball, table-tennis and badminton players in a club. N = {members who play netball}, T = {members who play table-tennis} and B = {members who play badminton}. Each member plays at least one of the three games.

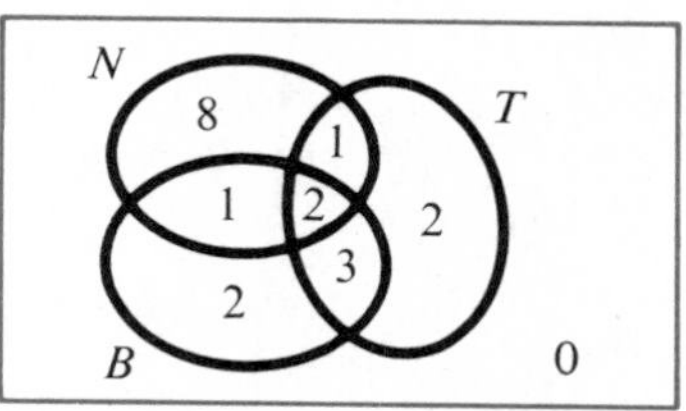

Fig. 13

(*a*) One of the members of the club is chosen at random. Find the probability that this member can play:

(i) badminton;
(ii) table-tennis but not badminton;
(iii) either table-tennis or netball (or both);
(iv) both netball and badminton but not table-tennis.

(*b*) If a netball player is chosen at random, what is the probability that the player can also play table-tennis?

8 Two dice are made in the shape of regular tetrahedra, one with faces numbered 1 to 4, the other with faces numbered 5 to 8 (see Figure 14). They are thrown at the same time and the score is calculated by adding the numbers on the faces on which they land.

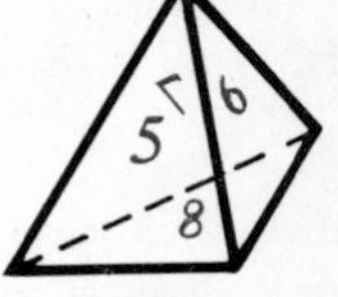

Fig. 14

Copy and complete the following table to give all the possible scores.

		Number on second die			
		5	6	7	8
Number	1	6			
on	2			9	
first	3				
die	4				

(*a*) Which is the most likely score?
(*b*) What is the probability of obtaining the highest possible score?
(*c*) What is the probability of scoring 7?
(*d*) What is the probability of scoring more than 9?
(*e*) What is the probability that the score is a prime number?

9 Two ordinary dice are thrown at the same time. Which of the following statements are true and which are false? If a statement is false, correct it.

(*a*) Throwing two sixes is less probable than throwing two fours.
(*b*) The probability that both dice show the same score is $\frac{1}{6}$.
(*c*) The most likely total score is 6.
(*d*) Throwing an even score and throwing an odd score are equally likely events.

(*Hint*: you may find it helpful to make a table which shows all the possible scores.)

10 Three unbiased coins are tossed. Find the probability that there will be one head and two tails.

11 A bag contains 20 discs: 12 yellow and 8 green. Two discs are taken out in succession and not replaced. By drawing a tree diagram, find the probability of taking two discs of different colours.

12 Assuming that the probability that a baby will be a boy is 0·6, calculate the probabilities that the first two children in a family will be:

(*a*) boys;
(*b*) of the same sex.

13 A bag contains 5 red discs, 4 blue discs and 3 green discs. Single discs are drawn at random out of the bag in succession and not replaced. Find the probability that 2 blues and a red, in that order, are drawn from the bag in in 3 draws.

14 A fairground stallholder earns his living by running a competition in which the public are invited to pay 10p for three table-tennis balls.

A competitor rolls each ball in turn down a chute and they then drop at random into one of three slots. The slots are numbered 1, 2 and 3, and a ball earns a number of points according to the slot it enters. Therefore the highest total a competitor can obtain is 9, $(3+3+3)$ and the lowest total is 3, $(1+1+1)$ (see Figure 15).

Certain total scores are rewarded with a prize worth 20p.

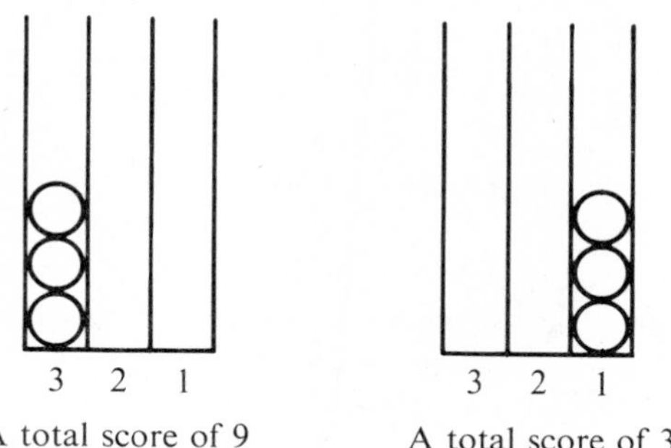

Fig. 15

(*a*) What is the probability that the first ball will enter the '3' slot?

(*b*) $3+3+1$ is one way of scoring seven with the three balls; $3+1+3$ is another. Altogether there are six ways of scoring a total of 7.

By any method, complete a copy of the following table:

Total score	9	8	7	6	5	4	3
Number of ways of scoring	1		6				1

(*c*) What is the probability of obtaining a total score of 6?

(*d*) What is the probability of obtaining a total score which is an even number?

(*e*) If you were the stallholder, for what total scores would you award prizes? Remember that you want to attract customers and also that you hope to make a living. Justify your answer.

15 From a Youth Club consisting of ten girls and twelve boys a Chairman and a Secretary are to be chosen at random. (No one person may hold both offices.)

(*a*) State the probability that a particular girl, Anne, will be chosen as Chairman.

(*b*) Copy and complete the tree diagram in Figure 16 by writing the appropriate probability on each of the branches. *A* stands for Anne being chosen, *B* for a boy, and *G* for a girl other than Anne. (Some probabilities may be zero.)

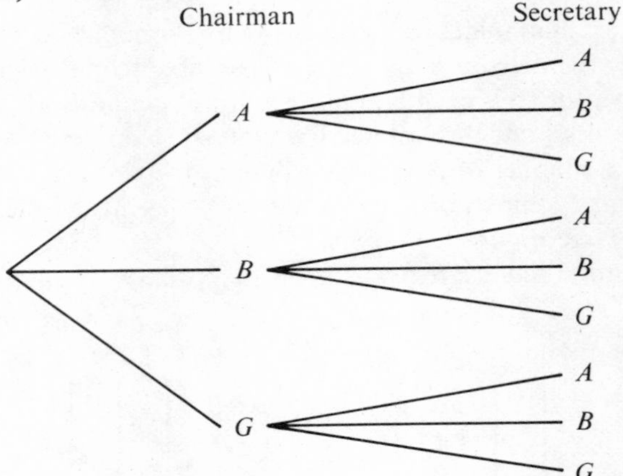

Fig. 16

(*c*) What are the probabilities, before any selections are made, that Anne (i) will be chosen as Secretary; (ii) will not be chosen for either office?

(*d*) A further rule is introduced to the effect that the Chairman and Secretary must be of opposite sexes. If the two names are chosen by drawing pieces of paper out of a hat, what is the probability that the second piece of paper drawn will be invalid under this rule?

16 A school blouse is made in three sizes, small, medium and large. In a class of 24 girls, five can wear the large size. Three of these five can also wear the medium size but none the small size. Altogether 18 of the class can wear the medium size, seven can wear the small size, and all can wear at least one size.

(*a*) Draw a Venn diagram to show this information.

(*b*) If one girl is chosen at random from the class, find the probability that she can wear

(i) a large size blouse;
(ii) a medium size but not a large size;
(iii) both a medium size and a small size.

(*c*) If a girl wearing a small size blouse is chosen, find the probability that she cannot wear a medium size blouse.

17 The pages of a book are numbered from 1 to 300.

(*a*) Copy and complete the following table, showing the frequencies of pages whose numbers have one, two and three digits:

Number of digits	1	2	3
Frequency			

(*b*) A page is chosen at random. Find the probability that its number will contain just two digits.

(*c*) A page is chosen at random from those with three-digit numbers. Find the probability that just one of its digits will be 0.

18 The roads shown in Figure 17 have no signposts. A motorist wishing to travel from A to D has no map, so he has to go by guesswork at each junction. He is twice as likely to go straight on, where that is possible, as he is to turn; and where both left and right turns are possible they are equally likely. He never turns back.

(*a*) State the following probabilities:

(i) that he will go from A to C;
(ii) that, if he goes via C, he will turn towards D.

(*b*) Calculate the probabilities that he will reach D:

(i) by the route ACD;
(ii) by one or other of the two sensible routes;
(iii) by the route $ACBCD$, assuming that he does not learn by his mistakes.

[*Warning*: read this question very carefully.]

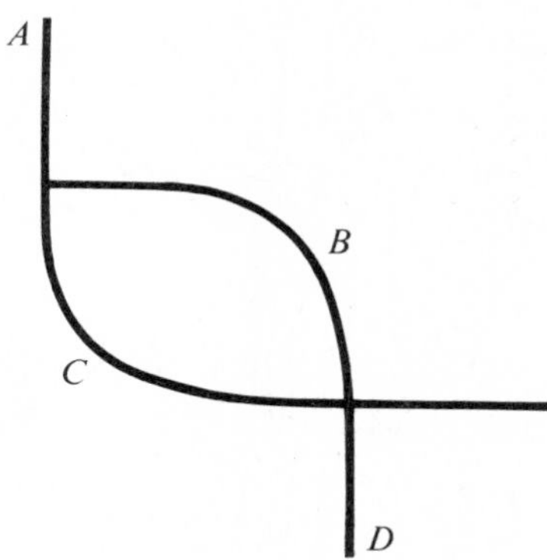

Fig. 17

19 (*a*) Two unbiased coins are tossed at the same time. By drawing a tree diagram, find the probability that they will both be the same (that is, either both heads or both tails).

(*b*) One unbiased coin is tossed together with a 'loaded' coin which has a probability of $\frac{2}{5}$ of landing heads and $\frac{3}{5}$ of landing tails. Find the probability that they will both be the same.

(*c*) Two coins, one with probability p, the other with probability q of landing heads, are tossed at the same time. Obtain an expression for the probability r that they will be the same. By multiplying out show that $2(p-\frac{1}{2})(q-\frac{1}{2})+\frac{1}{2}$ is equal to your expression for r.

If r is equal to $\frac{1}{2}$ state the value of $2(p-\frac{1}{2})(q-\frac{1}{2})$. What does this tell you about the coins?

20 A certain examination question has five parts, each requiring a single answer which must be either right or wrong. It is marked by giving 2 for each correct answer, $^{-}1$ for each incorrect answer, and 0 for each part which is not attempted.

(*a*) Write down the maximum and minimum marks for the whole question.

(*b*) State the possible ways in which 0 can be scored for the question as a whole.

(*c*) Show that just one of the totals between the maximum and the minimum is impossible to attain.

(*d*) If a candidate attempts all parts and is equally likely to be right or wrong in each part, what is the probability that he will score exactly 7?

Algebra

1 If A and B are two subsets of a universal set $\mathscr{E}$, where $n(A) = 16$, $n(B) = 9$ and $n(\mathscr{E}) = 20$, complete the inequalities: $\ldots \leqslant n(A \cap B) \leqslant \ldots$.

2 Write down the next term in the sequence $3\frac{1}{3}, 2\frac{1}{3}, 1\frac{1}{3}, \frac{1}{3}, \ldots$.

3 Express algebraically: x exceeds y by at least 2.

4 Express q in terms of p, given that $p = \sqrt{(q-5)}$.

5 P means 'Take the square of'; Q means 'Divide by 3'.

(i) State the value of $PQ(6)$.
(ii) What is $QP(x)$?

6 The function f is defined by $f: x \rightarrow 3x - 2$. Find:

(*a*) $f(2)$; (*b*) $f^{-1}(7)$; (*c*) $ff^{-1}(10)$.

7 $a * b$ denotes the remainder when the product $a \times b$ is divided by 11.

(i) Evaluate $4 * 5$.
(ii) Find an integer y such that $8 * y = 1$.

8 $f(x)$ is the greatest integer less than $\sqrt{x}$. Evaluate $f(f(120))$.

9 If $A = 3(p+5)$, express p in terms of A.

10 The operation $*$ is defined by $a * b = a(b+2)$.

(i) Solve the equation $6 * 3 = 3 * x$.
(ii) What can be said about a and b if

$$a * b = b * a?$$

(iii) Solve the equation $a * x = a$.
(iv) For what value of a is it impossible to solve the last equation?

11 f and g are the functions

$$f: x \rightarrow \frac{1}{x+3}, \qquad g: x \rightarrow \frac{1}{x} - 3.$$

Write down:

(i) $f(^-1)$; (ii) $g(\frac{1}{2})$; (iii) $gf(^-1)$; (iv) $fg(\frac{1}{6})$

What do your answers suggest about f and g?

12 Give the solution set for each of the following statements taking $\mathscr{E}$ as {Integers}.

(i) $(x+3)^2-(x-3)^2=12x$.
(ii) $(x+3)^2-(x-3)^2=0$.
(iii) $(x+3)^2+(x-3)^2=0$.

13 The operation $*$ is defined by $a*b=ab$, where $a, b \in \{^-2, 0, 2\}$.

(i) Construct the operation table for $*$.
(ii) State the identity element for $*$.
(iii) For each element, state its inverse or write 'None' as the case may be.

$*$	$^-2$	0	2
$^-2$			
0			
2			

(i)

Element	Inverse
$^-2$	
0	
2	

(iii)

14 Consider the sequence 2, 3, 5, 8, 12, ...

(i) Write down the next two terms.
(ii) Write down the first seven terms of the sequence whose nth term is $4+n(n-1)$.
(iii) State the relationship between the two sequences.
(iv) Deduce the 25th term of the first sequence.

15 (i) One factor of x^2+x-2 is $(x-1)$. State the other factor.
(ii) State the two possible values of x for which $x^2+x-2=0$.
(iii) State the value of x for which *both* $x^2+x-2=0$ *and* $x^2+5x+6=0$.

16 (i) The function $p: x \rightarrow \frac{1}{5}(x^3-1)$ can be expressed in the form hgf, where $f: x \rightarrow x^3$ and g and h are other simple functions. Express g and h in the form $g: x \rightarrow \ldots$ and $h: x \rightarrow \ldots$.
(ii) Express fgh in the form $fgh \rightarrow \ldots$.
(iii) Express in similar form the inverses of f, g, h and p.
(iv) If $A=\frac{1}{5}(B^3-1)$, express B in terms of A.

17 In this question $a \oplus b$ means $a+b+2$, and $a*b$ means $\dfrac{a-2b-2}{b+2}$.

(i) Show algebraically that $\oplus$ is associative and commutative.
(ii) (*a*) Evaluate $(25*7)*1$ and $25*(7*1)$.
(*b*) Evaluate $(^-1*0)$ and $(0*{}^-1)$.
What do these results show?

(iii) Find an identity element for $\oplus$ (i.e. a number x such that $a \oplus x = a$).

18 The functions p, q, r are defined by

$$p: x \to x,$$
$$q: x \to \frac{1}{x},$$
$$r: x \to 1 - x.$$

(i) Evaluate $qr(3)$, $qq(3)$, $rq(3)$, $rr(3)$, where, for instance, qr denotes the composite function formed by applying first r and then q.

(ii) From the above definition it follows that

$$rq: x \to 1 - \frac{1}{x}.$$

Express each of the functions qr, qq and rr in the form $x \to \ldots$, simplifying where possible.

(iii) Let s denote qr and t denote rq. Express each of the functions qs and rtq in terms of one or more of the functions p, q, r, giving each answer in the simplest form.

19 (i) State the solution set of the equation

$$(2x + 5)(x^2 - 2) = 0$$

when the universal set is (*a*) real numbers, (*b*) rational numbers, (*c*) integers.

(ii) State the solution sets of the inequality

$$x^2 - 2 \leqslant 0$$

in the same universal sets (*a*), (*b*) and (*c*).

(iii) Make a table showing the sign of $(2x + 5)(x^2 - 2)$ for $x = {}^{-}3, {}^{-}2, {}^{-}1, 0, 1, 2$ and sketch the graph of

$$y = (2x + 5)(x^2 - 2).$$

20 If x and y are numbers, $x * y$ represents the number $x + y + xy$.

(i) Solve the equation

$$x * 7 = (x * 3) + (x * 4).$$

(ii) If $x * y = 0$ express y in terms of x. State the value of x for which no value of y can be found.

(iii) Simplify $2 * (1 * x)$.

Computation

1 Multiply (i) 0·00012 by 0·04 and (ii) 13 000 by 700 giving both answers in standard index form.

2 Draw a Venn diagram to illustrate the sets

$\mathscr{E}$ = {natural numbers greater than 3}
P = {prime numbers}
Q = {even numbers}
T = {multiples of 3}.

Describe the set M which is such that $T \cap Q = M$.

3 See Figure 18:

(*a*) calculate the length of CD; (*b*) state the ratio of BD to AC.

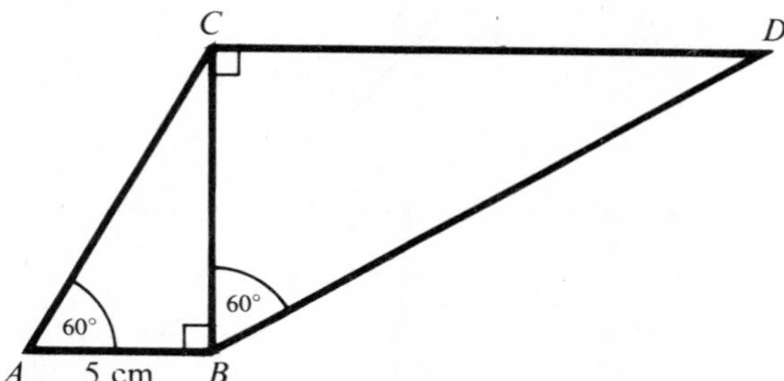

Fig. 18

4 Multiply 235_7 by seven, giving your answer

(i) as a base seven number;
(ii) as a number in the denary scale.

5 Which of the following are true?

(*a*) $0{\cdot}\dot{9} = 1$; (*b*) $0{\cdot}079 = 0{\cdot}08$ (to 2 S.F.);
(*c*) $\frac{5}{13} > \frac{6}{11}$; (*d*) $\sqrt{(31^2 + 27^2)} = 31 + 27$;
(*e*) $\sqrt{(31^2 \times 27^2)} = 31 \times 27$.

6 Work out (i) $\pi \times 6{\cdot}5^2$ and (ii) $\sqrt{\dfrac{50}{\pi}}$. Describe carefully, with reference to circles, what meaning could be given to each answer.

7 In a triangle ABC, $AB = AC$ and $AB:BC = 3:4$. If the perimeter of the triangle is 30 cm, find the length of AB.

8 A discount of 15 % off the marked price is allowed in a sale.

(*a*) What scale factor maps the marked price onto the actual selling price?
(*b*) If the marked price was 20 % above the cost price, give the scale factor which maps the cost price onto the marked price.
(*c*) In these circumstances give the scale factor which maps the cost price onto the actual selling price.

9 (*a*) Evaluate $142\,857 \times 7$.

(*b*) Work out (i) $142 + 857$ and (ii) $14 + 28 + 57$.
(*c*) Evaluate $32\,412_5 \times 12_5$.
(*d*) Establish for the product in (*c*), two results which correspond to those in (*b*) (i) and (*b*) (ii).

10 If $\mathscr{E} = \{\text{natural numbers}\}$, $A = \{\text{odd numbers}\}$ and $B = \{\text{multiples of } 5\}$, describe the sets

(a) $B \cap A'$; (b) $B' \cap A$.

11 Write down from your slide rule or tables a value for $\sqrt{65}$. Say which of the following can be written down without further use of slide rule or tables.

(a) $\sqrt{65\,000}$; (b) $\sqrt{0{\cdot}65}$; (c) $\sqrt{650}$; (d) $\sqrt{0{\cdot}0065}$.

12 Express (i) a and (ii) b, in Figure 19, in terms of h and θ.

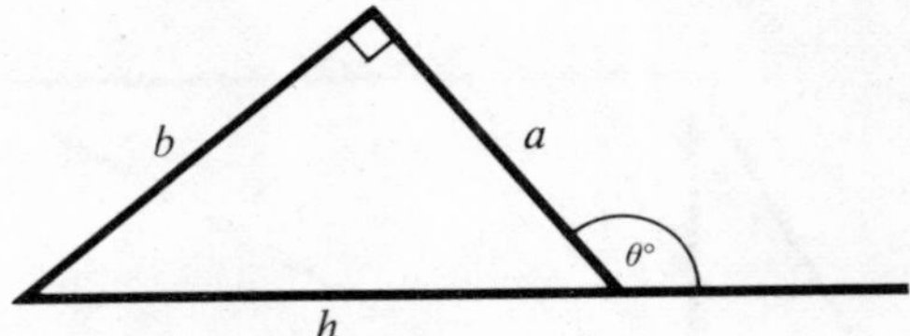

Fig. 19

13 The sector of the circle shown in Figure 20 is folded over to form a circular cone so that the arc AB becomes the circumference of the base. State the radius of the base circle.

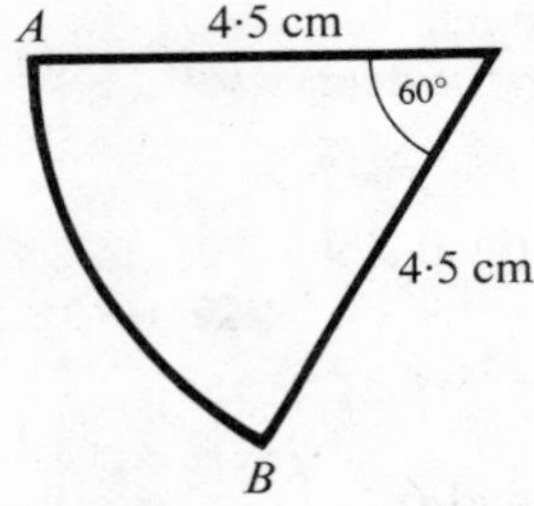

Fig. 20

14 The set of divisors of a natural number x is denoted by D_x and includes 1 and x. For example, $D_{12} = \{1, 2, 3, 4, 6, 12\}$.

(a) List the members of D_{24} and D_{60}.
(b) Find three values of x for which $n(D_x)$ is odd.
(c) What can you say in general about numbers which have an odd number of divisors?
(d) Find the *smallest* number x such that $n(D_x) = 10$.

15 (i) The distances between successive stations on a suburban railway line are given as 7 km, 8 km, 10 km and 6 km and the distance from the first station to the last as d km. All these distances are measured along the track and are given to the nearest kilometre, and you may assume that the platforms are short enough for their lengths to be neglected. State two inequalities satisfied by d, and explain briefly why it is not very likely that d is in fact equal to either the least or the greatest possible value.

(ii) Fares are calculated from the actual track distances at the rate of 1·1p/km and then rounded off to the nearest penny. The fare between the first two stations is 8p. Calculate to the nearest 0·1 km the least possible distance between them.

(iii) The fare for this journey was reduced to 7p and as a result the daily number of passengers, which had been about 2700 went up by about 20 per cent. Calculate approximately the resulting increase in takings.

16 (i) The diameter of a circle has length D. An arc of the circle has length πd. What fraction of the circumference is the arc?

(ii) An isolated hill standing on a plain has the shape of a perfect circular cone. A and B are points at the foot of the hill, A due north and B due south of the summit. The distance from A to B over the summit is 600 m while their direct distance through the hill is 540 m. Sketch the shape you would cut out to make a cardboard model of the hill, showing any necessary measurements. (If you show a distance, this should be given as an actual distance, not scaled.)

There is a path over the slope of the hill following the shortest route from A to B. Show this path on your sketch and find its length either by calculation or by an accurate drawing. Find also the shortest distance from the summit of the hill to the highest point of the path.

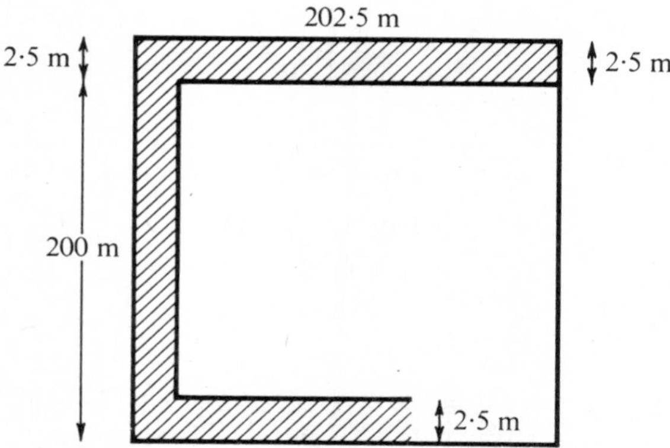

Fig. 21

17 (See Figure 21.)

(i) A square field has side 202·5 m. Find its area exactly.

(ii) A tractor pulls a mower 2·5 m wide once round the field just inside the boundary. Find the area of the grass mown. (Assume that none is missed at the corners.)

(iii) The process is now repeated for the square of grass remaining uncut. Find the area cut this time.

(iv) The tractor continues mowing until it reaches the middle. Show how, by finding in two different ways the total area cut, you could calculate the sum

$$2000 + 1950 + \cdots + 100 + 50.$$

18 A ship S, which is steaming at 12 knots on a bearing of 035° is 15 nautical miles west of a point B at noon. At the same time a ship T is at B, steaming due north. The paths of the ships cross at P.

(i) Show that the distance BP is 21·4 nautical miles correct to three significant figures.

(ii) Find the time at which S reaches P.

(iii) What limits must be imposed on T's speed if she is to be at least 1 nautical mile away when S reaches P?

19 Figure 22 shows a container in which cream is sold. It is 5 cm deep and has a circular top T of diameter 7 cm and a circular base B of diameter 5 cm. If its curved surface were continued downwards it would form a cone with vertex V as shown.

(i) By using a property of similar figures, show that the volume of the whole cone is approximately 2·74 times that of the smaller cone coming up to B.

(ii) Calculate the vertical height of B above V.

(iii) Obtain an expression (which should not be worked out) for the volume of the container.

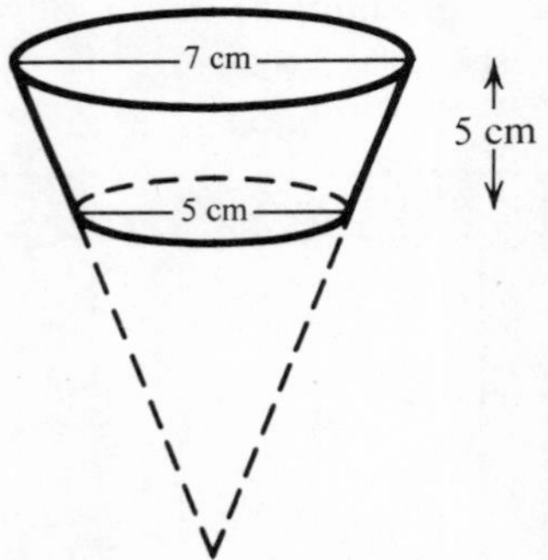

Fig. 22

20 Figure 23 represents a spool of recording tape. The shaded area with internal radius 3 cm and external radius 8 cm is occupied by tape.

(i) Find the shaded area in square centimetres.

(ii) If the tape is 730 metres long, calculate its thickness.

(iii) If half of the tape is unwound from the spool, calculate the external radius of the region still occupied by tape on the spool.

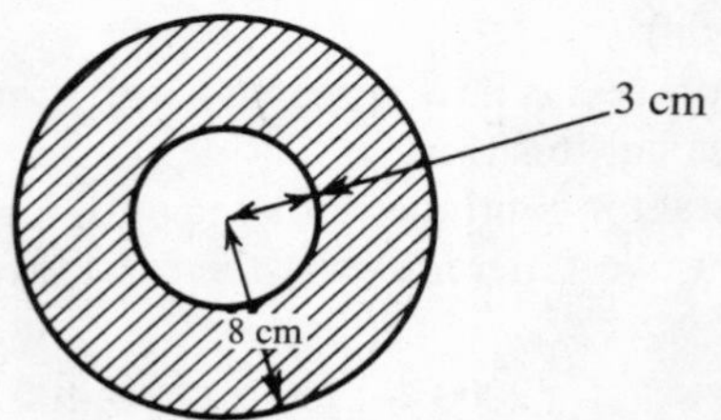

Fig. 23

Miscellaneous

1

1 Find the solution set of the equation $n^2 - 3n - 4 = 0$, where $\mathscr{E} = \{\text{integers}\}$.

2 In a class of 32 boys, 15 have brothers and 14 have sisters, 8 have neither brothers nor sisters. Draw a Venn diagram to illustrate these facts and find:

(*a*) the number in the class who have brothers but no sisters;
(*b*) the number in the class who have both brothers and sisters.

3 The square $OABC$ in which O is (0, 0), $A(2, 0)$, $B(2, 2)$, $C(0, 2)$ is mapped onto $OA'B'C'$ in which A' is (4, 2), $B'(6, 6)$, $C'(2, 4)$. Find the ratio of the area $OA'B'C'$ to area $OABC$.

4 State the number base in which each of the following correct multiplications is worked, and rewrite the calculation in base 10.

(*a*)
$$\begin{array}{r} 60 \\ +\ 33 \\ \hline 2420 \\ \hline \end{array}$$

(*b*)
$$\begin{array}{r} 44 \\ \times\ 30 \\ \hline 2420 \\ \hline \end{array}$$

5 $ABCDEF$ is a regular hexagon, and $ABXY$ is a square, these figures being on opposite sides of AB. Find:

(*a*) the size of angle XBC; (*b*) the size of angle CXY.

6 The numbers c, g, h and v (where $g \neq 0$) are connected by the formula

$$v^2 - gh = g\sqrt{(c^2 + h^2)}.$$

Express h in terms of v, g and c.

7 In the domain of positive real numbers

$$f: x \to \sqrt{x},$$
$$g: x \to \log x,$$
$$h: x \to 1/x.$$

(*a*) Find $f(2{\cdot}89)$.
(*b*) $f(x) = 8{\cdot}86$. Find x correct to 3 S.F.
(*c*) Find $g(7{\cdot}08)$ correct to 3 S.F.
(*d*) Evaluate $f(2{\cdot}89).g(7{\cdot}08).h(10)$ correct to 2 S.F.

8 If
$$\mathbf{a} = \begin{pmatrix} 2 \\ 3 \end{pmatrix}, \quad \mathbf{b} = \begin{pmatrix} ^-3 \\ 2 \end{pmatrix}, \quad \mathbf{c} = \begin{pmatrix} 9 \\ 7 \end{pmatrix},$$
find numbers m and n such that $m\mathbf{a} + n\mathbf{b} = \mathbf{c}$.

9 A five-sided polygon $ABCDE$ is symmetrical about the line $y = x$. A is (2, 0), C is (5, 5) and D is (3, 5). Write down the coordinates of B and of E. Calculate the area of the polygon.

10 A manufacturer of electric cookers sells them to an Electricity Board for £40 each, thus making a profit of 25 % on his cost price. The Electricity Board sells the cookers for £58 each. Find:

(*a*) the manufacturer's cost price;
(*b*) the percentage profit (based on the cost price to the Board) made by the Electricity Board on the sale of these cookers.

2

1 Find the value of $x^2 - y^2$ when $x = 3\frac{3}{8}$, $y = 2\frac{5}{8}$.

2 In a triangle ABC the foot D of the perpendicular from A to BC lies between B and C. The length of AD is 15 cm, of BD is 8 cm, and of AC is 25 cm.

(*a*) Calculate the length of DC.
(*b*) Calculate the length of AB.
(*c*) Calculate the size of angle ABD.
(*d*) Is triangle ABC right-angled? Justify your answer.

3 $R = \{\text{real numbers}\}$; $x \in R$, $y \in R$;
$P = \{(x, y): (y - x)(y - 2) = 0\}$;
$Q = \{(x, y): x^2 - xy = 0\}$.

Draw separate diagrams to illustrate the sets

(*a*) P; (*b*) Q; (*c*) $P \cup Q$; (*d*) $P \cap Q$.

4 A model is made of a house on a scale of 4 cm to 1 m. The height of the actual house is 10 m, its floor area is 120 m^2, and its cubic capacity is 320 m^3. Calculate the height, floor area and cubic capacity of the model in m, m^2 and m^3 respectively.

5 The side of a square is 10 cm correct to the nearest cm. Find the range of possible values of the area of the square.

6 A circle inscribed in a quadrilateral $ABCD$ touches the sides AB, BC, CD, DA at P, Q, R, S respectively.

(*a*) Show that the sum of the lengths of AB and DC is equal to the sum of the lengths of AD and BC.

(*b*) Given that SP is parallel to RQ, illustrate by means of a sketch that the figure is symmetrical about a certain line.

7 (*a*) Factorize $x^2 + 4x + 3$. (*b*) Solve the equation $x^2 + 4x + 3 = 0$.

8 A small swimming pool 10 m long and 3 m wide is filled to a mean depth of 2 m with sea water. The mass of 1 m^3 of sea water is 1030 kg. What is the mass (in kg) of sea water in the pool, correct to 1 S.F.?

9 If $f: x \to 3x$, $g: x \to x + 2$, $h: x \to x^2$, find fg and gf. Express $x \to 3(x + 2)^2$ in terms of f, g and h.

10 Two regular icosahedra each have their twenty faces numbered from 1 to 20. Both rest with one face in contact with a horizontal table. Find the probabilities that:

(*a*) similarly numbered faces are in contact with the table, and
(*b*) both faces in contact with the table are numbered 7.

3

1 Two dice are thrown simultaneously. Write down the probability that

(*a*) the sum of the pips is 12; (*b*) the sum of the pips is 5;
(*c*) the sum of the pips is 1; (*d*) the sum of the pips is 5 or 12.

2 If $$\begin{pmatrix}1 & 0\\ 2 & 1\end{pmatrix}\begin{pmatrix}x\\ y\end{pmatrix}=\begin{pmatrix}3\\ 8\end{pmatrix}, \quad \text{find } \begin{pmatrix}x\\ y\end{pmatrix}.$$

3 Find the solution set for the equation $(x+1)(x-2)(2x-3)=0$:

(*a*) in the set of positive rational numbers;
(*b*) in the set of integers.

4 Given $f\colon x \to 3\sin x^\circ$ has the domain $0 \leqslant x \leqslant 30$, find the range of the function.

5 A and B are two points in a plane, whose distance apart is 8 cm.

$$X=\{P\colon PA=5 \text{ cm}\} \quad \text{and} \quad Y=\{P\colon PB=r \text{ cm}\}.$$

What can be said about r if $X \cap Y$ has:

(*a*) 1 member; (*b*) 2 members; (*c*) no members?

6 In the quadrilateral $ABCD$, angle $BCD=90°$, $AB=AD$, angle $ABD=55°$, angle $DBC=45°$ and $BD=10$ cm. Calculate:

(*a*) the length of BC; (*b*) the length of AB; (*c*) the area of $ABCD$.

7 Consider the following four statements:

(*a*) $(x+1)^2=9$; (*b*) $(x+1)^2=(x-3)(x+5)+16$;
(*c*) $(x+1)^2=x(x+2)+5$; (*d*) $2(x+1)^2+13=x(2x-1)$.

One statement is never true, and another is true for only two values of x. Identify these two statements, and comment on the truth of the other two statements.

8 Three circles with centres P, Q, R have radii 1 cm, 2 cm, 3 cm respectively and touch each other externally. Prove that triangle PQR is right-angled.

9 $ABCD$ is a parallelogram whose diagonals intersect at E. M is the midpoint of DC. If $\mathbf{AB}=\mathbf{x}$ and $\mathbf{AD}=\mathbf{y}$, express in terms of $\mathbf{x}$ and $\mathbf{y}$ the vectors

(*a*) $\mathbf{AE}$; (*b*) $\mathbf{BD}$; (*c*) $\mathbf{MB}$.

10 An American book costs \$3.75. If the official rate of exchange is \$2.40 for £1, how much should the book cost in British money? (Give your answer correct to the nearest penny.)

4

1 Write down the latitude and longitude of the place on the Earth diametrically opposite to 50° N, 20° E.

2 $P = 65_8$, $Q = 302_4$, $R = 144_5$. Arrange the numbers P, Q, R in ascending order of magnitude.

3 Calculate $1110101_2 \div 1101_2$ giving your answer as a binary number.

4 A function is defined over the domain of positive real numbers by the equation $y = \sqrt{x}$. What are the images of $\frac{64}{169}$, 1·44 and 0·09 under the function?

5 Solve the equation $\dfrac{3x}{4} - \dfrac{x+1}{5} = \dfrac{3}{10}$.

6 $\mathbf{OA} = \begin{pmatrix}4\\7\end{pmatrix}$, $\mathbf{OB} = \begin{pmatrix}2\\12\end{pmatrix}$. If C is the mid-point of AB, find $\mathbf{OC}$ as a vector.

7 Make one alteration in each of the following sentences so that each is true.

(*a*) $6 \in \{x: x^2 + 5x - 6 = 0\}$.
(*b*) {squares} ∩ {rectangles} = {parallelograms}.
(*c*) $x = 3 \Leftrightarrow 4x^2 = 36$ for real values of x.

8 Given that $\frac{1}{3}(2x + 5) < 6$ and that $y = 4x - 3$, find in its simplest form the inequality which defines the set of possible values of y.

9 The point $P(8, 6)$ is reflected in the line $y = x$ to give the point Q, and P is also reflected in the line $y = {}^{-}x$ to give the point R. The mid-point of PQ is M and O is the origin. Find:

(*a*) the coordinates of Q; (*b*) the coordinates of R;
(*c*) the length of OM; (*d*) $\cos \angle MOQ$;
(*e*) the size of angle POR.

10 The functions f, g, h are defined on the set of real numbers between 0 and 180 (excluding 90) by the relations:

$$f(x) = \frac{1}{\cos x^\circ},$$
$$g(x) = \sin x^\circ,$$
$$h(x) = x^2.$$

(*a*) For what value of x is $f(x) = 2$?
(*b*) State two values of x such that $g(x) = 0{\cdot}837$.
(*c*) What is the range of g?
(*d*) Give the value of $h(2\sqrt{3}).f(50).g(130)$ correct to 2 S.F.

5

State the letters corresponding to correct answers.

1 The value of $\sqrt{0{\cdot}00038}$ is approximately:

(*a*) 0·02; (*b*) 0·06; (*c*) 0·006; (*d*) 0·002.

2 If the area of a lake is 8 km² and is represented by an area 2 cm² on a map, the representative fraction is

(*a*) $\frac{1}{4}$; (*b*) $\dfrac{1}{400\,000}$; (*c*) $\dfrac{1}{20\,000}$; (*d*) $\dfrac{1}{200\,000}$.

3 The equation of the line through ($^-1$, 4) with gradient 3 is:

(*a*) $x+3y=11$; (*b*) $y+3x=1$; (*c*) $x=3y-13$;
(*d*) $y-3x-7=0$.

4 The matrix $\begin{pmatrix} 1 & 0 \\ ^-1 & 1 \end{pmatrix}$ represents:

(*a*) a rotation; (*b*) a shear;
(*c*) a reflection; (*d*) none of these.

5 Figure 24 shows the speed–time graph of a particle moving for 20 seconds.

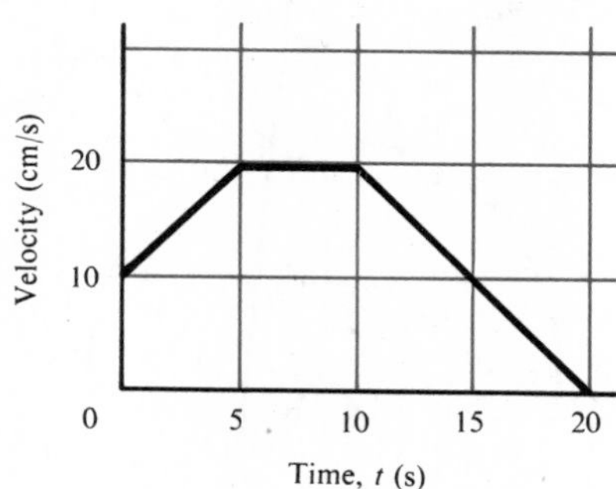

Fig. 24

(*a*) The initial speed is 10 cm/s;

(*b*) the total distance covered is 225 cm;

(*c*) the acceleration for $0<t<5$ is 2 cm/s²;

(*d*) the acceleration for $10<t<20$ is 1 cm/s².

6 AB is a vertical flagpole (see Figure 25). A person walks east to C from its base (a distance of 50 units) and then south to D such that $\angle DAC = 40°$. The angle of elevation of B from D is 30°.

The height of the flagpole is:

(*a*) $50 \cos 40° \tan 30°$; (*b*) $50 \sin 40° \tan 30°$;

(*c*) $\dfrac{50 \tan 30°}{\cos 40°}$; (*d*) $\dfrac{50}{\cos 40° \tan 30°}$.

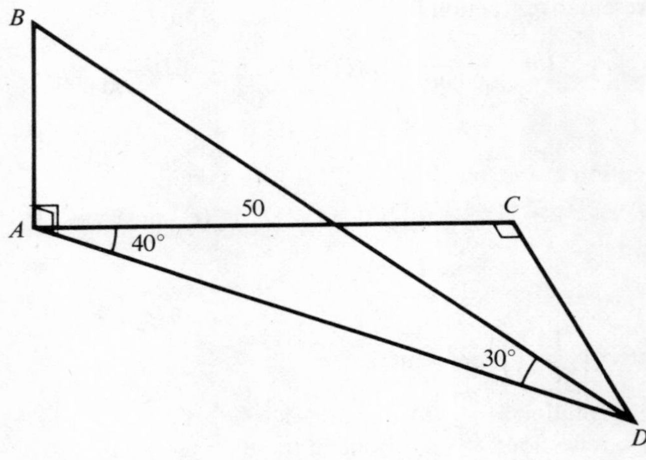

Fig. 25

7 (i) The number of elements in the solution set of $x^2 + 4x + 4 = (x + 2)^2$ is:

(*a*) 1; (*b*) 2; (*c*) none; (*d*) more than 2.

(ii) The number of elements in the solution set of $x^2 - 4x + 4 = (x + 2)^2$ is:

(*a*) 1; (*b*) 2; (*c*) none; (*d*) more than 2.

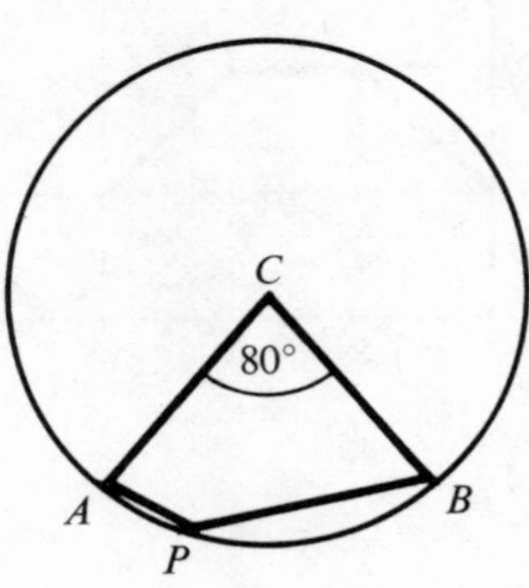

Fig. 26

8 In Figure 26, C is the centre of the circle. By joining CP and considering the two isosceles triangles ACP and BCP or by any other method, show that the size of $\angle APB$ is:

(*a*) 80°; (*b*) 140°; (*c*) 100°; (*d*) none of these.

9 The fraction represented by the recurring decimal $0{\cdot}\dot{2}\dot{7}$ is:

(*a*) $\frac{27}{100}$; (*b*) $\frac{26}{99}$; (*c*) $\frac{3}{11}$; (*d*) none of these.

10 A rough estimate of $(6{\cdot}04)^4$ is:

(*a*) 24; (*b*) 2000; (*c*) 1300; (*d*) 13 000.

6

State the letters corresponding to correct answers.

1 A stopwatch is known to gain by exactly 1 %. This means:

(*a*) after 99 s it reads 100 s; (*b*) after 100 s it reads 99 s;
(*c*) after 101 s it reads 100 s; (*d*) after 100 s it reads 101 s.

2 State for each of the following relations whether it is true:

(*a*) for all values of x; (*b*) for no values of x;
(*c*) for just one value of x; (*d*) for just two values of x;
(*e*) for more than two values of x (but not all).

(i) $x(x+3) = x(x+5)$;
(ii) $3(x+4) - (2+x) = 2(x+5)$;
(iii) $(x+1)(x+5) = (x+3)^2$.

3 The (x, y) plane is first reflected in the x-axis and is then rotated anti-clockwise through 90° about the origin. The point (3, 2) is transformed into:

(*a*) $(2, {}^-3)$; (*b*) $({}^-2, {}^-3)$; (*c*) $(2, 3)$;
(*d*) $({}^-2, 3)$; (*e*) none of these.

4 A and B are fixed points 5 cm apart in a given plane. The set of all points P in this plane such that the area of triangle APB is 10 cm² is:

(*a*) a straight line parallel to AB;
(*b*) a line segment parallel to AB;
(*c*) a pair of straight lines parallel to AB;
(*d*) a pair of line segments parallel to AB;
(*e*) none of these.

5 The shaded area in Figure 27 represents:

(*a*) $(C \cap A) \cap B$;
(*b*) $C \cup (A \cap B)$;
(*c*) $C \cap (A \cup B)$;
(*d*) $(C \cup B) \cap A$.

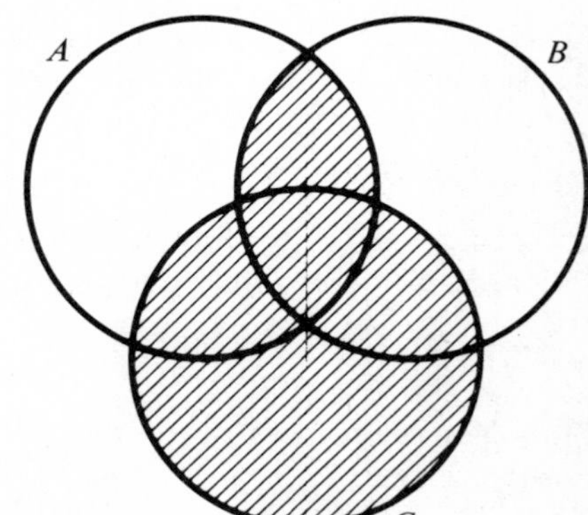

Fig. 27

6 The closest estimate of the value of $\dfrac{\pi^2 \times (14{\cdot}4)^4}{815}$ is:

(*a*) 20; (*b*) 100; (*c*) 500; (*d*) 750.

7 If $A(2, 3)$, $B(^-1, 2)$, $C(^-2, ^-1)$ are three vertices of a parallelogram $ABCD$, then D is:

(*a*) (1, 0); (*b*) ($^-1$, 4); (*c*) (0, 1); (*d*) none of these.

8 The transformation represented by the matrix $\begin{pmatrix} 0 & ^-1 \\ 1 & 0 \end{pmatrix}$ is:

(*a*) a $\frac{1}{4}$-turn about the origin; (*b*) a $\frac{1}{2}$-turn;
(*c*) a reflection in $y = 0$; (*d*) a reflection in $x = 0$.

9 The values of x and y which satisfy $(2x - 1)(y + 2) = 0$ are:

for x: (*a*) $\frac{1}{2}$; (*b*) $^-\frac{1}{2}$; (*c*) 2; (*d*) $^-2$;
for y: (*a*) $\frac{1}{2}$; (*b*) $^-\frac{1}{2}$; (*c*) 2; (*d*) $^-2$.

10 If $2x^2 + x - 3 = (x - p)(2x - q)$ then the values of p and q are:

for p: (*a*) 1; (*b*) $^-1$; (*c*) 3; (*d*) $^-3$;
for q: (*a*) 1; (*b*) $^-1$; (*c*) 3; (*d*) $^-3$.

7

State the letters corresponding to correct answers.

1 A(60° N, 20° W) and B(20° N, 160° W) are two points on the surface of the earth. An aeroplane is restricted to fly near the surface of the earth (in order to avoid detection) and along lines of latitude and longitude only. Which of the following is the shortest route from A to B?

(*a*) Westerly from A to longitude 160° W, then south to B;
(*b*) southerly from A to latitude 20° N, then west to B;
(*c*) northerly from A to the North Pole, then southerly along longitude 160° W to B;
(*d*) none of these (consistent with flight restrictions).

2 The measured diameter of a marble to 2 S.F. is 3·5 cm. The maximum possible volume in cm³ to 2 S.F. is:

(*a*) 20; (*b*) 21; (*c*) 22; (*d*) 23.

3 You are asked to draw a card from a traditionally composed pack of 52 cards. The probability that you draw:

(i) a heart is: (*a*) $\frac{1}{52}$; (*b*) $\frac{1}{4}$; (*c*) $\frac{1}{13}$; (*d*) none of these;
(ii) an ace is: (*a*) $\frac{1}{52}$; (*b*) $\frac{1}{4}$; (*c*) $\frac{1}{13}$; (*d*) none of these.

You are asked to throw a traditionally marked die. The probability that you throw

(iii) a 5 or a 6 is: (*a*) $\frac{1}{6}$; (*b*) $\frac{1}{3}$; (*c*) $\frac{1}{36}$; (*d*) none of these.

If you draw a card and throw a die, the probability that you obtain

(iv) both a spade and a 6 is:

(*a*) $\frac{1}{39}$; (*b*) $\frac{1}{24}$; (*c*) $\frac{1}{8}$; (*d*) none of these;

(v) either a jack or a 5 is:

(*a*) $\frac{10}{78}$; (*b*) $\frac{1}{24}$; (*c*) $\frac{5}{12}$; (*d*) none of these.

4 In Figure 28, O is the corner of a room such that OA, OB and OC are mutually perpendicular. $OA = 5$ m, $\angle OAB = 40°$, $\angle OBC = 56°$.
The length OB is:

(*a*) $5 \tan 40°$; (*b*) $5 \sin 40°$;

(*c*) $\dfrac{5}{\cos 40°}$; (*d*) none of these.

The length CB is:

(*a*) $\dfrac{5 \sin 40°}{\tan 56°}$; (*b*) $\dfrac{5 \tan 56°}{\cos 40°}$;

(*c*) $\dfrac{5 \sin 40°}{\sin 56°}$; (*d*) $\dfrac{5 \tan 40°}{\cos 56°}$.

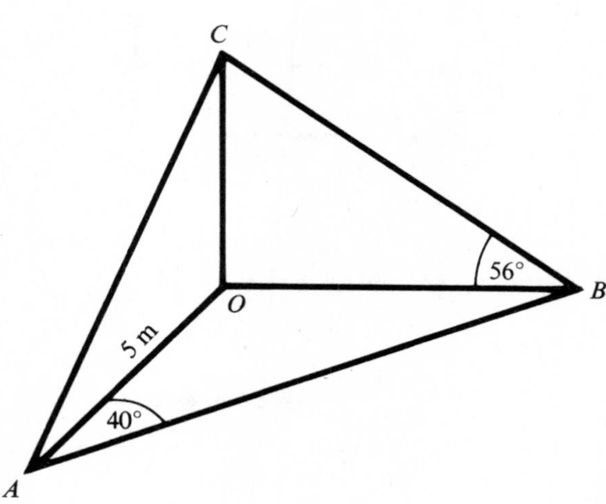

Fig. 28

5 If 10 km are represented by 2 cm on a map, then the area represented by 6 cm² is:

(*a*) 150 km²; (*b*) 90 km²;
(*c*) 30 km²; (*d*) none of these.

6 Figure 29 shows a pyramid standing on a cube:

(*a*) the angle between planes

OAB and $ABCD$ is $\angle OBC$;

(*b*) the angle between planes

CSP and $PQRS$ is $\angle BPQ$;

(*c*) the angle between line OP and plane $PQRS$ is $\angle OPQ$;

(*d*) the angle between lines OA and QR is $\angle DAO$.

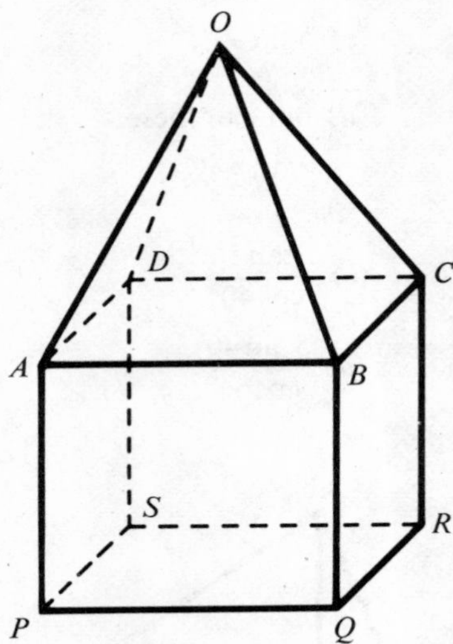

Fig. 29

7 $1{\cdot}5 \times 10^{-2}$ multiplied by 3×10^7 is:

(*a*) $4{\cdot}5 \times 10^{-14}$; (*b*) $4{\cdot}5 \times 10^9$;
(*c*) $4{\cdot}5 \times 10^{-7}$; (*d*) $4{\cdot}5 \times 10^5$.

8 (i) The next term in the sequence 1, 6, 13, 22, 33 is:

(*a*) 44; (*b*) 46;
(*c*) 52; (*d*) none of these.

(ii) The next term in the sequence 49, 31, 17, 7, 1 is:

(*a*) $^-1$; (*b*) $^-2$;
(*c*) $^-3$; (*d*) $\frac{1}{3}$.

9 The image of the point (1, 3) when reflected in the line $y = x + 2$ is:

(*a*) (1, 3); (*b*) (3, 1); (*c*) ($^-1$, 3); (*d*) (1, 1); (*e*) ($^-1$, $^-1$).

10 The inverse function of $f\colon \frac{x}{3} + 2$ is $f^{-1}\colon x \to$:

(*a*) $2 - \frac{x}{3}$; (*b*) $\frac{x}{3} - 2$; (*c*) $\frac{3}{x} - 2$; (*d*) $3(x-2)$; (*e*) $\frac{3}{x-2}$.

8

Say which statements are true and which are false.

1 (*a*) $\begin{pmatrix} 3 & 1 \\ 6 & 2 \end{pmatrix}\begin{pmatrix} 2 & ^-1 \\ ^-6 & 3 \end{pmatrix} = \begin{pmatrix} 0 & 0 \\ 0 & 0 \end{pmatrix}$;

(*b*) if **A** and **B** are 2×2 matrices, then it is impossible that $\mathbf{AB} = \mathbf{BA}$;

(*c*) $\begin{pmatrix} 4 & 1 \\ ^-3 & 0 \end{pmatrix}\begin{pmatrix} x \\ y \end{pmatrix} = \begin{pmatrix} 4x + y \\ ^-3x \end{pmatrix}$;

(*d*) $\begin{pmatrix} 0 & 1 \\ 1 & 0 \end{pmatrix}$ is the matrix of the transformation which reflects a point in the line $y = x$.

2 All the left-handed men in Ayton are members of the town police force. Then:

(*a*) all the Ayton police are left-handed;
(*b*) no Ayton man who is right-handed is a policeman;
(*c*) if a left-handed man is not a policeman, he does not live in Ayton.

3 Figure 30:

(*a*) $AC = BC \sin B$;
(*b*) $BC = \dfrac{BA}{\tan C}$;
(*c*) Area of triangle $ABC = \frac{1}{2}.CB.CA.\sin C$;
(*d*) $AD = AB \cos \angle BAD$.

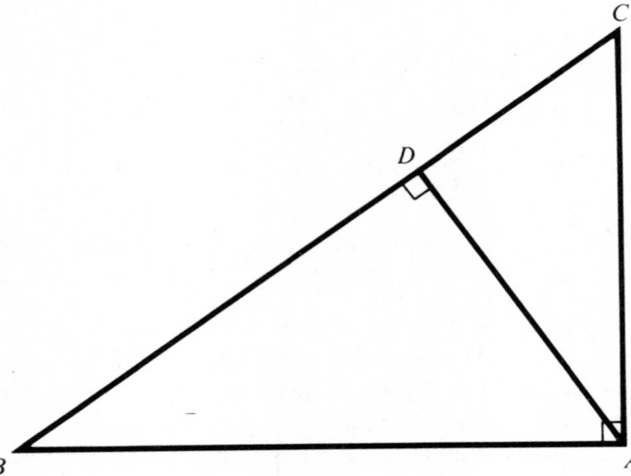

Fig. 30

4 (*a*) $\pi[6{\cdot}4^2 - 3{\cdot}6^2] = [\pi(6{\cdot}4 - 3{\cdot}6)][\pi(6{\cdot}4 + 3{\cdot}6)]$;

(*b*) $\dfrac{2x-2}{x} = \dfrac{2-2}{1} = \dfrac{0}{1} = 0$;

(*c*) $1011_2 \times 1011_2 = 10001111_2$;

(*d*) $S = \frac{4}{3}\pi r^3 \Leftrightarrow r^3 = \dfrac{3S}{4\pi} \Leftrightarrow r = \sqrt[3]{\left(\dfrac{3S}{4\pi}\right)}$.

5 (*a*) $x > 0 \Rightarrow x^2 > x$; (*b*) $a^2 - b^2 = (a - b)^2$;
(*c*) mn is even $\Leftrightarrow$ m and n are both even; (*d*) $\sqrt{2}$ is a rational number;
(*e*) if a quadrilateral has a centre of rotational symmetry, it is a rhombus.

6 If $x * y = x^2 + 2y^2$, then

(*a*) $4 * 2 = 32$; (*b*) $2 * 3 = 3 * 2$; (*c*) $1 * (2 * 1) = 73$;
(*d*) the operation $*$ is associative.

7 In Figure 31, $ABCD$ is a parallelogram. DQ is perpendicular to CB and CP is perpendicular to AB. Then

(*a*) $AP = 2$ cm; (*b*) $DQ = 8$ cm;
(*c*) triangle DCQ is similar to triangle CBP;
(*d*) triangle APQ is equal in area to triangle DCQ.

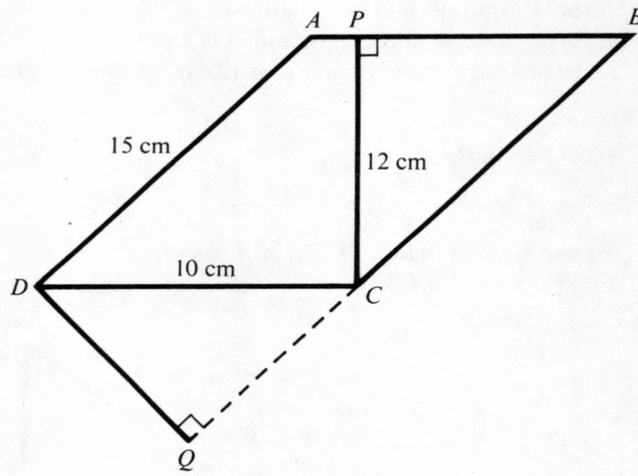

Fig. 31

8 (*a*) $\sqrt{6{\cdot}4} = 8$; (*b*) $\sqrt{0{\cdot}49} = 0{\cdot}7$; (*c*) $\sqrt{0{\cdot}009} = 0{\cdot}03$;
(*d*) $\sqrt{0{\cdot}016} = 0{\cdot}04$; (*e*) $\sqrt{2^9} = 2^3$; (*f*) $\sqrt{2^1} = 2^2$.

9 (*a*) In a single throw of two dice, it is twice as likely that you will throw either a 7 or an 11 as that you will throw a 2, a 3 or a 12.
(*b*) You are *more* likely to throw just one head in two tosses than to throw just two heads in four tosses.
(*c*) If I forecast the results (1, 2, X) of four football matches, the chance of my getting all four wrong is between $\frac{1}{5}$ and $\frac{1}{6}$. (Assume that each result is equally likely.)

10 $(5m - 2n)^2$ is equal to:

(*a*) $25m^2 - 4n^2$; (*b*) $5m^2 - 10mn + 2n^2$;
(*c*) $25m^2 - 20mn + 4n^2$; (*d*) $(2n - 5m)^2$;
(*e*) $25m^2 - 10mn + 4n^2$.

9

State the letters corresponding to correct answers.

1 The number of nautical miles separating the North and South Poles along the Greenwich Meridian is:

(*a*) 18 000; (*b*) 10 800; (*c*) 5400; (*d*) 21 600.

2 The length of the circumference of the line of latitude 60° S is, in nautical miles:

(*a*) 9000; (*b*) 5400; (*c*) 9350; (*d*) 10 800.

3 If

$$\mathbf{A}=\begin{pmatrix}1 & 2\\4 & 8\end{pmatrix} \text{ and } \mathbf{O}=\begin{pmatrix}0 & 0\\0 & 0\end{pmatrix},$$

then the equation $\mathbf{AX}=\mathbf{O}$ is satisfied by $\mathbf{X}$ equal to:

(*a*) $\begin{pmatrix}1 & 4\\2 & 8\end{pmatrix}$; (*b*) $\begin{pmatrix}1 & 2\\4 & 8\end{pmatrix}$; (*c*) $\begin{pmatrix}8 & ^-2\\^-4 & 1\end{pmatrix}$;

(*d*) $\begin{pmatrix}^-1 & ^-2\\^-4 & ^-8\end{pmatrix}$; (*e*) $\begin{pmatrix}6 & 2\\^-3 & ^-1\end{pmatrix}$.

4 The solution set of the equation $x^2+5x-6=0$ is:

(*a*) {3, 2}; (*b*) {3, ⁻2}; (*c*) {6, 1}; (*d*) {⁻6, 1}; (*e*) {6, ⁻1}.

5 (See Figure 32.) The length (in cm) of AB is:

(*a*) $\frac{1}{6}\cos 35°$; (*b*) $6\sin 35°$;

(*c*) $6\cos 35°$; (*d*) $\dfrac{6}{\sin 35°}$;

(*e*) $\dfrac{6}{\cos 35°}$.

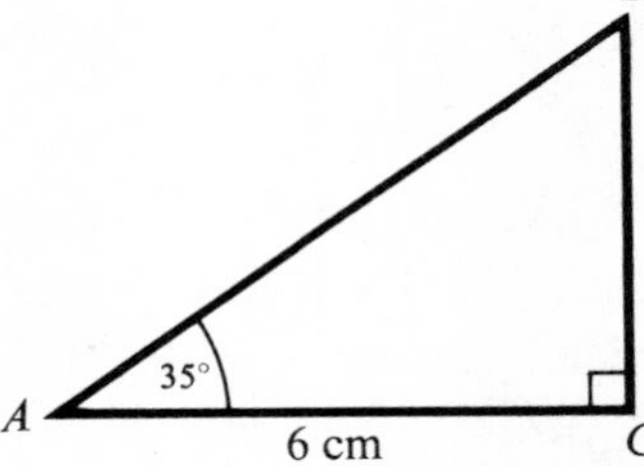

Fig. 41

6 State which has the largest area:

(*a*) a circle of radius 2 cm; (*b*) a square of side $3\frac{1}{2}$ cm;
(*c*) a triangle whose sides are 3 cm, 4 cm, 5 cm;
(*d*) a semi-circle of radius 4 cm.

7 The operation represented by the matrix $\begin{pmatrix}2 & 2\\0 & 2\end{pmatrix}$ is a combination of the two transformations:

(*a*) a translation; (*b*) a shear; (*c*) a reflection;
(*d*) a rotation; (*e*) an enlargement.

8 Two non-parallel coplanar lines meet in:

(*a*) 0 points; (*b*) 1 point;
(*c*) 2 points; (*d*) more than 2 points.

9 A new car is bought for £850. Its value decreases by £150 during the first year after purchase. Each year thereafter it loses 10 % of its value at the beginning of that year. An expression for its value in £ after n years ($n \geqslant 1$) is:

(a) $700 - (0{\cdot}1)^n . 700$;
(b) $70(10 - (n - 1))$;
(c) $850 \times (0{\cdot}9)^n$;
(d) $700 \times (0{\cdot}9)^{n-1}$.

10 If x and y are numbers, which of the following are true and which are false?

(a) $xy = 0 \Rightarrow (x = 0 \text{ and } y = 0)$.
(b) $(xy = 0 \text{ and } x \neq 0) \Rightarrow y = 0$.
(c) $x = 3 \Rightarrow x^2 = 3x$.
(d) $x^2 = 3x \Rightarrow x = 3$.

10

State the letters corresponding to the correct answers.

1 A shear is described by the matrix $\begin{pmatrix} 1 & 0 \\ 2 & 1 \end{pmatrix}$.

The invariant points of this shear lie on the line

(a) $x = 0$; (b) $y = 0$; (c) $2x = y$; (d) $y = 2$.

2 Given that $f: x \to \frac{3}{x}$ and $g: x \to \frac{x}{3}$, state whether it is true or false that

(a) $f^{-1} = f$; (b) $g^{-1} = g$; (c) $fg = gf$; (d) $(fg)^{-1} = g^{-1}f^{-1}$.

3 Which of the following statements are true and which are false?

(a) The cube root of 10 lies between 2·1 and 2·2.
(b) $\sqrt{(\sqrt{14})}$ is the cube root of 14.
(c) $\sqrt{2\frac{1}{4}}$ is a rational number.
(d) $\sqrt{12} = 4\sqrt{3}$.

4 Fred the mathematical fly crawls from X to Y in Figure 33 by the shortest route on the surface of the cuboid. The length of his journey is

(a) $(2 + \sqrt{34})$ cm; (b) $\sqrt{50}$ cm; (c) 10 cm; (d) $\sqrt{58}$ cm.

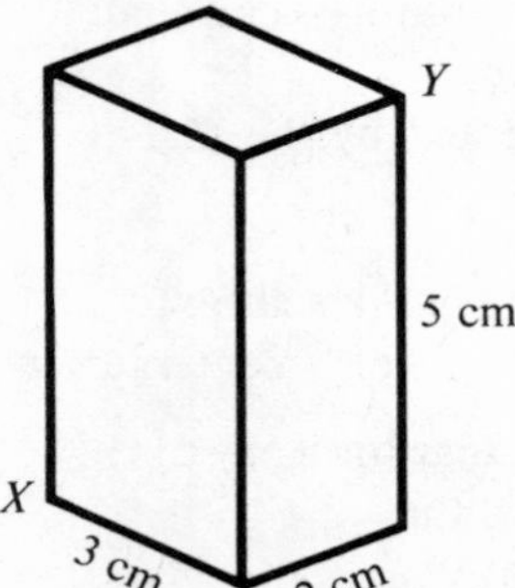

Fig. 33

5 Indicate which of the following regular polygons can be used by themselves to tessellate a plane, and which cannot.

(*a*) Equilateral triangles; (*b*) pentagons;
(*c*) hexagons; (*d*) octagons.

6 p, q, x are positive unequal numbers and $x \neq 1$. Indicate which of the following statements are true and which are false:

(*a*) $\dfrac{p+qx}{px+q} = 1$;

(*b*) $\dfrac{p^2-x^2}{p+x} = p-x$;

(*c*) $\dfrac{p^2+x^2}{p+x} = p+x$;

(*d*) $\dfrac{p/x+2}{q/x+3} = \dfrac{p+2x}{q+3x}$.

7 Which of the following belong, and which do not belong, to the set of points for which $2x - 3y < 10$?

(*a*) (4, 1); (*b*) (8, 2); (*c*) $(6, {}^-1)$; (*d*) $({}^-3, {}^-5)$.

8 Which of the following statements may be true, and which cannot possibly be true?

(*a*) $\tan x^\circ = 2{\cdot}1$; (*b*) $\sin x^\circ = 2{\cdot}9$;

(*c*) $\cos x^\circ = {}^-0{\cdot}6$; (*d*) $\dfrac{1}{\cos x^\circ} = 0{\cdot}5$.

9

x	5	15
y	6	$\frac{2}{3}$

The figures in the table are compatible with:

(*a*) $y \propto (1/x^2)$; (*b*) $y = {}^-\frac{8}{15}x + \frac{26}{3}$;
(*c*) $y = kx^2$; (*d*) none of these.

10 If $p: x \to 3x$ and $q: x \to x^2$, state which of the following are correct and which incorrect:

(*a*) $p(5) = \frac{3}{5}$;
(*b*) p and q are commutative;
(*c*) q^{-1} is always a function;
(*d*) $pq(x) = 3x^2$.

11

State the letters corresponding to the correct answers.

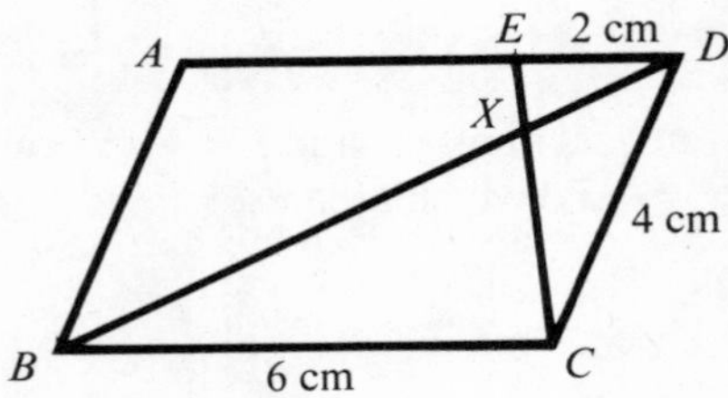

Fig. 34

1 $ABCD$ in Figure 34 is a parallelogram.

(i) The scale factor of the enlargement centre X of $\triangle EXD$ to $\triangle CXB$ is:

(*a*) $\frac{1}{4}$; (*b*) $\frac{1}{3}$; (*c*) 3; (*d*) $^-3$.

(ii) The area scale factor for EXD to CXB is:

(*a*) 4; (*b*) 3; (*c*) 9; (*d*) $^-9$.

2 If $5^x = 75$, the value of x is

(*a*) 15; (*b*) $2\frac{1}{2}$; (*c*) between 2 and 3; (*d*) none of those.

3 If
$$A = \{x: {}^-1 \leqslant x \leqslant 7\},$$
$$B = \{x: {}^-2 < x < 4\}$$
and
$$C = \{x: 0 < x < 5\}$$

then $A \cap C \cap B'$ is

(*a*) $\{x: 4 \leqslant x \leqslant 7\}$; (*b*) $\{x: 4 \leqslant x \leqslant 5\}$;
(*c*) $\{x: 0 < x \leqslant 4\}$; (*d*) $\{x: {}^-1 \leqslant x < 4\}$.

4 The point (3, 4) is mapped onto (5, 0) by an isometry that leaves the origin unmoved. The image of (0, 5) *may* be

(*a*) ($^-4$, 3); (*b*) (4, 3); (*c*) (4, $^-3$); (*d*) (2, 1).

5 If in the formula $H = PT^2$, P remains the same but T is increased by 10 % then H will be increased by

(*a*) 100 %; (*b*) 10 %; (*c*) 21 %; (*d*) 20 %.

6 If n is an integer such that

$$n < 7\sqrt{(0{\cdot}8)} < n + 1$$

then the value of n is:

(*a*) 2; (*b*) 5; (*c*) 6; (*d*) 7.

7 In a scale model of a building, the ground area is $\frac{1}{100}$th of the actual ground area. The ratio of the volume of the building to the volume of the model is:

(*a*) 100^3; (*b*) 10^3; (*c*) 100^{-3}; (*d*) 10^{-3}.

8 A set of inequalities which define the shaded area shown in Figure 35 (all the boundary is included) is:

(*a*) $0 \leqslant x \leqslant 1, \quad 0 \leqslant y \leqslant 1, \quad x + y \leqslant 2;$
(*b*) $0 \leqslant x \leqslant 1, \quad 0 \leqslant y \leqslant 1, \quad xy \leqslant 1;$
(*c*) $0 < x \leqslant 1, \quad 0 \leqslant y < 1, \quad x + y \geqslant 1;$
(*d*) $0 \leqslant x \leqslant 1, \quad 0 \leqslant y \leqslant 1, \quad x + y \geqslant 1.$

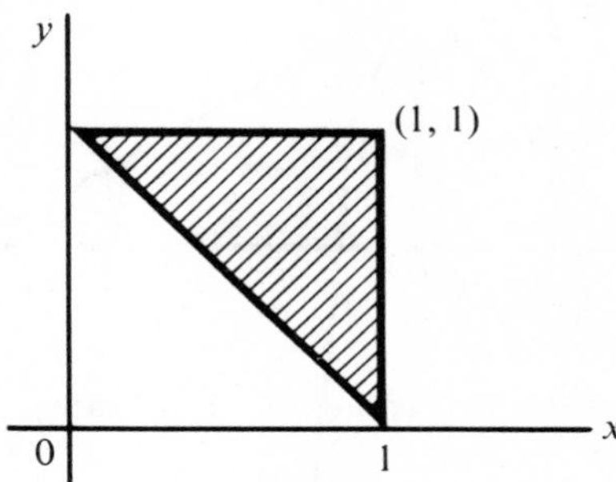

Fig. 35

9 The two circles in Figure 36 have centres A and B and touch at T. CD is a common tangent. $BT = 1$ cm, $TA = 2$ cm. If $\angle DCB = x°$, then:

(*a*) $\sin x° = \frac{1}{3}$; (*b*) $\tan x° = \frac{1}{3}$;
(*c*) $\sin x° = \frac{2}{3}$; (*d*) none of the above is true.

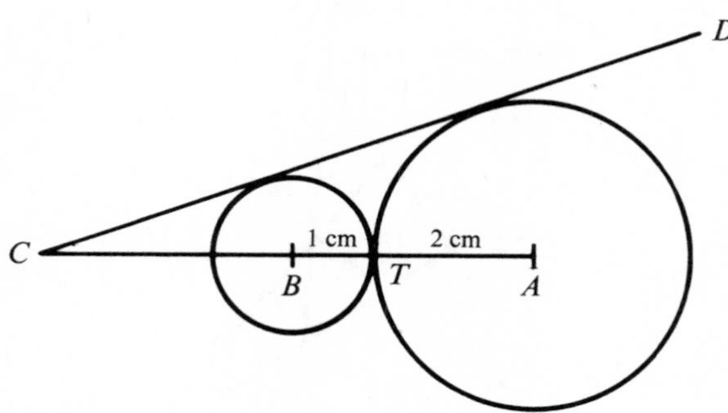

Fig. 36

10 In a certain triangle ABC, $AC^2 = AB^2 - BC^2$.
Which one or more of the following statements is (are) necessarily true?

(*a*) A is a right angle.
(*b*) B is a right angle.
(*c*) AC is the shortest side.
(*d*) AB is the longest side.

12

State the letters corresponding to the correct answers.

1 The length DC in Figure 37, in cm, is

(*a*) $5 \sin 40° \tan 30°$; (*b*) $5 \sin 40° \tan 60°$;
(*c*) $5 \cos 40° \cos 60°$; (*d*) $5 \cos 40° \tan 60°$.

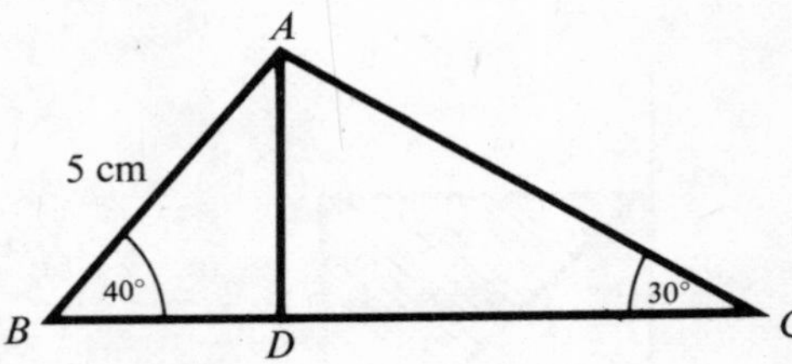

Fig. 37

2 State which of the following estimates of the value of

$$\tfrac{4}{3}\pi \frac{(18{\cdot}7)^3}{1728}$$

is the best: (*a*) 16; (*b*) 160; (*c*) 0·9; (*d*) 48.

3 The birth rate per 1000 of population is calculated correct to one decimal place. For a town with exactly 43 000 inhabitants the rate was stated to be 11·2 births per 1000. The maximum number of births consistent with this statement is

(*a*) 481; (*b*) 482; (*c*) 483; (*d*) 484.

4 In Figure 38 the angle BHC is equal to

(*a*) $A°$; (*b*) $2A°$; (*c*) $180° - A°$; (*d*) $360° - A°$.

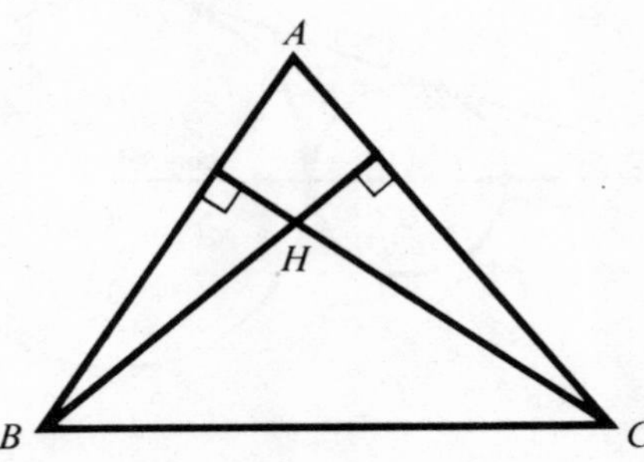

Fig. 38

5 In Figure 39 the shaded area in the Venn diagram represents

(*a*) $(C \cup A) \cap B$;
(*b*) $C \cap (A \cup B)$;
(*c*) $C \cup (A \cup B)$;
(*d*) $A \cap B \cap C$.

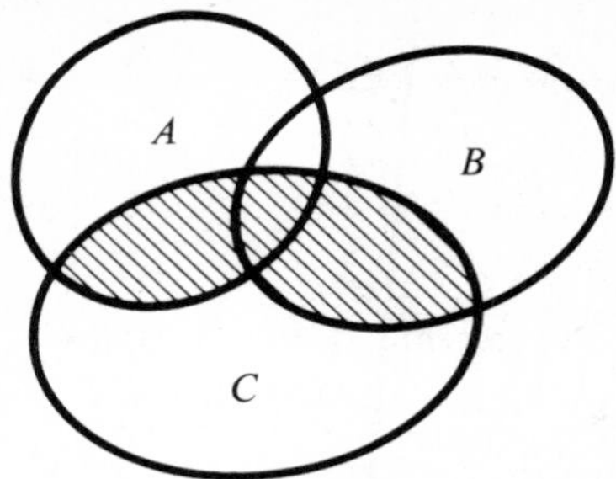

Fig. 39

6 In Figure 40 the length BC, in cm, is equal to:

(*a*) $10 \sin 40°$; (*b*) $20 \sin 20°$;
(*c*) $10 \tan 40°$; (*d*) $20 \tan 20°$.

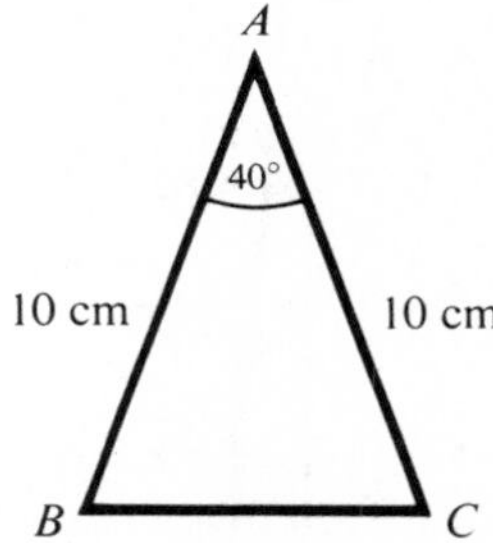

Fig. 40

7 Figure 41 shows part of a distance–time graph. Which of the following statements are true, which false?

(*a*) The body moves with speed 20 cm/s.
(*b*) The body starts with speed 4 cm/s.
(*c*) The body has no acceleration.
(*d*) The distance travelled in the first 15 s is 60 cm.

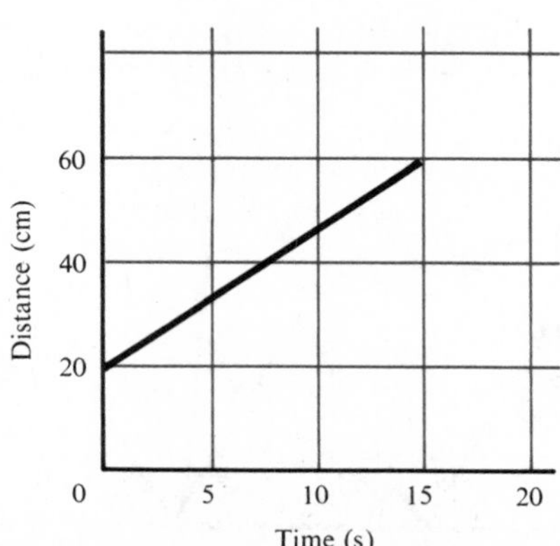

Fig. 41

8 The heights of 25 boys are measured to the nearest centimetre and are then grouped as follows:

Height (in cm)	151–155	156–160	161–165	166–170	171–175
Frequency	4	8	7	5	1

Which of the following statements may be true and which must be false?

(*a*) The modal class contains seven members.
(*b*) The median height is 158 cm.
(*c*) The median height is 161 cm.
(*d*) The range is 20 cm.

9 In Figure 42, P is the point (4, 2), 0 is the origin, **U** denotes reflection in the line $x = 1$. **V** denotes the half-turn about 0. Which of the following are true and which are false?

(*a*) **U**(P) is ($^-2$, 2);
(*b*) **VU**(P) is (2, $^-2$);
(*c*) **VU**(P) is **UV**(P);
(*d*) P is mapped onto **VU**(P) by a half-turn about (3, 0).

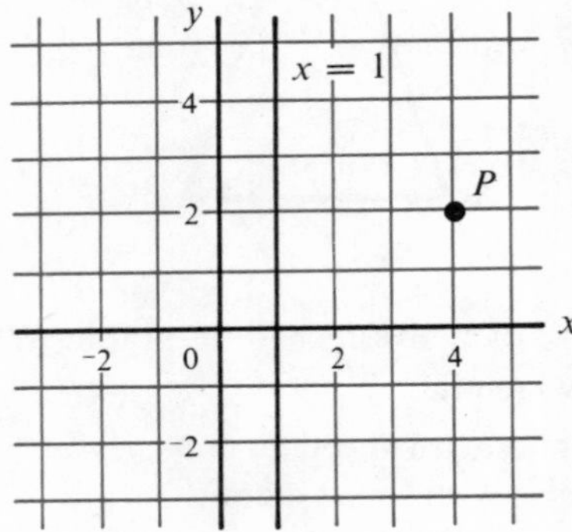

Fig. 42

10 A prism whose cross-section is an equilateral triangle with two differently marked corners lies on a table in position I (see Figure 43). It is rotated about the edge through P into position II, then about the edge through Q into position III. Which of the following transformations could describe the mapping I → III:

(*a*) a translation; (*b*) a reflection;
(*c*) a rotation; (*d*) none of these?

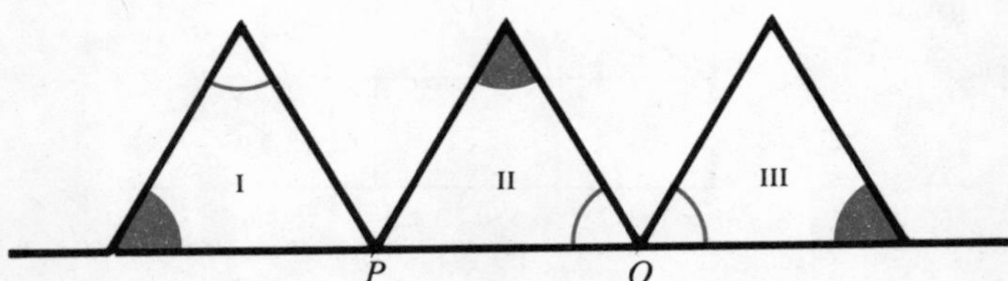

Fig. 43

Index

Index

Index

Page 29 Ex A No 2. Answer (graph) wrong. See notes on ans bk.

Page 34 (bottom). Ludicrous to get 9·8 m/s² from this sketch — 10 m/s² drawn.